Universitext

Daniel Huybrechts

Complex Geometry

An Introduction

 Springer

Daniel Huybrechts
Université Paris VII Denis Diderot
Institut de Mathématiques
2, place Jussieu
75251 Paris Cedex 05
France
e-mail: huybrech@math.jussieu.fr

Mathematics Subject Classification (2000): 14J32, 14J60, 14J81, 32Q15, 32Q20, 32Q25

Cover figure is taken from page 120.

Library of Congress Control Number: 2004108312

ISBN 3-540-21290-6 Springer Berlin Heidelberg New York

Springer is a part of Springer Science+Business Media
springeronline.com
© Springer-Verlag Berlin Heidelberg 2005
Printed in Germany

Cover design: Erich Kirchner, Heidelberg
Typesetting by the author using a Springer LaTeX macro package
Production: LE-TeX Jelonek, Schmidt & Vöckler GbR, Leipzig

Printed on acid-free paper   46/3142YL - 5 4 3 2 1 0

Preface

Complex geometry is a highly attractive branch of modern mathematics that has witnessed many years of active and successful research and that has recently obtained new impetus from physicists' interest in questions related to mirror symmetry. Due to its interactions with various other fields (differential, algebraic, and arithmetic geometry, but also string theory and conformal field theory), it has become an area with many facets. Also, there are a number of challenging open problems which contribute to the subject's attraction. The most famous among them is the Hodge conjecture, one of the seven one-million dollar millennium problems of the Clay Mathematics Institute. So, it seems likely that this area will fascinate new generations for many years to come.

Complex geometry, as presented in this book, studies the geometry of (mostly compact) complex manifolds. A complex manifold is a differentiable manifold endowed with the additional datum of a complex structure which is much more rigid than the geometrical structures in differential geometry. Due to this rigidity, one is often able to describe the geometry of complex manifolds in very explicit terms. E.g. the important class of projective manifolds can, in principle, be described as zero sets of polynomials.

Yet, a complete classification of all compact complex manifolds is too much to be hoped for. Complex curves can be classified in some sense (involving moduli spaces etc.), but already the classification of complex surfaces is tremendously complicated and partly incomplete.

In this book we will concentrate on more restrictive types of complex manifolds for which a rather complete theory is in store and which are also relevant in the applications. A prominent example are Calabi–Yau manifolds, which play a central role in questions related to mirror symmetry. Often, interesting complex manifolds are distinguished by the presence of special Riemannian metrics. This will be one of the central themes throughout this text. The idea is to study cases where the Riemannian and complex geometry on a differentiable manifold are not totally unrelated. This inevitably leads to

Kähler manifolds, and a large part of the book is devoted to techniques suited for the investigation of this prominent type of complex manifolds.

The book is based on a two semester course taught in 2001/2002 at the university of Cologne. It assumes, besides the usual facts from the theory of holomorphic functions in one variable, the basic notions of differentiable manifolds and sheaf theory. For the convenience of the reader we have summarized those in the appendices A and B. The aim of the course was to introduce certain fundamental concepts, techniques, and results in the theory of compact complex manifolds, without being neither too basic nor too sketchy.

I tried to teach the subject in a way that would enable the students to follow recent developments in complex geometry and in particular some of the exciting aspects of the interplay between complex geometry and string theory. Thus, I hope that the book will be useful for both communities, those readers aiming at understanding and doing research in complex geometry and those using mathematics and especially complex geometry in mathematical physics.

Some of the material was intended rather as an outlook to more specialized topics, and I have added those as appendices to the corresponding chapters. They are not necessary for the understanding of the subsequent sections.

I am aware of several shortcomings of this book. As I found it difficult to teach the deeper aspects of complex analysis to third-year students, the book cannot serve as an introduction to the fascinating program initiated by Siu, Demailly, and others, that recently has lead to important results in complex and algebraic geometry. So, for the analysis I have to refer to Demailly's excellent forthcoming (?) text book [35]. I also had to leave out quite a number of important tools, like higher direct image sheaves, spectral sequences, intermediate Jacobians, and others. The hope was to create a streamlined approach to some central results and so I did not want to enter too many of the promising side-roads. Finally, although relevant examples have been included in the text as well as in the exercises, the book does not discuss in depth any difficult type of geometry, e.g. Calabi–Yau or hyperkähler manifolds. But I believe that with the book at hand, it should not be too difficult to understand more advanced texts on special complex manifolds.

Besides Demailly's book [35], there are a number of text books on complex geometry, Hodge theory, etc. The classic [59] and the more recent one by Voisin [113] are excellent sources for more advanced reading. I hope that this book may serve as a leisurely introduction to those.

In the following, we will give an idea of the content of the book. For more information, the reader may consult the introductions at the beginning of each chapter.

Chapter 1 provides the minimum of the local theory needed for the global description of complex manifolds. It may be read along with the later chapters or worked through before diving into the general theory of complex manifolds beginning with Chapter 2.

Section 1.1 shows a way from the theory of holomorphic functions of one variable to the general theory of complex functions. Eventually, it would lead to the local theory of complex spaces, but we restrict ourselves to those aspects strictly necessary for the understanding of the rest of the book. The reader interested in this attractive combination of complex analysis and commutative algebra may consult [35] or any of the classics, e.g. [57, 64].

Section 1.2 is a lesson in linear algebra and as such rather elementary. We start out with a real vector space endowed with a scalar product and the additional datum of an almost complex structure. We shall investigate what kind of structure is induced on the exterior algebra of the vector space. I tried to present the material with some care in order to make the reader feel comfortable when later, in the global context, the machinery is applied to compact Kähler manifolds.

Section 1.3 proves holomorphic versions of the Poincaré lemma and is supposed to accustom the reader to the yoga of complex differential forms on open sets of $\mathbb{C}^n$.

With **Chapter 2** the story begins. Sections 2.1 and 2.2 deal with complex manifolds and holomorphic vector bundles, both holomorphic analogues of the corresponding notions in real differential geometry. But a few striking differences between the real and the complex world will become apparent right away. The many concrete examples of complex manifolds are meant to motivate the discussion of the more advanced techniques in the subsequent chapters.

Section 2.3 illuminates the intimate relation between complex codimension one submanifolds (or, more generally, divisors) and holomorphic line bundles with their global sections. This builds the bridge to classical algebraic geometry, e.g. Veronese and Segre embedding are discussed. The section ends with a short discussion of the curve case.

Section 2.4 is devoted to the complex projective space $\mathbb{P}^n$, a universal object in complex (algebraic) geometry comparable to spheres in the real world. We describe its tangent bundle by means of the Euler sequence and certain tautological line bundles. A discussion of the Riemannian structure of $\mathbb{P}^n$ (e.g. the Fubini–Study metric) is postponed until Section 3.1.

Section 2.5 provides an example of the universal use of the projective space. It explains a complex surgery, called blow-up, which modifies a given complex manifold along a distinguished complex submanifold, replacing a point by a projective space. Apart from its importance in the birational classification of complex manifolds, blow-ups will turn out to be of use in the proof of the Kodaira embedding theorem in Section 5.2.

Section 2.6 interprets complex manifolds as differentiable manifolds together with an additional linear datum (an almost complex structure) satisfying an integrability condition. Here, the linear algebra of Section 1.2 comes in handy. The crucial Newlander–Nierenberg theorem, asserting the equivalence of the two points of view, is formulated but not proved.

Chapter 3 is devoted to (mostly compact) Kähler manifolds. The existence of a Kähler metric on a compact complex manifold has far reaching consequences for its cohomology. Behind most of the results on Kähler manifolds one finds the so-called Kähler identities, a set of commutator relations for the various differential and linear operators. They are the topic of Section 3.1.

In Section 3.2, Hodge theory for compact manifolds is used to pass from arbitrary forms to harmonic forms and eventually to cohomology classes. This immediately yields central results, like Serre duality and, in Section 3.3, Lefschetz decomposition.

Section 3.3 also explains how to determine those classes in the second cohomology $H^2(X)$ of a compact Kähler manifold X that come from holomorphic line bundles. This is the Lefschetz theorem on $(1,1)$-classes. A short introduction to the hoped for generalization to higher degree cohomology classes, i.e. the Hodge conjecture, ends this section.

There are three appendices to Chapter 3. Appendix 3.A proves the formality of compact Kähler manifolds, a result that interprets the crucial $\partial\bar{\partial}$-lemma of Section 3.2 homologically. Appendix 3.B is a first introduction to some mathematical aspects of supersymmetry. The cohomological structures encountered in the bulk of the chapter are formalized by the notion of a Hodge structure. Appendix 3.C collects a few basic notions and explains how they fit in our context.

Chapter 4 provides indispensable tools for the study of complex manifolds: connections, curvature, and Chern classes. In contrast to previous sections, we will not just study complex manifolds and their tangent bundles but broaden our perspective by considering arbitrary holomorphic vector bundles. However, we will not be in the position to undertake an indepth analysis of all fundamental questions. E.g. the question whether there exist holomorphic vector bundles besides the obvious ones on a given manifold (or holomorphic structures on a given complex vector bundle) will not be addressed. This is partially due to the limitations of the book, but also to the state of the art. Only for curves and projective surfaces the situation is fairly well understood (see [70]).

In the appendices to Chapter 4 we discuss the interplay of complex and Riemannian geometry. Appendix 4.A tries to clarify the relation between the Levi-Civita connection and the Chern connection on a Kähler manifold. The concept of holonomy, well known in classical Riemannian geometry, allows to view certain features in complex geometry from a slightly different angle. Appendix 4.B outlines some basic results about Kähler–Einstein and Hermite–Einstein metrics. Before, the hermitian structure on a holomorphic vector bundle was used as an auxiliary in order to apply Hodge theory, etc. Now, we ask whether canonical hermitian structures, satisfying certain compatibility conditions, can be found.

In order to illustrate the power of cohomological methods, we present in **Chapter 5** three central results in complex algebraic geometry. Except for the

Hirzebruch–Riemann–Roch theorem, complete proofs are given, in particular for Kodaira's vanishing and embedding theorems. The latter one determines which compact complex manifolds can be embedded into a projective space. All three results are of fundamental importance in the global theory of complex manifolds.

Chapter 6 is relevant to the reader interested in Calabi–Yau manifolds and mirror symmetry. It is meant as a first encounter with deformation theory, a notoriously difficult and technical subject. In Section 6.1 we leave aside convergence questions and show how to study deformations by a power series expansion leading to the Maurer–Cartan equation. This approach can successfully be carried out for compact Kähler manifolds with trivial canonical bundle (Calabi–Yau manifolds) due to the Tian–Todorov lemma. Section 6.2 surveys the more abstract parts of deformation theory, without going into detail. The appendix to this chapter is very much in the spirit of appendix 3.A. Here, the content of Section 6.1 is put in the homological language of Batalin–Vilkovisky algebras, a notion that has turned out to be useful in the construction of Frobenius manifolds and in the formulation of mirror symmetry.

In general, all results are proved except for assertions presented as 'theorems', indicating that they are beyond the scope of this book, and a few rather sketchy points in the various appendices to the chapters. Certain arguments, though, are relegated to the exercises, not because I wanted to leave unpleasant bits to the reader, but because sometimes it is just more rewarding performing a computation on ones own.

Acknowledgement: I learned much of the material from the two classics [8, 59] and from my teacher H. Kurke. Later, the interplay of algebraic geometry and gauge theory as well as the various mathematical aspects of mirror symmetry have formed my way of thinking about complex geometry. The style of the presentation has been influenced by stimulating discussions with D. Kaledin, R. Thomas, and many others over the last few years.

I want to thank G. Hein, M. Nieper-Wißkirchen, D. Ploog, and A. Schmidt, who read preliminary versions of the text and came up with long lists of comments, corrections, and suggestions. Due to their effort, the text has considerably improved.

Paris, June 2004   *Daniel Huybrechts*

Contents

Local Theory

This chapter consists of three sections. Section 1.1 collects the principal facts from the theory of holomorphic functions of several variables. We assume that the reader has mastered the theory of holomorphic functions of one variable, but the main results shall be briefly recalled.

Section 1.2 is pure linear algebra. The reader may skip this part, or skim through it, and come back to it whenever feeling uncomfortable about certain points in the later chapters. I tried to present the material with great care. In particular, the interplay between the Hodge and Lefschetz operators is explained with all the details.

In Section 1.3 the techniques of the previous two sections are merged. The reader will be introduced to the theory of complex differential forms on open subsets of $\mathbb{C}^n$. This gives him the opportunity to do some explicit calculations before these notions will be reconsidered in the global context. The central result of this section is the $\bar{\partial}$-Poincaré lemma.

1.1 Holomorphic Functions of Several Variables

Let us first recall some basic facts and definitions from the theory of holomorphic functions of one variable. For proofs and further discussion the reader may consult any textbook on the subject, e.g. [98].

Let $U \subset \mathbb{C}$ be an open subset. A function $f : U \to \mathbb{C}$ is called *holomorphic* if for any point $z_0 \in U$ there exists a ball $B_\varepsilon(z_0) \subset U$ of radius $\varepsilon > 0$ around z_0 such that f on $B_\varepsilon(z_0)$ can be written as a convergent power series, i.e.

$$f(z) = \sum_{n=0}^{\infty} a_n(z - z_0)^n \text{ for all } z \in B_\varepsilon(z_0). \tag{1.1}$$

There are equivalent definitions of holomorphicity. The most important one uses the *Cauchy–Riemann equations*. Let us denote the real and imaginary part of $z \in \mathbb{C}$ by x respectively y. Thus, f can be regarded as a complex function $f(x, y)$ of two real variables x and y. Furthermore, f can be written in the form $f(x, y) = u(x, y) + iv(x, y)$, where $u(x, y)$ and $v(x, y)$ denote the real and imaginary part of f, respectively. Then one shows that f is holomorphic if and only if u and v are continuously differentiable and

$$\frac{\partial u}{\partial x} = \frac{\partial v}{\partial y}, \quad \frac{\partial u}{\partial y} = -\frac{\partial v}{\partial x}. \tag{1.2}$$

In other words, the derivative of f has to be complex linear. Let us introduce the differential operators

$$\frac{\partial}{\partial z} := \frac{1}{2} \left(\frac{\partial}{\partial x} - i \frac{\partial}{\partial y} \right) \quad \text{and} \quad \frac{\partial}{\partial \bar{z}} := \frac{1}{2} \left(\frac{\partial}{\partial x} + i \frac{\partial}{\partial y} \right). \tag{1.3}$$

The notation is motivated by the properties $\frac{\partial}{\partial z}(z) = \frac{\partial}{\partial \bar{z}}(\bar{z}) = 1$ and $\frac{\partial}{\partial \bar{z}}(z) = \frac{\partial}{\partial z}(\bar{z}) = 0$. Then, the Cauchy–Riemann equations (1.2) can be rewritten as $\frac{\partial f}{\partial \bar{z}} = 0$. This is easy if one uses $f = u + iv$. It might be instructive to do the same calculation for f written as the vector $\binom{u}{v}$ and $\frac{\partial}{\partial \bar{z}} = \frac{1}{2}(\frac{\partial}{\partial x} + \binom{0 \ -1}{1 \ \ 0}\frac{\partial}{\partial y})$.

As the transition from the real partial differentials $\frac{\partial}{\partial x}, \frac{\partial}{\partial y}$ to the complex partial differentials $\frac{\partial}{\partial z}, \frac{\partial}{\partial \bar{z}}$ is a crucial point, let us discuss this a little further. Consider a differentiable map $f : U \subset \mathbb{C} = \mathbb{R}^2 \to \mathbb{C} = \mathbb{R}^2$. Its differential $df(z)$ at a point $z \in U$ is an $\mathbb{R}$-linear map between the tangent spaces $df(z) : T_z\mathbb{R}^2 \to T_{f(z)}\mathbb{R}^2$. Writing the complex coordinate on the left hand side as $z = x + iy$ and on the right hand side as $w = r + is$ the two tangent spaces can be given canonical bases $\langle \partial/\partial x, \partial/\partial y \rangle$ and $\langle \partial/\partial r, \partial/\partial s \rangle$, respectively. With respect to these the differential $df(z)$ is given by the real Jacobian

$$J_\mathbb{R}(f) = \begin{pmatrix} \dfrac{\partial u}{\partial x} & \dfrac{\partial u}{\partial y} \\[2mm] \dfrac{\partial v}{\partial x} & \dfrac{\partial v}{\partial y} \end{pmatrix},$$

where $f = u + iv$ as before, i.e. $u = r \circ f$ and $v = s \circ f$.

After extending $df(z)$ to a $\mathbb{C}$-linear map $df(z)_\mathbb{C} : T_z\mathbb{R}^2 \otimes \mathbb{C} \to T_{f(z)}\mathbb{R}^2 \otimes \mathbb{C}$, we may choose different bases $\langle \frac{\partial}{\partial z} = \frac{1}{2}(\frac{\partial}{\partial x} - i\frac{\partial}{\partial y}), \frac{\partial}{\partial \bar{z}} = \frac{1}{2}(\frac{\partial}{\partial x} + i\frac{\partial}{\partial y}) \rangle$ and correspondingly for the right hand side. With respect to those $df(z)$ is given by the matrix

$$\begin{pmatrix} \dfrac{\partial f}{\partial z} & \dfrac{\partial f}{\partial \bar{z}} \\[2mm] \dfrac{\partial \bar{f}}{\partial z} & \dfrac{\partial \bar{f}}{\partial \bar{z}} \end{pmatrix}.$$

E.g. the vector $\frac{\partial}{\partial z}$ is sent to the vector $\frac{\partial f}{\partial z} \cdot \frac{\partial}{\partial w} + \frac{\partial \bar{f}}{\partial z} \cdot \frac{\partial}{\partial \bar{w}}$. For the chain rule it would be more natural to change the order of $\frac{\partial}{\partial w}$ and $\frac{\partial f}{\partial z}$ (and of $\frac{\partial}{\partial \bar{w}}$ and $\frac{\partial \bar{f}}{\partial z}$). In the following, we will use that for any function f one has $\frac{\partial \bar{f}}{\partial \bar{z}} = \overline{\left(\frac{\partial f}{\partial z}\right)}$. If f is holomorphic, then $\frac{\partial f}{\partial \bar{z}} = \frac{\partial \bar{f}}{\partial z} = 0$ and thus $df(z)$ in the new base is given by the diagonal matrix

$$\begin{pmatrix} \dfrac{\partial f}{\partial z} & 0 \\[2ex] 0 & \dfrac{\partial \bar{f}}{\partial \bar{z}} \end{pmatrix}.$$

Holomorphicity of f is also equivalent to the *Cauchy integral formula*. More precisely, a function $f : U \to \mathbb{C}$ is holomorphic if and only if f is continuously differentiable and for any $B_\varepsilon(z_0) \subset U$ the following formula holds true

$$f(z_0) = \frac{1}{2\pi i} \int_{\partial B_\varepsilon(z_0)} \frac{f(z)}{z - z_0} dz. \tag{1.4}$$

Actually, the formula holds true for any continuous function $f : \overline{B_\varepsilon(z_0)} \to \mathbb{C}$ which is holomorphic in the interior. Let us remind that the Cauchy integral formula is used to prove the existence of a power series expansion of any function satisfying the Cauchy–Riemann equations. (If f is just continuous, one only has $f(z_0) = (1/2\pi i) \lim_{\varepsilon \to 0} \int_{\partial B_\varepsilon(z_0)} f(z)/(z - z_0) dz$.)

The following list collects a few well-known facts, which will be important for our purposes.

Maximum principle. Let $U \subset \mathbb{C}$ be open and connected. If $f : U \to \mathbb{C}$ is holomorphic and non-constant, then $|f|$ has no local maximum in U. If U is bounded and f can be extended to a continuous function $f : \overline{U} \to \mathbb{C}$, then $|f|$ takes its maximal values on the boundary ∂U.

Identity theorem. If $f, g : U \to \mathbb{C}$ are two holomorphic functions on a connected open subset $U \subset \mathbb{C}$ such that $f(z) = g(z)$ for all z in a non-empty open subset $V \subset U$, then $f = g$. There are stronger versions of the identity theorem, but in this form it immediately generalizes to higher dimensions.

Riemann extension theorem. Let $f : B_\varepsilon(z_0) \setminus \{z_0\} \to \mathbb{C}$ be a bounded holomorphic function. Then f can be extended to a holomorphic function $f : B_\varepsilon(0) \to \mathbb{C}$.

Riemann mapping theorem. Let $U \subset \mathbb{C}$ be a simply connected proper open subset. Then U is biholomorphic to the unit ball $B_1(0)$, i.e. there exists a bijective holomorphic map $f : U \to B_1(0)$ such that its inverse f^{-1} is also holomorphic.

Liouville. Every bounded holomorphic function $f : \mathbb{C} \to \mathbb{C}$ is constant. In particular, there is no biholomorphic map between $\mathbb{C}$ and a ball $B_\varepsilon(0)$ with $\varepsilon < \infty$. This is a striking difference to the real situation and will cause a different concept of locality for complex manifolds than the one we are used to from real differential geometry.

Residue theorem. Let $f : B_\varepsilon(0) \setminus \{0\} \to \mathbb{C}$ be a holomorphic function. Then f can be expanded in a Laurent series $f(z) = \sum_{n=-\infty}^{\infty} a_n z^n$ and the coefficient a_{-1} is given by the residue formula $a_{-1} = (1/2\pi i) \int_{|z|=\varepsilon/2} f(z) dz$. The residue theorem is usually applied to more general situations where the function f has several isolated singularities in a connected open subset and the integral is taken over a closed contractible path surrounding the singularities.

The notion of a holomorphic function of one variable can be extended in two different ways. Firstly, one can consider functions of several variables $\mathbb{C}^n \to \mathbb{C}$ and, secondly, functions that take values in $\mathbb{C}^n$. As a basis for the topology in higher dimensions we will usually take the *polydiscs* $B_\varepsilon(w) = \{z \mid |z_i - w_i| < \varepsilon_i\}$, where $\varepsilon := (\varepsilon_1, \ldots, \varepsilon_n)$.

Definition 1.1.1 Let $U \subset \mathbb{C}^n$ be an open subset and let $f : U \to \mathbb{C}$ be a continuously differentiable function. Then f is said to be *holomorphic* if the Cauchy–Riemann equations (1.2) holds for all coordinates $z_i = x_i + iy_i$, i.e.

$$\frac{\partial u}{\partial x_i} = \frac{\partial v}{\partial y_i}, \quad \frac{\partial u}{\partial y_i} = -\frac{\partial v}{\partial x_i}, \quad i = 1, \ldots, n. \tag{1.5}$$

(It should be clear that i appears with two different meanings here, as an index and as $\sqrt{-1}$. This is a bit unfortunate, but it will always be clear which one is meant.)

By definition, a continuous(ly differentiable) function f is holomorphic if the induced functions

$$U \cap \{(z_1, \ldots, z_{i-1}, z, z_{i+1}, \ldots, z_n) \mid z \in \mathbb{C}\} \to \mathbb{C}$$

are holomorphic for all choices of i and fixed $z_1, \ldots, z_{i-1}, z_{i+1}, \ldots, z_n \in \mathbb{C}$. Introducing

$$\frac{\partial}{\partial z_i} := \frac{1}{2}\left(\frac{\partial}{\partial x_i} - i\frac{\partial}{\partial y_i}\right) \quad \text{and} \quad \frac{\partial}{\partial \bar{z}_i} := \frac{1}{2}\left(\frac{\partial}{\partial x_i} + i\frac{\partial}{\partial y_i}\right),$$

(1.5) can be rewritten as

$$\frac{\partial f}{\partial \bar{z}_i} = 0 \quad \text{for} \quad i = 1, \ldots, n. \tag{1.6}$$

Sometimes all these equations together are written as $\bar{\partial} f = 0$. Later in Section 1.3, a precise meaning will be given to this equation.

The comparison between real and complex Jacobian can be carried over to several variables. This will be discussed shortly.

But before, let us discuss the Cauchy integral formula for functions of several variables and a few central results.

Proposition 1.1.2 *Let* $f : \overline{B_\varepsilon(w)} \to \mathbb{C}$ *be a continuous function such that* f *is holomorphic with respect to every single component* z_i *in any point of* $B_\varepsilon(w)$. *Then for any* $z \in B_\varepsilon(w)$ *the following formula holds true*

$$f(z) = \frac{1}{(2\pi i)^n} \int_{|\xi_i - w_i| = \varepsilon_i} \frac{f(\xi_1, \ldots, \xi_n)}{(\xi_1 - z_1) \ldots (\xi_n - z_n)} d\xi_1 \ldots d\xi_n. \qquad (1.7)$$

Proof. Repeated application of the Cauchy integral formula in one variable yields

$$f(z) = \frac{1}{(2\pi i)^n} \int_{|\xi_1 - w_1| = \varepsilon_1} \cdots \int_{|\xi_n - w_n| = \varepsilon_n} \frac{f(\xi_1, \ldots, \xi_n)}{(\xi_1 - z_1) \ldots (\xi_n - z_n)} d\xi_n \ldots d\xi_1.$$

Since the integrand is continuous on the boundary of $B_\varepsilon(w)$, the iterated integral can be replaced by the multiple integral. This proves the assertion. $\square$

The proposition can easily be applied to show that any continuous(!) function on an open subset $U \subset \mathbb{C}^n$ with the property that the function is holomorphic with respect to any single coordinate is holomorphic itself (Osgood's Lemma, cf. [64]). Clearly the integrand in the above integral is holomorphic as a function of $\xi = (\xi_1, \ldots, \xi_n)$.

As in the one-dimensional case, the integral formula (1.7) can be used to write down a power series expansion of any holomorphic function $f : U \to \mathbb{C}$. More precisely, for any $w \in U$ there exists a polydisc $B_\varepsilon(w) \subset U \subset \mathbb{C}^n$ such that the restriction of f to $B_\varepsilon(w)$ is given by a power series

$$\sum_{i_1, \ldots, i_n = 0}^{\infty} a_{i_1 \ldots i_n} (z_1 - w_1)^{i_1} \ldots (z_n - w_n)^{i_n},$$

with

$$a_{i_1 \ldots i_n} = \frac{1}{i_1! \ldots i_n!} \cdot \frac{\partial^{i_1 + \ldots + i_n} f}{\partial z_1^{i_1} \ldots \partial z_n^{i_n}}.$$

From the above list the maximum principle, the identity theorem, and the Liouville theorem generalize easily to the higher dimensional situation. A version of the Riemann extension theorem holds true, although the proof needs some work. The Riemann mapping theorem definitely fails (see Exercise 1.1.16). There are also some new unexpected features in dimension > 1, e.g. Hartogs' theorem (see Proposition 1.1.4).

Often the holomorphicity of a function of several variables is shown by representing the function as an integral, using residue theorem or Cauchy integral formula, of a function which is known to be holomorphic. For later use we state this principle as a separate lemma.

Lemma 1.1.3 *Let $U \subset \mathbb{C}^n$ be an open subset and let $V \subset \mathbb{C}$ be an open neighbourhood of the boundary of $B_\varepsilon(0) \subset \mathbb{C}$. Assume that $f : V \times U \to \mathbb{C}$ is a holomorphic function. Then*

$$g(z) := g(z_1, \ldots, z_n) := \int_{|\xi| = \varepsilon} f(\xi, z_1, \ldots, z_n) d\xi$$

is a holomorphic function on U.

Proof. Let $z \in U$. If $|\xi| = \varepsilon$ then there exists a polydisc $B_{\delta(\xi)}(\xi) \times B_{\delta'(\xi)}(z) \subset V \times U$ on which f has a power series expansion.

Since $\partial B_\varepsilon(0)$ is compact, we can find a finite number of points $\xi_1, \ldots, \xi_k \in \partial B_\varepsilon(0)$ and positive real numbers $\delta(\xi_1), \ldots, \delta(\xi_k)$ such that

$$\bigcup \big(\partial B_\varepsilon(0) \cap B_{\delta(\xi_i)/2}(\xi_i) \big) \quad \text{is a disjoint union}$$

and

$$\partial B_\varepsilon(0) = \bigcup \left(\partial B_\varepsilon(0) \cap \overline{B_{\delta(\xi_i)/2}(\xi_i)} \right).$$

Hence, $g(z) = \int_{|\xi| = \varepsilon} f(\xi, z_1, \ldots, z_n) d\xi = \sum_{i=1}^k \int_{|\xi| = \varepsilon, |\xi_i - \xi| < \delta(\xi_i)/2} f d\xi$. Each summand is holomorphic, as the power series expansion of f converges uniformly on $\overline{B_{\delta(\xi_i)/2}(\xi_i)}$ and thus commutes with the integral. $\square$

The next result is only valid in dimension at least two.

Proposition 1.1.4 (Hartogs' theorem) *Suppose $\varepsilon = (\varepsilon_1, \ldots, \varepsilon_n)$ and $\varepsilon' = (\varepsilon'_1, \ldots, \varepsilon'_n)$ are given such that for all i one has $\varepsilon'_i < \varepsilon_i$. If $n > 1$ then any holomorphic map $f : B_\varepsilon(0) \setminus \overline{B_{\varepsilon'}(0)} \to \mathbb{C}$ be can can be uniquely extended to a holomorphic map $f : B_\varepsilon(0) \to \mathbb{C}$.*

Proof. We may assume that $\varepsilon = (1, \ldots, 1)$. Moreover, there exists $\delta > 0$ such that the open subset $V := \{z \mid 1 - \delta < |z_1| < 1, |z_{i \neq 1}| < 1\} \cup \{z \mid 1 - \delta < |z_2| < 1, |z_{i \neq 2}| < 1\}$ is contained in the complement of $B_{\varepsilon'}(0)$.

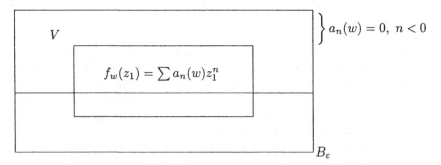

In particular, f is holomorphic on V. Thus, for any $w := (z_2, \ldots, z_n)$ with $|z_i| < 1$ this yields a holomorphic function $f_w(z_1) := f(z_1, z_2, \ldots, z_n)$ on the annulus $1 - \delta < |z_1| < 1$. Let $f_w(z_1) = \sum_{n=-\infty}^{\infty} a_n(w) z_1^n$ be the Laurent series of this function. Then $a_n(w) = (1/2\pi i) \int_{|\xi|=1-\delta/2} (f_w(\xi)/\xi^{n+1}) d\xi$ and by Lemma 1.1.3 this function is holomorphic for w in the unit polydisc of $\mathbb{C}^{n-1}$.

On the other hand, the function $z_1 \mapsto f_w(z_1)$ is holomorphic on the unit disc for fixed w such that $1 - \delta < |z_2| < 1$. Thus, $a_n(w) = 0$ for $n < 0$ and $1 - \delta < |z_2| < 1$. By the identity principle this yields $a_n(w) \equiv 0$ for $n < 0$. But then we can define the holomorphic extension $\tilde{f}$ of f by the power series $\sum_{n=0}^{\infty} a_n(w) z_1^n$. This series converges uniformly, as the $a_n(w)$ are holomorphic and thus attain their maximum at the boundary. So the convergence of the Laurent series on the annulus yields the uniform convergence everywhere. Clearly, the holomorphic function given by this series glues with f to give the desired holomorphic function.   □

Of course, the theorem definitely fails for holomorphic functions of one variable. The informal reason for the theorem is that the singularities of f are given by the vanishing of a holomorphic function. But the zero set of a single holomorphic function would 'stick out' of the smaller disc. As usual the result can easily be generalized to other situations, e.g. when the discs are not centered at the origin.

Next we will prove the Weierstrass preparation theorem (WPT) which is an important technical tool in the theory of functions of several complex variables. Let $f : B_\varepsilon(0) \to \mathbb{C}$ be a holomorphic function on the polydisc $B_\varepsilon(0)$. For any $w = (z_2, \ldots, z_n)$ we denote by $f_w(z_1)$ the function $f(z_1, z_2, \ldots, z_n)$. We will show that all the zeros of f are caused by a factor of f which has the form of a Weierstrass polynomial.

Definition 1.1.5 A *Weierstrass polynomial* is a polynomial in z_1 of the form

$$z_1^d + \alpha_1(w) z_1^{d-1} + \ldots + \alpha_d(w)$$

where the coefficients $\alpha_i(w)$ are holomorphic functions on some small disc in $\mathbb{C}^{n-1}$ vanishing at the origin.

Before stating the result, let us recall that any holomorphic function $f(z)$ in one variable with a zero of order d at the origin can be written as $z^d \cdot h(z)$ with $h(0) \neq 0$. If we let this decomposition depend on extra parameters, then the polynomial z^d becomes an arbitrary polynomial of degree d whose coefficients depend on the parameter. This is due to the fact that a zero of order d of $f_0(z_1)$ might deform to a collection of zeros of $f_w(z_1)$ whose orders sum up to d.

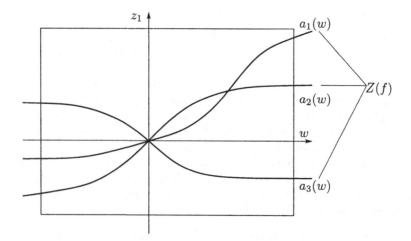

Proposition 1.1.6 (Weierstrass preparation theorem) *Let $f : B_\varepsilon(0) \to \mathbb{C}$ be a holomorphic function on the polydisc $B_\varepsilon(0)$. Assume $f(0) = 0$ and $f_0(z_1) \not\equiv 0$. Then there exists a Weierstrass polynomial $g(z_1, w) = g_w(z_1)$ and a holomorphic function h on some smaller polydisc $B_{\varepsilon'}(0) \subset B_\varepsilon(0)$ such that $f = g \cdot h$ and $h(0) \neq 0$. The Weierstrass polynomial g is unique.*

Proof. Since $f_0 \not\equiv 0$, we find $\varepsilon_1 > 0$ such that f_0 in the closure of the disc of radius ε_1 vanishes only in 0. Let d be its multiplicity. Next we choose $\varepsilon_2, \ldots, \varepsilon_n > 0$ such that $f(z_1, \ldots, z_n) \neq 0$ for $|z_1| = \varepsilon_1$ and $|z_i| < \varepsilon_i$, $i = 2, \ldots, n$.

Let $a_1(w), \ldots, a_d(w)$ be the zeros of $f_w(z_1)$ in the disc of radius ε_1, where any zero occurs as often as determined by its multiplicity. In particular, $a_1(0) = \ldots = a_d(0) = 0$. At this point we do not know yet that the number of zeros is constant, i.e. that d does not depend on w. The argument will be given below.

The polynomial $g_w(z_1) := \prod_{i=1}^d (z_1 - a_i(w))$ has the same zeros as f_w (again with multiplicities). Thus, for fixed w the function $h_w(z_1) := f_w(z_1)/g_w(z_1)$ is holomorphic in z_1. It remains to show that the functions $g_w(z_1)$ and $h(z_1, w) = h_w(z_1)$ are holomorphic in w.

In order to see this, one first notes that the coefficients of the polynomial $g_w(z_1)$ can be written as polynomials in the expressions $\sum_{i=1}^d a_i(w)^k$, $k = 1, \ldots, d$. Thus, $g - w(z_1)$ is holomorphic in w if these sums are holomorphic in w. The latter can be seen by applying the residue theorem to the function $z_1^k f_w'(z_1)/f_w(z_1)$.

Let $f_w(\xi) = \sum_{j=m}^\infty \alpha_j (\xi - a)^j$ be the power series expansion of f_w in a zero a. Then $f_w'(\xi) = \sum_{j=m}^\infty j \cdot \alpha_j (\xi - a)^{j-1}$. Moreover, $\xi^k = a^k + ka^{k-1}(\xi - a) + \ldots$. We leave it to the reader to verify that this immediately yields

$$\operatorname{Res}_{\xi=a} \xi^k \frac{f_w'(\xi)}{f_w(\xi)} = ma^k.$$

Thus, we have

$$\sum_{i=1}^{d} a_i(w)^k = \frac{1}{2\pi i} \int_{|\xi|=\varepsilon_1} \xi^k \frac{f_w'(\xi)}{f_w(\xi)} d\xi.$$

The left hand side is a holomorphic function in w (use Lemma 1.1.3) and thus $g(z_1, w) := g_w(z_1)$ is holomorphic in $z_1, \ldots, z_n$. Note that for $k = 0$ the left hand side is just the number of zeros of f_w counted with multiplicities. So this integer ($= d$) depends holomorphically on w and therefore does not depend on w at all.

The complement of the set $\{(z_1, w) \mid z_1 = a_i(w) \text{ for some } i\}$ contains a neighbourhood of $\{(z_1, w) \mid |z_1| = \varepsilon_1, |z_{i>1}| < \varepsilon_i\}$. Thus, the Cauchy integral formula $h(z_1, w) = (1/2\pi i) \int_{|\xi|=\varepsilon_1} h(\xi, w)/(\xi - z_1) d\xi$ and the holomorphicity of f/g show that h is holomorphic everywhere (again by Lemma 1.1.3).

The uniqueness of the Weierstrass polynomial g is clear: Since $h(0) \neq 0$, we can assume that h does not vanish anywhere and thus f_w and g_w have the same zero sets. But the only Weierstrass polynomial with this property is the polynomial constructed above. $\qquad\square$

As a short hand, we will denote the *zero set* of a holomorphic function f by $Z(f)$, i.e. $Z(f) := \{z \mid f(z) = 0\}$.

Proposition 1.1.7 (Riemann extension theorem) *Let f be a holomorphic function on an open subset $U \subset \mathbb{C}^n$. If $g : U \setminus Z(f) \to \mathbb{C}$ is holomorphic and locally bounded near $Z(f)$, then g can uniquely be extended to a holomorphic function $\tilde{g} : U \to \mathbb{C}$.*

Proof. Before launching into the proof let us consider the following special case: $n = 2$ and $f(z) = z_1$. Then $g_{z_2}(z_1) := g(z_1, z_2)$ is a bounded holomorphic function on a punctured disc in the complex plane. Thus, by the usual Riemann extension theorem one finds an extension of g_{z_2} to a holomorphic function on the whole disc. It then remains to show that all these functions glue together.

For the general case we may assume that $U = B_\varepsilon(0)$ and that $Z(f)$ does not contain the intersection of U with the line $\{(z_1, 0, \ldots, 0) \mid z_1 \in \mathbb{C}\}$. Furthermore, we can restrict to the case that the restriction f_0 of f to this line vanishes only in the origin. Thus, $f_0(z_1) \neq 0$ for $|z_1| = \varepsilon_1/2$ and modifying $\varepsilon_2, \ldots, \varepsilon_n > 0$ if necessary we can also assume that $f(z) \neq 0$ if $|z_1| = \varepsilon_1/2$ and $|z_i| < \varepsilon_i/2$ for $i = 2, \ldots, n$. In other words, for any w with $|z_i| < \varepsilon_i/2$, $i = 2, \ldots, n$ the function f_w has no zeros at the boundary of $B_{\varepsilon_1/2}(0)$.

By assumption, the restriction g_w of g to $B_{\varepsilon_1/2}(0) \setminus Z(f_w)$ is bounded and can thus be extended to a holomorphic function $\tilde{g}_w$ on $B_{\varepsilon_1/2}(0)$. By the Cauchy integral formula this extension is given by

$$\tilde{g}_w(z_1) = \frac{1}{2\pi i} \int_{\partial B_{\varepsilon_1/2}(0)} \frac{g_w(\xi)}{\xi - z_1} d\xi.$$

Since $f_w(\xi) \neq 0$ for any $\xi \in \partial B_{\varepsilon_1/2}(0)$, the integrand is holomorphic in (z_1, w). By Lemma 1.1.3 this shows that $\tilde{g}(z_1, w) := \tilde{g}_w(z_1)$ is holomorphic in (z_1, w). This yields the desired holomorphic extension of g. □

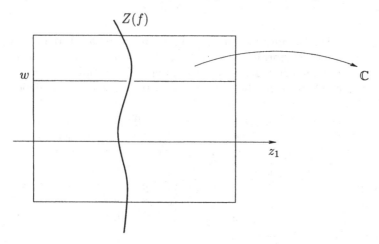

The result also proves that a locally bounded holomorphic function on the complement of a *thin* subset can be uniquely extended. Here, a thin subset is a subset which locally is contained in the zero set of a non-trivial holomorphic function.

Extending the notion of holomorphicity to functions f that take values in $\mathbb{C}^n$ is purely formal. Similarly to the definition of a differentiable function with values in $\mathbb{R}^n$ this is done as follows.

Definition 1.1.8 Let $U \subset \mathbb{C}^m$ be an open subset. A function $f : U \to \mathbb{C}^n$ is called *holomorphic* if all coordinate functions $f_1, \ldots, f_n$ are holomorphic functions $U \to \mathbb{C}$.

In analogy to the one-dimensional case one says that a map $f : U \to V$ between two open subsets $U, V \subset \mathbb{C}^n$ is *biholomorphic* if and only if f is bijective and holomorphic and its inverse $f^{-1} : V \to U$ is holomorphic as well.

Definition 1.1.9 Let $U \subset \mathbb{C}^m$ be an open subset and let $f : U \to \mathbb{C}^n$ be a holomorphic map. The (complex) *Jacobian* of f at a point $z \in U$ is the matrix

$$J(f)(z) := \left(\frac{\partial f_i}{\partial z_j}(z) \right)_{\substack{1 \leq i \leq n \\ 1 \leq j \leq m}}.$$

A point $z \in U$ is called *regular* if $J(f)(z)$ is surjective. If every point $z \in f^{-1}(w)$ is regular then w is called a *regular value*.

As in the one-dimensional case, it is useful to relate the complex Jacobian $J(f)$ to the real one. This goes as follows. The differentiable map

$f : U \subset \mathbb{C}^m = \mathbb{R}^{2m} \to \mathbb{C}^n = \mathbb{R}^{2n}$ induces for $z \in U$ the $\mathbb{R}$-linear map $df(z) :$ $T_z\mathbb{R}^{2m} \to T_{f(z)}\mathbb{R}^{2n}$. With respect to the bases $\langle \frac{\partial}{\partial x_1}, \ldots, \frac{\partial}{\partial x_m}, \frac{\partial}{\partial y_1}, \ldots, \frac{\partial}{\partial y_m} \rangle$ and $\langle \frac{\partial}{\partial r_1}, \ldots, \frac{\partial}{\partial r_n}, \frac{\partial}{\partial s_1}, \ldots, \frac{\partial}{\partial s_n} \rangle$ the linear map $df(z)$ is given by the real Jacobian

$$
J_{\mathbb{R}}(f) = \begin{pmatrix} \left(\dfrac{\partial u_i}{\partial x_j} \right)_{i,j} & \left(\dfrac{\partial u_i}{\partial y_j} \right)_{i,j} \\[2ex] \left(\dfrac{\partial v_i}{\partial x_j} \right)_{i,j} & \left(\dfrac{\partial v_i}{\partial y_j} \right)_{i,j} \end{pmatrix}.
$$

The $\mathbb{C}$-linear extension $df(z)_{\mathbb{C}} : T_z\mathbb{R}^{2m} \otimes \mathbb{C} \to T_{f(z)}\mathbb{R}^{2n} \otimes \mathbb{C}$ with respect to the basis $\langle \frac{\partial}{\partial z_1}, \ldots, \frac{\partial}{\partial z_m}, \frac{\partial}{\partial \bar{z}_1}, \ldots, \frac{\partial}{\partial \bar{z}_m} \rangle$ and $\langle \frac{\partial}{\partial w_1}, \ldots, \frac{\partial}{\partial w_n}, \frac{\partial}{\partial \bar{w}_1}, \ldots, \frac{\partial}{\partial \bar{w}_n} \rangle$ is given by

$$
\begin{pmatrix} \left(\dfrac{\partial f_i}{\partial z_j} \right)_{i,j} & \left(\dfrac{\partial f_i}{\partial \bar{z}_j} \right)_{i,j} \\[2ex] \left(\dfrac{\partial \bar{f}_i}{\partial z_j} \right)_{i,j} & \left(\dfrac{\partial \bar{f}_i}{\partial \bar{z}_j} \right)_{i,j} \end{pmatrix} \quad \text{and for } f \text{ holomorphic by} \quad \begin{pmatrix} J(f) & 0 \\ 0 & \overline{J(f)} \end{pmatrix}.
$$

In particular, for a holomorphic function f one has $\det J_{\mathbb{R}}(f) = \det J(f) \cdot \det \overline{J(f)} = |\det J(f)|^2$, which is non-negative.

In analogy to the implicit function theorem and the inverse function theorem, for real functions (or rather using these two results), one has the following standard facts.

Proposition 1.1.10 (Inverse function theorem) *Let $f : U \to V$ be a holomorphic map between two open subsets $U, V \subset \mathbb{C}^n$. If $z \in U$ is regular then there exist open subsets $z \in U' \subset U$ and $f(z) \in V' \subset V$ such that f induces a biholomorphic map $f : U' \to V'$.*

Proposition 1.1.11 (Implicit function theorem) *Let $U \subset \mathbb{C}^m$ be an open subset and let $f : U \to \mathbb{C}^n$ be a holomorphic map, where $m \geq n$. Suppose $z_0 \in U$ is a point such that*

$$
\det \left(\frac{\partial f_i}{\partial z_j}(z_0) \right)_{1 \leq i,j \leq n} \neq 0.
$$

Then there exist open subsets $U_1 \subset \mathbb{C}^{m-n}, U_2 \subset \mathbb{C}^n$ and a holomorphic map $g : U_1 \to U_2$ such that $U_1 \times U_2 \subset U$ and $f(z) = f(z_0)$ if and only if $g(z_{n+1}, \ldots, z_m) = (z_1, \ldots, z_n)$.

Proof. Using the relation between complex and real Jacobian explained above, one finds that z is regular if and only if $\det J_{\mathbb{R}}(f)(z) \neq 0$, i.e. z is a regular point for the underlying real map. Thus, the real inverse function theorem applies and one finds a continuously differentiable inverse $f^{-1} : V' \to U' \subset U$ of f. It suffices to show that f^{-1} satisfies the Cauchy–Riemann equation. Clearly, $\partial(f^{-1} \circ f)/\partial \bar{z}_j = 0$ and, since on the other hand f is holomorphic,

$$
\begin{aligned}
\frac{\partial(f^{-1} \circ f)}{\partial \bar{z}_j} &= \sum_{k=1}^{n} \frac{\partial f^{-1}}{\partial w_k} \cdot \frac{\partial f_k}{\partial \bar{z}_j} + \frac{\partial f^{-1}}{\partial \bar{w}_k} \cdot \frac{\partial \bar{f}_k}{\partial \bar{z}_j} \\
&= \sum_{k=1}^{n} \frac{\partial f^{-1}}{\partial \bar{w}_k} \cdot \frac{\partial \bar{f}_k}{\partial \bar{z}_j}.
\end{aligned}
$$

Thus, $\left(\frac{\partial f^{-1}}{\partial \bar{w}_k}\right)_k \cdot \overline{J(f)} = 0$. Since $\det J_{\mathbb{R}}(f) \neq 0$ on U', this yields $\frac{\partial f^{-1}}{\partial \bar{w}_k} = 0$ on V' for all k, i.e. f^{-1} is holomorphic.

One could also simply use the fact that f^{-1} is holomorphic if and only if $d(f^{-1})$ is complex linear, where $T_x\mathbb{C}^n = T_x\mathbb{R}^{2n}$ is endowed with the natural complex structure. Since $d(f^{-1}) = (df)^{-1}$, this follows from the assumption that df is complex linear.

For the implicit function theorem we again use the real version of the theorem. Thus, the asserted function g exists, and it remains to show that g is holomorphic. Clearly, $\frac{\partial}{\partial \bar{z}_j}(f_i(g(z_{n+1}, \ldots, z_m), z_{n+1}, \ldots, z_m)) = 0$ for $n+1 \leq j \leq m$. On the other hand, the holomorphicity of f yields

$$
\begin{aligned}
&\frac{\partial}{\partial \bar{z}_j} \left(f_i(g(z_{n+1}, \ldots, z_m), z_{n+1}, \ldots, z_m)\right) \\
&= \frac{\partial f_i}{\partial \bar{z}_j}(g(z_{n+1}, \ldots, z_m), z_{n+1}, \ldots, z_m) + \sum_{k=1}^{n} \frac{\partial f_i}{\partial z_k} \cdot \frac{\partial g_k}{\partial \bar{z}_j} + \frac{\partial f_i}{\partial z_k} \cdot \frac{\partial \bar{g}_k}{\partial \bar{z}_j} \\
&= \sum_{k=1}^{n} \frac{\partial f_i}{\partial z_k} \cdot \frac{\partial g_k}{\partial \bar{z}_j}.
\end{aligned}
$$

Thus, $\left(\frac{\partial f_i}{\partial z_j}\right)_{1 \leq i,j \leq n} \cdot \left(\frac{\partial g_k}{\partial \bar{z}_j}\right)_{k=1,\ldots,n} = 0$ for all j. This implies $\partial g_k/\partial \bar{z}_j = 0$ for $1 \leq k, j \leq n$. Hence, the g_k are all holomorphic. $\qquad\square$

The proofs of the following facts are literally the same as in the real situation. They are left to the reader.

Corollary 1.1.12 *Let $U \subset \mathbb{C}^m$ be an open subset and let $f : U \to \mathbb{C}^n$ be a holomorphic map. Assume that $z_0 \in U$ such that $\mathrm{rk}(J(f)(z_0))$ is maximal.*

i) If $m \geq n$ then there exists a biholomorphic map $h : V \to U'$, where U' is an open subset of U containing z_0, such that $f(h(z_1, \ldots, z_m)) = (z_1, \ldots, z_n)$ for all $(z_1, \ldots, z_m) \in V$.

ii) If $m \leq n$ then there exists a biholomorphic map $g : V \to V'$, where V is an open subset of $\mathbb{C}^n$ containing $f(z_0)$, such that $g(f(z)) = (z_1, \ldots, z_m, 0, \ldots, 0)$. $\qquad\square$

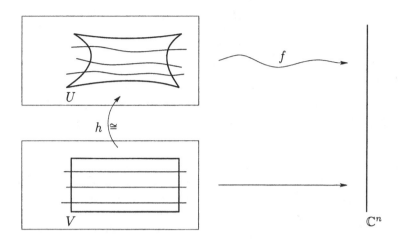

Contrary to the real world, a holomorphic map is biholomorphic if and only if it is bijective. So the regularity of the Jacobian follows from the bijectivity. This is

Proposition 1.1.13 *Let $f : U \to V$ be a bijective holomorphic map between two open subsets $U, V \subset \mathbb{C}^n$. Then for all $z \in U$ one has $\det J(f)(z) \neq 0$. In particular, f is biholomorphic.*

Proof. The proof proceeds by induction. For $n = 1$ this is standard, but for completeness sake we recall the argument. Suppose f' has a zero. After a suitable coordinate change, we can assume $f(0) = f'(0) = 0$. Then the power series expansion of f has the form $f(z) = z^d h(z)$, where $d \geq 2$ and $h(0) \neq 0$. In a small neighbourhood of 0 we can consider a d-th root $\sqrt[d]{h(z)}$. Then $w := z\sqrt[d]{h(z)}$ is a local coordinate. With respect to this coordinate f has the form $f(w) = w^d$. Hence, f is not injective. Contradiction.

Now let n be arbitrary and assume that the assertion is proven for all $k < n$. Let $z \in U$ such that $\det J(f)(z) = 0$. We will show that this implies $J(f)(z) = 0$. Suppose $\mathrm{rk}(J(f)(z)) = k \geq 1$. We may assume that $\left(\frac{\partial f_i}{\partial z_j}(z) \right)_{1 \leq i,j \leq k}$ is non-singular. By the inverse function theorem, $\tilde{z}_i := f_i(z)$ for $i = 1, \ldots, k$ and $\tilde{z}_i := z_i$ for $i = k+1, \ldots, n$ form a local coordinate system around $z \in U$. Clearly, f maps $U' = \{\tilde{z} \mid \tilde{z}_i = 0 \text{ for } i = 1, \ldots, k\} \cap U$ bijectively onto $\{w \mid w_i = 0 \text{ for } i = 1, \ldots, k\} \cap V$. But the Jacobian of the restriction of f to U' is singular at z. This contradicts the induction hypothesis. Therefore, $k = 0$. In other words, whenever the Jacobian $J(f)(z)$ is singular it vanishes completely.

Let $\det J(f)(z) = 0$ and assume that z is a regular point of the holomorphic function $\det J(f) : U \to \mathbb{C}$. A neighbourhood W of z in the fibre of the function $\det J(f)$ over $0 \in \mathbb{C}$ is biholomorphic to an open subset of $\mathbb{C}^{n-1}$ (Proposition 1.1.11), and for $n > 1$ this has positive dimension. This yields a function $f|_W : W \to \mathbb{C}^n$ with vanishing Jacobian everywhere, which is, therefore, constant. This contradicts the injectivity of f.

The proof is not complete yet as we still have to show that there always exists a regular point of $\det J(f)$ in the fibre over 0. We will come back to this later (see Remark 1.1.20). □

It is convenient to work throughout with the stalk of the sheaf of holomorphic functions. This way we avoid shrinking open neighbourhoods explicitly again and again. Let us introduce the necessary notations.

Definition 1.1.14 By $\mathcal{O}_{\mathbb{C}^n}$ we denote the *sheaf of holomorphic functions* on $\mathbb{C}^n$. Thus, for any open subset $U \subset \mathbb{C}^n$ the space of sections $\mathcal{O}_{\mathbb{C}^n}(U)$ of this sheaf over U is the set of all holomorphic functions $f : U \to \mathbb{C}$. The *stalk* $\mathcal{O}_{\mathbb{C}^n,z}$ of $\mathcal{O}_{\mathbb{C}^n}$ at a point $z \in \mathbb{C}^n$ is the set of all germs (U, f), where U is an arbitrary small open neighbourhood of z and f is a holomorphic function on U. (Compare this with the general definition of the stalk of a sheaf given in Appendix B.)

Clearly, two stalks at different points are isomorphic. Note that the partial derivatives $\frac{\partial}{\partial z_i}$ define $\mathbb{C}$-linear endomorphisms of $\mathcal{O}_{\mathbb{C}^n,0}$. We can also still speak about the zero set $Z(f)$ of an element $f \in \mathcal{O}_{\mathbb{C}^n,0}$, which here just means the germ of it (see Definition 1.1.21 or trust your own intuition for the moment). The ring $\mathcal{O}_{\mathbb{C}^n,0}$ is local and its maximal ideal $\mathfrak{m}$ consists of all functions that vanish in 0. In other words, the set of units $\mathcal{O}_{\mathbb{C}^n,0}^*$ consists of all functions f with $f(0) \neq 0$.

Using these notation the WPT can be rephrased by saying that after an appropriate coordinate choice any function $f \in \mathcal{O}_{\mathbb{C}^n,0}$ can be uniquely written as $f = g \cdot h$, where $h \in \mathcal{O}_{\mathbb{C}^n,0}$ is a unit and $g \in \mathcal{O}_{\mathbb{C}^{n-1},0}[z_1]$ is a Weierstrass polynomial. The WPT implies the following

Proposition 1.1.15 *The local ring $\mathcal{O}_{\mathbb{C}^n,0}$ is a UFD.*

Before giving the proof, let us recall the necessary definitions from commutative algebra.

Definition 1.1.16 Let R be an integral domain, i.e. R has no zero divisors. An element $f \in R$ is *irreducible* if it cannot be written as the product of two non-units in R. An integral domain is called *unique factorization domain* (UFD) if every element can be written as a product of irreducible ones and if the factors are unique up to reordering and multiplication with units.

For the proof we will need one basic fact from algebra: If R is a UFD, then also the polynomial ring $R[x]$ is a UFD (Gauss Lemma).

Proof. We prove the assumption by induction on n. For $n = 0$ the ring $\mathcal{O}_{\mathbb{C}^n,0} = \mathbb{C}$ is a field and thus a UFD. Suppose that $\mathcal{O}_{\mathbb{C}^{n-1},0}$ is a UFD. If $f \in \mathcal{O}_{\mathbb{C}^n,0}$ we can choose coordinates such that the WPT can be applied. Thus $f = g \cdot h$, where $g \in \mathcal{O}_{\mathbb{C}^{n-1},0}[z_1]$ is a Weierstrass polynomial and h is a unit in $\mathcal{O}_{\mathbb{C}^n,0}$. As an element of $\mathcal{O}_{\mathbb{C}^{n-1},0}[z_1]$ the Weierstrass polynomial g can be uniquely written as a product of irreducible elements $g_i \in \mathcal{O}_{\mathbb{C}^{n-1},0}[z_1]$. All that is left to show is that any irreducible factor g_i is also irreducible as an element in $\mathcal{O}_{\mathbb{C}^n,0}$.

Let us first show that any Weierstrass polynomial can be written as a product of irreducible Weierstrass polynomials (not every irreducible factor g_i need to be a Weierstrass polynomial). Assume that a Weierstrass polynomial $g \in \mathcal{O}_{\mathbb{C}^{n-1},0}[z_1]$ can be written as the product of non-units $g_i \in \mathcal{O}_{\mathbb{C}^{n-1},0}[z_1]$. Consider the decomposition of g_i according to the WPT $g_i = \tilde{g}_i \cdot h_i$ with Weierstrass polynomials $\tilde{g}_i \in \mathcal{O}_{\mathbb{C}^{n-1},0}[z_1]$. Note that since g is a Weierstrass polynomial, all the factors g_i are non-trivial on the z_1-line and thus satisfy the hypothesis of the Weierstrass preparation theorem. Then $g = \prod \tilde{g}_i \cdot \prod h_i$ and by the uniqueness assertion of the WPT $g = \prod \tilde{g}_i$ and $\prod h_i = 1$. A priori, the $\tilde{g}_i$ need not be irreducible in $\mathcal{O}_{\mathbb{C}^{n-1},0}[z_1]$ as the h_i are just units in $\mathcal{O}_{\mathbb{C}^n,0}$. But, since the degree of g as a polynomial in z_1 is finite, repeating the process leads to a decomposition, where either the $\tilde{g}_i$ are irreducible Weierstrass polynomials or elements in $\mathcal{O}_{\mathbb{C}^{n-1},0}$. To the latter we can apply the induction hypothesis.

We conclude by showing that any irreducible Weierstrass polynomial g is actually irreducible as an element of $\mathcal{O}_{\mathbb{C}^n,0}$. Suppose $g = f_1 \cdot f_2$, where $f_1, f_2 \in \mathcal{O}_{\mathbb{C}^n,0}$ are non-units. We apply the WPT to both functions. Hence $f_i = g_i \cdot h_i$ for $i = 1, 2$ and thus $g = (g_1 \cdot g_2) \cdot (h_1 \cdot h_2)$. By the uniqueness part of the WPT, this yields $g = g_1 \cdot g_2$, which contradicts the irreducibility of g as an element of $\mathcal{O}_{\mathbb{C}^{n-1},0}[z_1]$. $\square$

It will be important later on to know that the ring of holomorphic functions in the origin is noetherian. For this we will need the Weierstrass division theorem.

Proposition 1.1.17 (Weierstrass division theorem) *Let $f \in \mathcal{O}_{\mathbb{C}^n,0}$ and let $g \in \mathcal{O}_{\mathbb{C}^{n-1},0}[z_1]$ be a Weierstrass polynomial of degree d. Then there exist $r \in \mathcal{O}_{\mathbb{C}^{n-1},0}[z_1]$ of degree $< d$ and $h \in \mathcal{O}_{\mathbb{C}^n,0}$ such that $f = g \cdot h + r$. The functions h and r are uniquely determined.*

Before giving the proof let us consider the case $n = 1$. If $f(z) = \sum_{j=0}^{\infty} a_j z^j$ and $g = z^d$ the assertion is equivalent to the obvious equation $f = (z^d)(\sum_{j=d}^{\infty} a_j z^{j-d}) + \sum_{j=0}^{d-1} a_j z^j$, i.e. $h := \sum_{j=d}^{\infty} a_j z^{j-d}$ and $r := \sum_{j=0}^{d-1} a_j z^j$. Since h is allowed to be a power series, the claimed division by g is still possible when g is not just z^d, but an arbitrary polynomial in z.

Proof. The uniqueness is easy. Assume $f = g \cdot h_1 + r_1 = g \cdot h_2 + r_2$. Then $r_1 - r_2 = g \cdot (h_2 - h_1)$. For $w = (z_2, \ldots, z_n)$ we consider the function $g_w(z_1) := g(z_1, z_2, \ldots, z_n)$. Since g is a Weierstrass polynomial, g_w for $w = (0, \ldots, 0)$ has a zero of order d. Thus, for any small w the polynomial g_w has d zeros (counted with multiplicities) close to 0. Then the same must hold for $(r_1 - r_2)_w$. But the latter is a polynomial of degree $< d$. Thus $(r_1 - r_2)_w \equiv 0$ for generic w and hence $r_1 = r_2$.

To prove the existence of h and r we first define h by

$$h(z_1, \ldots, z_n) := \frac{1}{2\pi i} \int_{|\xi|=\varepsilon_1} \frac{f_w(\xi)}{g_w(\xi) \cdot (\xi - z_1)} d\xi,$$

where $|z_1| < \varepsilon_1$. For ε_i, $i = 1, \ldots, n$, small enough we may assume that $g_w(\xi) \neq 0$ for any $|\xi| = \varepsilon_1$ and $|z_i| < \varepsilon_i$ for $i = 2, \ldots, n$. Thus, h is holomorphic by Lemma 1.1.3. All that remains to show is that $r := f - gh$ is a polynomial in $\mathcal{O}_{\mathbb{C}^{n-1},0}[z_1]$ of degree $< d$. This is shown by the following straightforward computation.

$$
\begin{aligned}
&r(z_1, \ldots, z_n) \\
&= \frac{1}{2\pi i} \int_{|\xi|=\varepsilon_1} \frac{f_w(\xi)}{\xi - z_1} d\xi - \frac{g_w(z_1)}{2\pi i} \int_{|\xi|=\varepsilon_1} \frac{f_w(\xi)}{g_w(\xi) \cdot (\xi - z_1)} d\xi \\
&= \frac{1}{2\pi i} \int_{|\xi|=\varepsilon_1} \frac{f_w(\xi) \cdot (g_w(\xi) - g_w(z_1))}{g_w(\xi) \cdot (\xi - z_1)} d\xi \\
&= \frac{1}{2\pi i} \int_{|\xi|=\varepsilon_1} \frac{f_w(\xi)}{g_w(\xi)} \cdot \left(\frac{(\xi^d - z_1^d) + \alpha_1(w) \cdot (\xi^{d-1} - z_1^{d-1}) + \ldots}{(\xi - z_1)} \right) d\xi \\
&= \frac{1}{2\pi i} \int_{|\xi|=\varepsilon_1} \frac{f_w(\xi)}{g_w(\xi)} \cdot (z_1^{d-1} \beta_1(\xi, w) + z_1^{d-2} \beta_2(\xi, w) + \ldots) \, d\xi.
\end{aligned}
$$

Here, the $\alpha_i(w)$ are the coefficients of $g \in \mathcal{O}_{\mathbb{C}^{n-1},0}[z_1]$ and the $\beta_i(\xi, w)$ are determined by them. The integrand in the last integral is a polynomial in z_1 of order $< d$ whose coefficients are holomorphic functions in (ξ, w).   □

Proposition 1.1.18 *The local UFD $\mathcal{O}_{\mathbb{C}^n,0}$ is noetherian.*

Proof. We have to show that any ideal in $\mathcal{O}_{\mathbb{C}^n,0}$ is finitely generated. We use again induction on n. The case $n = 0$ is trivial, as any field is noetherian.

Next, we assume that $\mathcal{O}_{\mathbb{C}^{n-1},0}$ is noetherian. Then, also the polynomial ring $\mathcal{O}_{\mathbb{C}^{n-1},0}[x]$ is noetherian (cf. [2]). Let $I \subset \mathcal{O}_{\mathbb{C}^n}$ be a non-trivial ideal and choose $0 \neq f \in I$. Changing coordinates if necessary, we can assume that the WPT can be applied, i.e. $f = g \cdot h$ with $g \in \mathcal{O}_{\mathbb{C}^{n-1},0}[z_1]$ a Weierstrass polynomial and $h \in \mathcal{O}_{\mathbb{C}^n,0}$ a unit. Hence $g \in I$. The ideal $I \cap \mathcal{O}_{\mathbb{C}^{n-1},0}[z_1]$ in $\mathcal{O}_{\mathbb{C}^{n-1},0}[z_1]$ is generated by finitely many elements $g_1, \ldots, g_k$.

For any other $\tilde{f} \in I$ the Weierstrass division theorem yields $\tilde{f} = g \cdot \tilde{h} + r$ for some $r \in \mathcal{O}_{\mathbb{C}^{n-1},0}[z_1]$. Since $\tilde{f}, g \cdot \tilde{h} \in I$, also $r \in I$ and therefore $r \in I \cap \mathcal{O}_{\mathbb{C}^{n-1},0}[z_1]$. Thus, $\tilde{f} = g \cdot \tilde{h} + \sum_{i=1}^{k} a_i \cdot g_i$. This shows that I is finitely generated by the elements $g, g_1, \ldots, g_k$.   □

Corollary 1.1.19 *Let $g \in \mathcal{O}_{\mathbb{C}^n,0}$ be an irreducible function. If $f \in \mathcal{O}_{\mathbb{C}^n,0}$ vanishes on $Z(g)$, then g divides f.*

Proof. By the WPT (Proposition 1.1.6) we may assume that $g \in \mathcal{O}_{\mathbb{C}^{n-1},0}[z_1]$ is a Weierstrass polynomial of degree d. By the Weierstrass division theorem (Proposition 1.1.17) one finds $h \in \mathcal{O}_{\mathbb{C}^n,0}$ and $r \in \mathcal{O}_{\mathbb{C}^{n-1},0}[z_1]$ of degree $< d$ such that $f = g \cdot h + r$. By assumption r_w vanishes on the zero set of g_w. If the zeros of g_w for generic w had multiplicity one, then this would yield $r_w \equiv 0$, as r_w is of degree $< d$. All we have to show is that the set of $w \in \mathbb{C}^{n-1}$ such that g_w has zeros with multiplicity > 1 is contained in the zero set of a non-trivial holomorphic function $\gamma \in \mathcal{O}_{\mathbb{C}^{n-1},0}$. Once γ is found, one concludes by using the easy fact that the complement of the zero set of γ is dense (cf. Exercise 1.1.8).

Since g is irreducible and $\frac{\partial g}{\partial z_1}$ is of degree $d - 1$, there exist elements $h_1, h_2 \in \mathcal{O}_{\mathbb{C}^{n-1},0}[z_1]$ and $0 \neq \gamma \in \mathcal{O}_{\mathbb{C}^{n-1},0}$ such that $h_1 \cdot g + h_2 \cdot \frac{\partial g}{\partial z_1} = \gamma$ (Gauss lemma). If g_w has a zero ξ of multiplicity > 1, then $\gamma(\xi) = h_1(\xi, w) \cdot g_w(\xi) + h_2(\xi, w) \cdot \frac{\partial g_w}{\partial z_1}(\xi) = 0$. $\square$

Remark 1.1.20 Let $g \in \mathcal{O}_{\mathbb{C}^n,0}$ be irreducible. The arguments above show that the holomorphic function $g : U \to \mathbb{C}$ admits regular points in the fibre over $0 \in \mathbb{C}$. Indeed, if $g = \tilde{g} \cdot h$ is the decomposition according to the WPT, then locally $g^{-1}(0) = \tilde{g}^{-1}(0)$. Since $\frac{\partial g}{\partial z_i}(z) = \frac{\partial \tilde{g}}{\partial z_i}(z)h(z)$ for all $z \in g^{-1}(0)$, one knows that in $g^{-1}(0)$ the g-regular points coincide with the $\tilde{g}$-regular points. Thus, we may assume that g is an irreducible Weierstrass polynomial. Then, as above, one proves that $\frac{\partial g}{\partial z_1}$ cannot be trivial on $Z(g)$. Thus, g-regular points exist.

Let us now consider an arbitrary holomorphic function $g : U \to \mathbb{C}$. A priori, the fibre $Z(g) = g^{-1}(0)$ might not contain any g-regular point (e.g. $g = z^2$). Fix a point $z \in g^{-1}(0)$. For simplicity assume $z = 0$. Write $g = \prod g_i^{n_i}$, where the $g_i \in \mathcal{O}_{\mathbb{C}^n}$ are relatively prime and irreducible. Hence, $Z(g) = \bigcup Z(g_i^{n_i}) = \bigcup Z(g_i)$. Note that two zero sets $Z(g_i)$ might intersect each other, but according to the corollary they cannot be contained in each other for $i \neq j$. Now we can apply what was explained before to each $Z(g_i)$. So, $Z(g)$ might not contain any g-regular point, but every component $Z(g_i)$ contains g_i-regular points.

Let us apply this to the situation considered in the proof of Proposition 1.1.13. There we wanted to show that the fibre of the holomorphic map $g = \det J(f) : U \to \mathbb{C}$ contains a positive dimensional W. If there exist g-regular points in the fibre $g^{-1}(0)$, then W exists by the implicit function theorem 1.1.11. Using the above arguments we see that any component $Z(g_i)$, where $g = \prod g_i^{n_i}$ is the prime factor decomposition of g, contains such a set W. This completes the proof of Proposition 1.1.13.

By now, the reader should be convinced that working with the local ring $\mathcal{O}_{\mathbb{C}^n,0}$ has many advantages when one is interested only in local properties of holomorphic functions. In some of the above arguments we treated elements $f \in \mathcal{O}_{\mathbb{C}^n,0}$ as honest functions and associated to them their zero sets $Z(f)$. In the following we will make this more rigorous. This leads us to the notion of germs of analytic varieties in a point of $\mathbb{C}^n$. As usual, we will choose the origin as the distinguished point.

Definition 1.1.21 The *germ* of a set in the origin $0 \in \mathbb{C}^n$ is given by a subset $X \subset \mathbb{C}^n$. Two subsets $X, Y \subset \mathbb{C}^n$ define the same germ if there exists an open neighbourhood $0 \in U \subset \mathbb{C}^n$ with $U \cap X = U \cap Y$.

Sometimes, one also writes $(X, 0)$ for a germ of a set in the origin. Let $f \in \mathcal{O}_{\mathbb{C}^n, 0}$. Then we denote by $Z(f)$ the germ of the zero set of f, i.e. if f is represented by a holomorphic function $f : U \to \mathbb{C}$, then $Z(f)$ is represented by the zero set of this holomorphic function. Clearly, the germ $Z(f)$ does not depend on the chosen representative of f. If f is a unit in $\mathcal{O}_{\mathbb{C}^n, 0}$, then $Z(f)$ is the empty set. Analogously, one defines $Z(f_1, \ldots, f_k) := Z(f_1) \cap \ldots \cap Z(f_k)$ and more generally $Z(A)$ as $\bigcap_{f \in A} Z(f)$ for a finite subset $A \subset \mathcal{O}_{\mathbb{C}^n, 0}$. Note that intersection, union, and inclusion for germs of sets are well-defined notions.

Definition 1.1.22 A germ $X \subset \mathbb{C}^n$ in 0 is called *analytic* if there exist elements $f_1, \ldots, f_k \in \mathcal{O}_{\mathbb{C}^n, 0}$, such that X and $Z(f_1, \ldots, f_k)$ define the same germ.

Here is a global version of this definition:

Definition 1.1.23 Let $U \subset \mathbb{C}^n$ be an open subset. An *analytic subset* of U is a closed subset $X \subset U$ such that for any $x \in X$ there exists an open neighbourhood $x \in V \subset U$ and holomorphic functions $f_1, \ldots, f_k : V \to \mathbb{C}$ such that $X \cap V = \{z \mid f_1(z) = \ldots = f_k(z) = 0\}$.

Obviously, any analytic set X defines an analytic germ in any point $z \in X$. We will mostly stick to the purely local situation, as we cannot become honestly global until we will have introduced the notion of a complex manifold.

Definition 1.1.24 Let $X \subset \mathbb{C}^n$ be a germ in the origin. Then $I(X)$ denotes the set of all elements $f \in \mathcal{O}_{\mathbb{C}^n, 0}$ with $X \subset Z(f)$.

Lemma 1.1.25 *For any germ $X \subset \mathbb{C}^n$ the set $I(X) \subset \mathcal{O}_{\mathbb{C}^n, 0}$ is an ideal. If $(A) \subset \mathcal{O}_{\mathbb{C}^n, 0}$ denotes the ideal generated by a subset $A \subset \mathcal{O}_{\mathbb{C}^n, 0}$, then $Z(A) = Z((A))$ and $Z(A)$ is analytic.*

Proof. All assertions are easily verified. Except, perhaps, the last one. Here one has to use the fact that $\mathcal{O}_{\mathbb{C}^n, 0}$ is noetherian, i.e. any ideal is generated by finitely many elements. $\qquad\qquad\square$

Lemma 1.1.26 *If $X_1 \subset X_2$, then $I(X_2) \subset I(X_1)$. If $I_1 \subset I_2$, then $Z(I_2) \subset Z(I_1)$. For any analytic germ X one has $Z(I(X)) = X$. For any ideal $I \subset \mathcal{O}_{\mathbb{C}^n, 0}$ one has $I \subset I(Z(I))$.*

Proof. The first two assertions are obvious. Clearly, $X \subset Z(I(X))$. On the other hand, there exist elements $f_1, \ldots, f_k \in \mathcal{O}_{\mathbb{C}^n,0}$ with $X = Z(f_1, \ldots, f_k)$. Then, $f_1, \ldots, f_k \in I(X)$ and thus $Z(I(X)) \subset X = Z(f_1, \ldots, f_k)$. Hence, $X = Z(I(X))$. The last assertion is again trivial. $\square$

Also note that $Z(I \cdot J) = Z(I) \cup Z(J)$ and $Z(I + J) = Z(I) \cap Z(J)$, for two ideals $I, J \subset \mathcal{O}_{\mathbb{C}^n,0}$.

Definition 1.1.27 An analytic germ is *irreducible* if the following condition is satisfied: Let $X = X_1 \cup X_2$, where X_1 and X_2 are analytic germs. Then $X = X_1$ or $X = X_2$.

This property translates easily into an algebraic property of the associated ideal:

Lemma 1.1.28 *An analytic germ X is irreducible if and only if $I(X) \subset \mathcal{O}_{\mathbb{C}^n,0}$ is a prime ideal.*

Proof. If X is irreducible and $f_1 \cdot f_2 \in I(X)$, then $X = (X \cap Z(f_1)) \cup (X \cap Z(f_2))$ is a union of analytic germs. Thus, $X = X \cap Z(f_i)$ for $i = 1$ or $i = 2$. Hence, at least one of the functions f_1 or f_2 vanishes on X and, therefore, is contained in the ideal $I(X)$.

Conversely, let $I(X)$ be a prime ideal and let $X = X_1 \cup X_2$ with X_i analytic. If $f_i \in I(X_i)$, $i = 1, 2$, then $f_1 \cdot f_2 \in I(X)$. Hence, $f_1 \in I(X)$ or $f_2 \in I(X)$. Thus, it suffices to show that if $X \neq X_1$ and $X \neq X_2$, then there exist elements $f_1 \in I(X_1) \setminus I(X)$ and $f_2 \in I(X_2) \setminus I(X)$. This follows immediately from Lemma 1.1.26. $\square$

Lemma 1.1.28 also shows that for $f \in \mathcal{O}_{\mathbb{C}^n,0}$ the zero set $Z(f)$ is irreducible if and only if there exists an irreducible $g \in \mathcal{O}_{\mathbb{C}^n,0}$ such that $f = g^k$ for some k. Indeed, if $f = g^k$ with g irreducible, then $Z(f) = Z(g)$ and if $h \in I(Z(g))$ then g divides h by Lemma 1.1.19. This yields $I(Z(g)) = (g)$ and thus $Z(f)$ is irreducible. Conversely, if $f = \prod g_i^{n_i}$, then $Z(f) = \bigcup Z(g_i)$, which cannot be irreducible except for the case $f = g^k$ with g irreducible.

Recall that the *radical* of an ideal $I \subset \mathcal{O}_{\mathbb{C}^n,0}$ is the ideal $\sqrt{I}$ of all elements $f \in \mathcal{O}_{\mathbb{C}^n,0}$ such that $f^k \in I$ for some $k > 0$. One easily proves that $\sqrt{I} \subset I(Z(I))$, but the other inclusion also holds true. This is the content of the following deep theorem.

Proposition 1.1.29 (Nullstellensatz) *If $I \subset \mathcal{O}_{\mathbb{C}^n,0}$ is any ideal, then $\sqrt{I} = I(Z(I))$.*

The assertion is easily reduced to the case of prime ideals as follows. From commutative algebra one knows that $\sqrt{I}$ is the intersection of all prime ideals $\mathfrak{p}$ containing I (cf. [2]). Thus, it suffices to show that $I(Z(I)) \subset \mathfrak{p}$ for all prime ideals $I \subset \mathfrak{p}$. If we know already that $\mathfrak{p} = \sqrt{\mathfrak{p}} = I(Z(\mathfrak{p}))$ for all prime ideals $\mathfrak{p}$, this follows from $Z(\mathfrak{p}) \subset Z(I)$.

In order to prove the assertion for prime ideals we need an important result on the structure of irreducible analytic germs which could be considered as a far reaching generalization of the WPT (Proposition 1.1.6). As the formulation of this theorem is already quite involved and its proof is rather lengthy, we will not attempt to present it here. We will state the following theorem without proof. It should however be noted that we actually present only two aspects of the full result.

Theorem 1.1.30 *Let $X \subset \mathbb{C}^n$ be an irreducible analytic germ defined by a prime ideal $\mathfrak{p} \subset \mathcal{O}_{\mathbb{C}^n,0}$. Then one can find a coordinate system*

$$(z_1, \ldots, z_{n-d}, z_{n-d+1}, \ldots, z_n)$$

such that the projection $(z_1, \ldots, z_n) \mapsto (z_{n-d+1}, \ldots, z_n)$ induces a surjective map (of germs!) $\pi : X \to \mathbb{C}^d$ and such that the induced ring homomorphism $\mathcal{O}_{\mathbb{C}^d,0} \to \mathcal{O}_{\mathbb{C}^n,0}/\mathfrak{p}$ is a finite integral ring extension.

Before applying this general result to the proof of the Nullstellensatz, let us explain that for principal prime ideals $\mathfrak{p}$ these two statements follow directly from the WPT. Let $\mathfrak{p} = (g)$ with g irreducible. Clearly, we may assume that g is a Weierstrass polynomial. Let $f \in \mathcal{O}_{\mathbb{C}^n,0}$ be arbitrary. By the Weierstrass division theorem one has $f = g \cdot h + r$ with $r \in \mathcal{O}_{\mathbb{C}^{n-1},0}[z_1]$ has degree $< e := \deg(g)$ and thus the induced element $\bar{f}$ in $\mathcal{O}_{\mathbb{C}^n,0}/\mathfrak{p}$ is contained in $\sum_{i=0}^e z_1^i \cdot \mathcal{O}_{\mathbb{C}^{n-1},0}/\mathfrak{p}$. Hence, $\mathcal{O}_{\mathbb{C}^n,0}/\mathfrak{p}$ is a finite $\mathcal{O}_{\mathbb{C}^{n-1},0}$-module. The surjectivity of $Z(g) \to \mathbb{C}^{n-1}$ follows from the fact that the polynomial g_w has a zero for any w close to 0.

Proof of the Nullstellensatz. We have seen already that it suffices to prove the assertion for prime ideals. Now let us apply Theorem 1.1.30. Since $\mathfrak{p} \subset I(Z(\mathfrak{p}))$ holds for trivial reasons (see Lemma 1.1.26), we only have to ensure that any $f \in I(Z(\mathfrak{p}))$ is contained in $\mathfrak{p}$. For an appropriate coordinate system $(z_1, \ldots, z_n)$ the induced element $\bar{f} \in \mathcal{O}_{\mathbb{C}^n,0}/\mathfrak{p}$ satisfies an irreducible algebraic equation $\bar{f}^k + a_1 \bar{f}^{k-1} + \ldots + a_k = 0$ over $\mathcal{O}_{\mathbb{C}^d,0}$, i.e. $a_i = a_i(w) \in \mathcal{O}_{\mathbb{C}^d,0}$, where $w = (z_{n-d+1}, \ldots, z_n)$. Since f vanishes along $Z(\mathfrak{p})$, the 0-th coefficient a_k does as well. As $Z(\mathfrak{p}) \to \mathbb{C}^d$ is surjective, this yields $a_k = 0$. Hence, the above algebraic equation cannot be irreducible except for $k = 1$. Therefore, $\bar{f} = 0$ and, thus, $f \in \mathfrak{p}$.   □

Theorem 1.1.30 also shows that the following definition of the dimension of an analytic germ coincides with the geometric intuition.

Definition 1.1.31 Let X be an irreducible analytic germ defined by a prime ideal $\mathfrak{p} \subset \mathcal{O}_{\mathbb{C}^n,0}$. Then the *dimension* of X is defined by $\dim(X) = d$, where d is as in Theorem 1.1.30.

Using commutative algebra one verifies that this definition does not depend on any choice. In fact, $\dim(X) = \dim(\mathcal{O}_{\mathbb{C}^n,0}/\mathfrak{p})$. An arbitrary analytic germ is of dimension d if all its irreducible components are of the same dimension d.

Remark 1.1.32 Let $X \subset \mathbb{C}^n$ be an irreducible analytic germ of codimension one, i.e. $\dim(X) = n - 1$. Then the prime ideal $\mathfrak{p}$ defining X is of height one, i.e. the only other prime ideal contained in $\mathfrak{p}$ is (0). A basic result in commutative algebra tells us, that any prime ideal of height one in a UFD is principal. Therefore, $\mathfrak{p} = (f)$ for some irreducible $f \in \mathcal{O}_{\mathbb{C}^n,0}$. In other words, any irreducible analytic germ of codimension one is defined by one irreducible holomorphic function. More generally, any analytic germ of codimension one is the zero set of a single holomorphic function.

In the theory of functions of one variable a meromorphic function f on an open subset $U \subset \mathbb{C}$ is a holomorphic function defined on the complement of a discrete set of points $S \subset U$ such that f has poles of finite order in all points of S. Then one shows that locally around any point of S the function can be written as the quotient of two holomorphic functions. The latter description generalizes to the notion of a meromorphic function of several variables.

Definition 1.1.33 Let $U \subset \mathbb{C}^n$ be open. A *meromorphic* function f on U is a function on the complement of a nowhere dense subset $S \subset U$ with the following property: There exist an open cover $U = \bigcup U_i$ and holomorphic functions $g_i, h_i : U_i \to \mathbb{C}$ with $h_i|_{U_i \setminus S} \cdot f|_{U_i \setminus S} = g_i|_{U_i \setminus S}$.

We denote by $K(U)$ the set of all meromorphic functions on U. One easily checks that $K(U)$ is a field if U is connected. Let f be a meromorphic function on U, i.e. $f \in K(X)$. Then for any $z \in U$ the meromorphic function f in a neighbourhood of z is given by $\frac{g}{h}$ with $g, h \in \mathcal{O}_{\mathbb{C}^n,z}$. If we assume that g and h are chosen to be relatively prime, then they are unique up to units. Hence, the zero set and the pole set of a meromorphic function are well-defined.

Definition 1.1.34 Let f be a meromorphic function on an open subset $U \subset \mathbb{C}^n$. Then the *zero set* $Z(f) \subset U$ of f and the *pole set* $P(f) \subset U$ are the analytic sets that in every point $z \in U$ are given by $Z(g)$ respectively $Z(h)$, where f on an open neighbourhood of z is given by $\frac{g}{h}$ with $g, h \in \mathcal{O}_{\mathbb{C}^n,z}$ relatively prime.

Already for meromorphic functions of one variable this definition is not satisfactory. We know that we have to count zeros and poles with their multiplicities in order to obtain sensitive information. The following local result enables us to define the irreducible decomposition of the zero set $Z(f)$ with multiplicities. In particular, we will be able to define the divisor of a meromorphic function globally (Definition 2.3.8).

Proposition 1.1.35 *Let $f \in \mathcal{O}_{\mathbb{C}^n,0}$ be irreducible. Then for sufficiently small ε and $z \in B_\varepsilon(0)$ the induced element $f \in \mathcal{O}_{\mathbb{C}^n,z}$ is irreducible.*

If $f, g \in \mathcal{O}_{\mathbb{C}^n,0}$ are relatively prime, then they are relatively prime in $\mathcal{O}_{\mathbb{C}^n,z}$ for z in a sufficiently small neighbourhood of 0.

Proof. We may assume that $f \in \mathcal{O}_{\mathbb{C}^{n-1},0}[z_1]$ is a Weierstrass polynomial. Suppose that f as an element of $\mathcal{O}_{\mathbb{C}^n,z}$ is reducible. Then $f = f_1 \cdot f_2$ with $f_i \in \mathcal{O}_{\mathbb{C}^n,z}$ non-units, i.e. $f_1(z) = f_2(z) = 0$. Thus, $\frac{\partial f}{\partial z_1}(z) = \frac{\partial f_1}{\partial z_1}(z) \cdot f_2(z) + f_1(z) \cdot \frac{\partial f_2}{\partial z_1}(z) = 0$.

Thus, the set of points $z \in B_\varepsilon(0)$ where f as an element of $\mathcal{O}_{\mathbb{C}^n,z}$ is reducible is contained in the analytic set $Z(f, \frac{\partial f}{\partial z_1})$. We have to show that this set is a proper subset of $Z(f)$. If not, then $\frac{\partial f}{\partial z_1}$ would vanish on $Z(f)$. Since f is irreducible, we can apply Corollary 1.1.19. This yields a contradiction for degree reasons.

The second assertion is proved as follows. We may assume that f and g are Weierstrass polynomials in $\mathcal{O}_{\mathbb{C}^{n-1},0}[z_1]$. Then, f and g are relatively prime as elements in $\mathcal{O}_{\mathbb{C}^n,0}$ if and only if they are relatively prime in $\mathcal{O}_{\mathbb{C}^{n-1},0}[z_1]$. By the Gauss lemma the polynomials f and g are relatively prime if and only if there exist polynomials $h_1, h_2 \in \mathcal{O}_{\mathbb{C}^{n-1},0}[z_1]$ such that $0 \neq \gamma = h_1 \cdot f + h_2 \cdot g \in \mathcal{O}_{\mathbb{C}^{n-1},0}$. This immediately proves the assertion, where the open neighbourhood of the origin is given by the open subset where γ, h_1, h_2, f, and g are defined and γ does not vanish. $\qquad\square$

The next proposition will be used later in the proof of Siegel's theorem 2.1.9 which says that the transcendence degree of the function field of a compact complex manifold is at most the dimension of the manifold.

Proposition 1.1.36 (Schwarz lemma) *Let $\varepsilon := (\delta, \ldots, \delta)$ and let f be a holomorphic function on an open neighbourhood of the closure of the polydisc $\overline{B_\varepsilon(0)}$. Assume that f vanishes of order k at the origin, i.e. in the power series expansion non-trivial monomials of degree $< k$ do not occur.*

If $|f(z)|$ for $z \in \overline{B_\varepsilon(0)}$ can be bounded from above by C, then

$$|f(z)| \leq C \left(\frac{|z|}{\delta} \right)^k$$

for all $z \in \overline{B_\varepsilon(0)}$.

Proof. Fix $0 \neq z \in B_\delta(0)$ and define a holomorphic function g_z of one variable as follows: For $w \leq \delta$ one sets

$$g_z(w) := w^{-k} f \left(w \cdot \frac{z}{|z|} \right).$$

Then $|g_z(w)| \leq \delta^{-k} C$ for $|w| = \delta$. The maximum principle implies that $|g_z(w)| \leq \delta^{-k} C$ for $|w| \leq \delta$. Hence,

$$|z|^{-k} |f(z)| = |g_z(|z|)| \leq \delta^{-k} C.$$

This yields the assertion. $\qquad\square$

Exercises

1.1.1   Show that every holomorphic map $f : \mathbb{C} \to \mathbb{H} := \{z \mid \mathrm{Im}(z) > 0\}$ is constant.

1.1.2   Show that real and imaginary part u respectively v of a holomorphic function $f = u + iv$ are harmonic, i.e. $\sum_i \frac{\partial^2 u}{\partial x_i^2} + \sum_i \frac{\partial^2 u}{\partial y_i^2} = 0$ and similarly for v.

1.1.3   Deduce the maximum principle and the identity theorem for holomorphic functions of several variables from the corresponding one-dimensional results.

1.1.4   Prove the chain rule $\frac{\partial (f \circ g)}{\partial z} = \frac{\partial f}{\partial w} \cdot \frac{\partial g}{\partial z} + \frac{\partial f}{\partial \bar{w}} \frac{\partial \bar{g}}{\partial z}$ and its analogue for $\partial / \partial \bar{z}$. Use this to show that the composition of two holomorphic functions is holomorphic.

1.1.5   Deduce the implicit function theorem for holomorphic functions $f : U \to \mathbb{C}$ from the Weierstrass preparation theorem.

1.1.6   Consider the function $f : \mathbb{C}^2 \to \mathbb{C}$, $(z_1, z_2) \mapsto z_1^3 z_2 + z_1 z_2 + z_1^2 z_2^2 + z_2^2 + z_1 z_2^3$ and find an explicit decomposition $f = h \cdot g_w$ as claimed by the WPT.

1.1.7   State and prove the product formula for $\partial / \partial z$ and $\partial / \partial \bar{z}$. Show that the product $f \cdot g$ of two holomorphic functions f and g is holomorphic and that $1/f$ is holomorphic on the complement of the zero set $Z(f)$.

1.1.8   Let $U \subset \mathbb{C}^n$ be open and connected. Show that for any non-trivial holomorphic function $f : U \to \mathbb{C}$ the complement $U \setminus Z(f)$ of the zero set of f is connected and dense in U.

1.1.9   Let $U \subset \mathbb{C}^n$ be open and connected. Show that the set $K(U)$ of meromorphic functions on U is a field. What is the relation between $K(U)$ and the quotient field of $\mathcal{O}_{\mathbb{C}^n, z}$ for $z \in U$?

1.1.10   Let $U := B_\varepsilon(0) \subset \mathbb{C}^n$ and consider the ring $\mathcal{O}(U)$ of holomorphic functions on U. Show that $\mathcal{O}(U)$ is naturally contained in $\mathcal{O}_{\mathbb{C}^n, 0}$. What is the relation between the localization of $\mathcal{O}(U)$ at the prime ideal of all functions vanishing at the origin and $\mathcal{O}_{\mathbb{C}^n, 0}$? Is this prime ideal maximal?

1.1.11   The notion of irreducibility for analytic germs generalizes in a straightforward way to the corresponding notion for analytic sets $X \subset \mathbb{C}^n$. Give an example of an irreducible analytic set that does not define irreducible analytic germs at every point and of an analytic set whose induced germs are all irreducible, but the set is not.

1.1.12   Let $U \subset \mathbb{C}^n$ be an open subset and let $f : U \to \mathbb{C}$ be holomorphic. Show that for $n \geq 2$ the zero set $Z(f)$ cannot consist of a single point. Analogously, show that for a holomorphic function $f : \mathbb{C}^n \to \mathbb{C}$, $n \geq 2$ and $w \in \mathrm{Im}(f)$ there exists $z \in f^{-1}(w)$ such that $\|z\| \gg 0$.

1.1.13   Show that the product of two analytic germs is in a natural way an analytic germ.

1.1.14   Let $X \subset \mathbb{C}^n$ be an irreducible analytic set of dimension d. A point $x \in X$ is called *singular* if X cannot be defined by $n - d$ holomorphic functions locally around x for which x is regular. Then the set of singular points $X_{\mathrm{sing}} \subset X$ is empty or an analytic subset of dimension $< d$. Although the basic idea behind this result is very simple, its complete proof is rather technical (see [58, Ch. 6]). Try to prove the fact in easy cases, e.g. when X is defined by a single holomorphic function.

If x is a regular point, i.e. $x \in X_{\mathrm{reg}} := X \setminus X_{\mathrm{sing}}$, the $n-d$ holomorphic functions defining X near x can be completed to a local coordinate system.

1.1.15   Consider the holomorphic map $f : \mathbb{C} \to \mathbb{C}^2$, $z \mapsto (z^2 - 1, z^3 - z)$. Is the image an analytic set?

1.1.16   The aim of this exercise is to establish the theorem of Poincaré stating that the polydisc $B_{(1,1)}(0) \subset \mathbb{C}^2$ and the unit disc $D := \{ z \in \mathbb{C}^2 \mid \|z\| < 1 \}$ are not biholomorphic. (Thus the Riemann mapping theorem does not generalize to higher dimensions. We refer to [93] for details.)
(a) Recall the description of the group of automorphisms of the unit disc in the complex plane. Show that the group of unitary matrices of rank two is a subgroup of the group of biholomorphic maps of D which leave the origin fixed.
(b) Show that for any $z \in B_{(1,1)}(0)$ there exists a biholomorphic map $f : B_{(1,1)}(0) \to B_{(1,1)}(0)$ with $f(z) = 0$.
(c)* Show that group of biholomorphic maps of $B_{(1,1)}(0)$ which leave invariant the origin is abelian.
(d) Show that D and $B_{(1,1)}(0)$ are not biholomorphic.

1.1.17   Let $X \subset \mathbb{C}^n$ be an analytic subset. Show that locally around any point $x \in X$ the regular part X_{reg} has finite volume. (Hint: Use Theorem 1.1.30.) This will be needed later when we integrate differential forms over singular subvarieties.

1.1.18   Let $f : U \to V$ be holomorphic and let $X \subset V$ be an analytic set. Show that $f^{-1}(X) \subset V$ is analytic. What is the relation between the irreducibility of X and $f^{-1}(X)$?

1.1.19   Let $I \subset \mathcal{O}_{\mathbb{C}^2,0}$ be the ideal generated by $z_1^2 - z_2^3 + z_1$ and $z_1^4 - 2z_1 z_2^3 + z_1^2$. Describe $\sqrt{I}$.

1.1.20   Let $U \subset \mathbb{C}^n$ be an open subset and $f : U \setminus \mathbb{C}^{n-2} \to \mathbb{C}$ a holomorphic map. Show that there exists a unique holomorphic extension $\tilde{f} : U \to \mathbb{C}$ of f.

Comment: We have presented the absolute minimum of the local theory that is needed for the understanding of certain points in the later chapters. There are many excellent text books on the subject, e.g. [35, 58, 64].

1.2 Complex and Hermitian Structures

In this section, which is essentially a lesson in linear algebra, we shall study additional structures on a given real vector space, e.g. scalar products and (almost) complex structures. They induce linear operators on the exterior algebra (Hodge, Lefschetz, etc.), and we will be interested in the interaction between these operators.

In the following, V shall denote a finite-dimensional real vector space.

Definition 1.2.1 An endomorphism $I : V \to V$ with $I^2 = -\mathrm{id}$ is called an *almost complex structure* on V.

Clearly, if I is an almost complex structure then $I \in \mathrm{Gl}(V)$. If V is the real vector space underlying a complex vector space then $v \mapsto i \cdot v$ defines an almost complex structure I on V. The converse holds true as well:

Lemma 1.2.2 *If I is an almost complex structure on a real vector space V, then V admits in a natural way the structure of a complex vector space.*

Proof. The $\mathbb{C}$-module structure on V is defined by $(a + ib) \cdot v = a \cdot v + b \cdot I(v)$, where $a, b \in \mathbb{R}$. The $\mathbb{R}$-linearity of I and the assumption $I^2 = -\mathrm{id}$ yield $((a + ib)(c + id)) \cdot v = (a + ib)((c + id) \cdot v)$ and in particular $i(i \cdot v) = -v$. $\square$

Thus, almost complex structures and complex structures are equivalent notions for vector spaces. In particular, an almost complex structure can only exist on an even dimensional real vector space.

Corollary 1.2.3 *Any almost complex structure on V induces a natural orientation on V.*

Proof. Using the lemma, the assertion reduces to the statement that the real vector space $\mathbb{C}^n$ admits a natural orientation. We may assume $n = 1$ and use the orientation given by the basis $(1, i)$. The orientation is well-defined, as it does not change under $\mathbb{C}$-linear automorphisms. $\square$

For a real vector space V the complex vector space $V \otimes_{\mathbb{R}} \mathbb{C}$ is denoted by $V_{\mathbb{C}}$. Thus, the real vector space V is naturally contained in the complex vector space $V_{\mathbb{C}}$ via the map $v \mapsto v \otimes 1$. Moreover, $V \subset V_{\mathbb{C}}$ is the part that is left invariant under complex conjugation on $V_{\mathbb{C}}$ which is defined by $\overline{(v \otimes \lambda)} := v \otimes \overline{\lambda}$ for $v \in V$ and $\lambda \in \mathbb{C}$.

Suppose that V is endowed with an almost complex structure I. Then we will also denote by I its $\mathbb{C}$-linear extension to an endomorphism $V_{\mathbb{C}} \to V_{\mathbb{C}}$. Clearly, the only eigenvalues of I on $V_{\mathbb{C}}$ are $\pm i$.

Definition 1.2.4 Let I be an almost complex structure on a real vector space V and let $I : V_{\mathbb{C}} \to V_{\mathbb{C}}$ be its $\mathbb{C}$-linear extension. Then the $\pm i$ eigenspaces are denoted $V^{1,0}$ and $V^{0,1}$, respectively, i.e.

$$V^{1,0} = \{v \in V_{\mathbb{C}} \mid I(v) = i \cdot v\} \text{ and } V^{0,1} = \{v \in V_{\mathbb{C}} \mid I(v) = -i \cdot v\}.$$

Lemma 1.2.5 *Let V be a real vector space endowed with an almost complex structure I. Then*

$$V_{\mathbb{C}} = V^{1,0} \oplus V^{0,1}.$$

Complex conjugation on $V_{\mathbb{C}}$ induces an $\mathbb{R}$-linear isomorphism $V^{1,0} \cong V^{0,1}$.

Proof. Since $V^{1,0} \cap V^{0,1} = 0$, the canonical map

$$V^{1,0} \oplus V^{0,1} \longrightarrow V_{\mathbb{C}}$$

is injective. The first assertion follows from the existence of the inverse map

$$v \longmapsto \tfrac{1}{2}\left(v - iI(v)\right) \oplus \tfrac{1}{2}(v + iI(v)) .$$

For the second assertion we write $v \in V_{\mathbb{C}}$ as $v = x + iy$ with $x, y \in V$. Then $\overline{(v - iI(v))} = (x - iy + iI(x) + I(y)) = (\bar{v} + iI(\bar{v}))$. Hence, complex conjugation interchanges the two factors. $\qquad\square$

One should be aware of the existence of two almost complex structures on $V_{\mathbb{C}}$. One is given by I and the other one by i. They coincide on the subspace $V^{1,0}$ but differ by a sign on $V^{0,1}$. Obviously, $V^{1,0}$ and $V^{0,1}$ are complex subspaces of $V_{\mathbb{C}}$ with respect to both almost complex structures. In the sequel, we will always regard $V_{\mathbb{C}}$ as the complex vector space with respect to i. The $\mathbb{C}$-linear extension of I is the additional structure that gives rise to the above decomposition. If $V^{1,0}$ and $V^{0,1}$ are considered with the complex structure i, then the compositions $V \subset V_{\mathbb{C}} \to V^{1,0}$ and $V \subset V_{\mathbb{C}} \to V^{0,1}$ are complex linear respectively complex antilinear. Here, V is endowed with the almost complex structure I.

Lemma 1.2.6 *Let V be a real vector space endowed with an almost complex structure I. Then the dual space $V^* = \mathrm{Hom}_{\mathbb{R}}(V, \mathbb{R})$ has a natural almost complex structure given by $I(f)(v) = f(I(v))$. The induced decomposition on $(V^*)_{\mathbb{C}} = \mathrm{Hom}_{\mathbb{R}}(V, \mathbb{C}) = (V_{\mathbb{C}})^*$ is given by*

$$(V^*)^{1,0} = \{f \in \mathrm{Hom}_{\mathbb{R}}(V, \mathbb{C}) \mid f(I(v)) = if(v)\} = (V^{1,0})^*$$
$$(V^*)^{0,1} = \{f \in \mathrm{Hom}_{\mathbb{R}}(V, \mathbb{C}) \mid f(I(v)) = -if(v)\} = (V^{0,1})^*.$$

Also note that $(V^*)^{1,0} = \mathrm{Hom}_{\mathbb{C}}((V, I), \mathbb{C})$. $\qquad\square$

If V is a real vector space of dimension d, the natural decomposition of its exterior algebra is of the form

$$\bigwedge{}^* V = \bigoplus_{k=0}^{d} \bigwedge{}^k V.$$

Analogously, $\bigwedge^* V_{\mathbb{C}}$ denotes the exterior algebra of the complex vector space $V_{\mathbb{C}}$, which decomposes as

$$\bigwedge{}^* V_{\mathbb{C}} = \bigoplus_{k=0}^{d} \bigwedge{}^k V_{\mathbb{C}}. \tag{1.8}$$

Moreover, $\bigwedge^* V_{\mathbb{C}} = \bigwedge V \otimes_{\mathbb{R}} \mathbb{C}$ and $\bigwedge^* V$ is the real subspace of $\bigwedge^* V_{\mathbb{C}}$ that is left invariant under complex conjugation.

If V is endowed with an almost complex structure I, then its real dimension d is even, say $d = 2n$, and $V_{\mathbb{C}}$ decomposes as above $V_{\mathbb{C}} = V^{1,0} \oplus V^{0,1}$ with $V^{1,0}$ and $V^{0,1}$ complex vector spaces of dimension n.

Definition 1.2.7 One defines

$$\bigwedge{}^{p,q} V := \bigwedge{}^p V^{1,0} \otimes_{\mathbb{C}} \bigwedge{}^q V^{0,1},$$

where the exterior products of $V^{1,0}$ and $V^{0,1}$ are taken as exterior products of complex vector spaces. An element $\alpha \in \bigwedge^{p,q} V$ is of bidegree (p,q).

Proposition 1.2.8 *For a real vector space V endowed with an almost complex structure I one has:*

i) $\bigwedge^{p,q} V$ *is in a canonical way a subspace of $\bigwedge^{p+q} V_{\mathbb{C}}$.*

ii) $\bigwedge^k V_{\mathbb{C}} = \bigoplus_{p+q=k} \bigwedge^{p,q} V.$

iii) *Complex conjugation on $\bigwedge^* V_{\mathbb{C}}$ defines a ($\mathbb{C}$-antilinear) isomorphism $\bigwedge^{p,q} V \cong \bigwedge^{q,p} V$, i.e. $\overline{\bigwedge^{p,q} V} = \bigwedge^{q,p} V$.*

iv) *The exterior product is of bidegree $(0,0)$, i.e. $(\alpha, \beta) \mapsto \alpha \wedge \beta$ maps $\bigwedge^{p,q} V \times \bigwedge^{r,s} V$ to the subspace $\bigwedge^{p+r,q+s} V$.*

Proof. Let $v_1, \ldots, v_n \in \bigwedge^{1,0} V = V^{1,0}$ and $w_1, \ldots, w_n \in \bigwedge^{0,1} V = V^{0,1}$ be $\mathbb{C}$-basis. Then $v_{J_1} \otimes w_{J_2} \in \bigwedge^{p,q} V$ with $J_1 = \{i_1 < \ldots < i_p\}$ and $J_2 = \{j_1 < \ldots < j_q\}$ form a basis of $\bigwedge^{p,q} V$.

This shows i) and ii). Here, one could as well use the general fact that any direct sum decomposition $V_{\mathbb{C}} = W_1 \oplus W_2$ induces a direct sum decomposition $\bigwedge^k V_{\mathbb{C}} = \bigoplus_{p+q=k} \bigwedge^p W_1 \otimes \bigwedge^q W_2$.

Since complex conjugation is multiplicative, i.e. $\overline{w_1 \wedge w_2} = \overline{w_1} \wedge \overline{w_2}$, assertion iii) follows from $\overline{V^{1,0}} = V^{0,1}$. The last assertion holds again true for any decomposition $V_{\mathbb{C}} = W_1 \oplus W_2$. $\qquad \square$

Any vector $v \in V_{\mathbb{C}}$ can be written as $v = x + iy$ with $x, y \in V$. Assume that $z_i = \frac{1}{2}(x_i - iy_i) \in V^{1,0}$ is a $\mathbb{C}$-basis of $V^{1,0}$ with $x_i, y_i \in V$. Since $I(z_i) = iz_i$, one finds $y_i = I(x_i)$ and $x_i = -I(y_i)$. Moreover, $x_i, y_i \in V$ form a real basis of V and, therefore, a basis of the complex vector space $V_{\mathbb{C}}$. A natural basis of the complex vector space $V^{0,1}$ is then provided by $\bar{z}_i = \frac{1}{2}(x_i + iy_i)$.

Conversely, if $v \in V$, then $\frac{1}{2}(v - iI(v)) \in V^{1,0}$. Therefore, if $\langle x_i, y_i := I(x_i) \rangle$ is a basis of the real vector space V, then $z_i = \frac{1}{2}(x_i - iy_i)$ is a basis of the complex vector space $V^{1,0}$. With these notations one has the following

Lemma 1.2.9 *For any* $m \leq \dim_{\mathbb{C}} V^{1,0}$ *one has*

$$(-2i)^m (z_1 \wedge \bar{z}_1) \wedge \ldots \wedge (z_m \wedge \bar{z}_m) = (x_1 \wedge y_1) \wedge \ldots \wedge (x_m \wedge y_m).$$

For $m = \dim_{\mathbb{C}} V^{1,0}$, *this defines a positive oriented volume form for the natural orientation of* V *(cf. Corollary 1.2.3).*

Proof. This is a straightforward calculation using induction on m. $\square$

There is an analogous formula for the dual basis. Let $\langle x^i, y^i \rangle$ be the basis of V^* dual to $\langle x_i, y_i \rangle$. Then, $z^i = x^i + iy^i$ and $\bar{z}^i = x^i - iy^i$ are the basis of $V^{1,0*}$ and $V^{0,1*}$ dual to $\langle z_i \rangle$ respectively $\langle \bar{z}_i \rangle$. The above formula yields

$$\left(\frac{i}{2} \right)^m (z^1 \wedge \bar{z}^1) \wedge \ldots \wedge (z^m \wedge \bar{z}^m) = (x^1 \wedge y^1) \wedge \ldots \wedge (x^m \wedge y^m).$$

Note that $I(x^i) = -y^i$ and $I(y^i) = x^i$. We tacitly use the natural isomorphism $\bigwedge^k V^* \cong (\bigwedge^k V)^*$ given by $(\alpha_1 \wedge \ldots \wedge \alpha_k)(v_1 \wedge \ldots \wedge v_k) = \det(\alpha_j(v_j))_{i,j}$.

Definition 1.2.10 With respect to the direct sum decompositions (1.8) and ii) of Proposition 1.2.8 one defines the natural projections

$$\Pi^k : \bigwedge{}^* V_{\mathbb{C}} \longrightarrow \bigwedge{}^k V_{\mathbb{C}} \quad \text{and} \quad \Pi^{p,q} : \bigwedge{}^* V_{\mathbb{C}} \longrightarrow \bigwedge{}^{p,q} V.$$

Furthermore, $\mathbf{I} : \bigwedge^* V_{\mathbb{C}} \to \bigwedge^* V_{\mathbb{C}}$ is the linear operator that acts on $\bigwedge^{p,q} V$ by multiplication with i^{p-q}, i.e.

$$\mathbf{I} = \sum_{p,q} i^{p-q} \cdot \Pi^{p,q}.$$

The operator Π^k does not depend on the almost complex structure I, but the operators $\mathbf{I}$ and $\Pi^{p,q}$ certainly do. Note that $\mathbf{I}$ is the multiplicative extension of the almost complex structure I on $V_{\mathbb{C}}$, but $\mathbf{I}$ is not an almost complex structure. Since I is defined on the real vector space V, also $\mathbf{I}$ is an endomorphism of the real exterior algebra $\bigwedge^* V$.

We denote the corresponding operators on the dual space $\bigwedge^* V_{\mathbb{C}}^*$ also by Π^k, $\Pi^{p,q}$, respectively $\mathbf{I}$. Note that $\mathbf{I}(\alpha)(v_1, \ldots, v_k) = \alpha(\mathbf{I}(v_1), \ldots, \mathbf{I}(v_k))$ for $\alpha \in \bigwedge^k V_{\mathbb{C}}^*$ and $v_i \in V_{\mathbb{C}}$.

Let $(V, \langle \ , \ \rangle)$ be a finite-dimensional euclidian vector space, i.e. V is a real vector space and $\langle \ , \ \rangle$ is a positive definite symmetric bilinear form.

Definition 1.2.11 An almost complex structure I on V is *compatible* with the scalar product $\langle \ , \ \rangle$ if $\langle I(v), I(w) \rangle = \langle v, w \rangle$ for all $v, w \in V$, i.e. $I \in O(V, \langle \ , \ \rangle)$.

Before considering the general situation, let us study the two-dimensional case, where scalar products and almost complex structures are intimately related. It turns out that these two notions are almost equivalent. This definitely fails in higher dimensions.

Example 1.2.12 Let V be a real vector space of dimension two with a fixed orientation. If $\langle\ ,\ \rangle$ is a scalar product, then there exists a natural almost complex structure I on V associated to it which is defined as follows: For any $0 \neq v \in V$ the vector $I(v) \in V$ is uniquely determined by the following three conditions: $\langle v, I(v)\rangle = 0$, $\|I(v)\| = \|v\|$, and $\{v, I(v)\}$ is positively oriented. Equivalently, I is the rotation by $\pi/2$. Thus, $I^2 = -\mathrm{id}$, i.e. I is an almost complex structure. One also sees that $I \in \mathrm{SO}(V)$ and, thus, I is compatible with $\langle\ ,\ \rangle$.

Two scalar products $\langle\ ,\ \rangle$ and $\langle\ ,\ \rangle'$ are called *conformal equivalent* if there exists a (positive) scalar λ with $\langle\ ,\ \rangle' = \lambda \cdot \langle\ ,\ \rangle$. Clearly, two conformally equivalent scalar products define the same almost complex structure. Conversely, for any given almost complex structure I there always exists a scalar product $\langle\ ,\ \rangle$ to which I is associated.

In this way one obtains a bijection between the set of conformal equivalence classes of scalar products on the two-dimensional oriented vector space V and the set of almost complex structures that induce the given orientation:

$$\{\langle\ ,\ \rangle\}_{/\sim_{\mathrm{conf}}} \longleftrightarrow \{I \in \mathrm{Gl}(V)_+ \mid I^2 = -\mathrm{id}\}.$$

Let us now come back to an euclidian vector space $(V, \langle\ ,\ \rangle)$ of arbitrary dimension endowed with a compatible almost complex structure I.

Definition 1.2.13 The *fundamental form* associated to $(V, \langle\ ,\ \rangle, I)$ is the form

$$\omega := -\langle(\ ), I(\ )\rangle = \langle I(\ ), (\ )\rangle.$$

Lemma 1.2.14 *Let $(V, \langle\ ,\ \rangle)$ be an euclidian vector space endowed with a compatible almost complex structure. Then, its fundamental form ω is real and of type $(1,1)$, i.e. $\omega \in \bigwedge^2 V^* \cap \bigwedge^{1,1} V^*$.*

Proof. Since

$$\langle v, I(w)\rangle) = \langle I(v), I(I(w))\rangle = -\langle I(v), w\rangle = -\langle w, I(v)\rangle$$

for all $v, w \in V$, the form ω is alternating, i.e. $\omega \in \bigwedge^2 V^*$.
   Since

$$(\mathbf{I}\omega)(v, w) = \omega(\mathbf{I}(v), \mathbf{I}(w)) = \langle I(I(v)), I(w)\rangle = \omega(v, w),$$

one finds $\mathbf{I}(\omega) = \omega$, i.e. $\omega \in \bigwedge^{1,1} V^*_{\mathbb{C}}$. $\square$

Note that two of the three structures $\{\langle\ ,\ \rangle, I, \omega\}$ determine the remaining one.

Following a standard procedure, the scalar product and the fundamental form are encoded by a natural hermitian form.

Lemma 1.2.15 *Let* $(V, \langle \ , \ \rangle)$ *be an euclidian vector space endowed with a compatible complex structure. The form* $(\ , \ ) := \langle \ , \ \rangle - i \cdot \omega$ *is a positive hermitian form on* (V, I).

Proof. The form $(\ , \ )$ is clearly $\mathbb{R}$-linear and $(v, v) = \langle v, v \rangle > 0$ for $0 \neq v \in V$. Moreover, $(v, w) = \overline{(w, v)}$ and

$$
\begin{aligned}
(I(v), w) &= \langle I(v), w \rangle - i \cdot \omega(I(v), w) \\
&= \langle I(I(v)), I(w) \rangle + i \cdot \langle v, w \rangle \\
&= i \cdot (i \cdot \langle v, I(w) \rangle + \langle v, w \rangle) = i \cdot (v, w).
\end{aligned}
$$

$\square$

One also considers the extension of the scalar product $\langle \ , \ \rangle$ to a positive definite hermitian form $\langle \ , \ \rangle_{\mathbb{C}}$ on $V_{\mathbb{C}}$. This is defined by

$$
\langle v \otimes \lambda, w \otimes \mu \rangle_{\mathbb{C}} := (\lambda \overline{\mu}) \cdot \langle v, w \rangle
$$

for $v, w \in V$ and $\lambda, \mu \in \mathbb{C}$.

Lemma 1.2.16 *If* $(V, \langle \ , \ \rangle)$ *is an euclidian vector space with a compatible almost complex structure* I. *Then* $V_{\mathbb{C}} = V^{1,0} \oplus V^{0,1}$ *is an orthogonal decomposition with respect to the hermitian product* $\langle \ , \ \rangle_{\mathbb{C}}$.

Proof. Let $v - iI(v) \in V^{1,0}$ and $w + iI(w) \in V^{0,1}$ with $v, w \in V$. Then an easy calculation shows $\langle v - iI(v), w + iI(w) \rangle_{\mathbb{C}} = 0$. $\square$

Let us now study the relation between $(\ , \ )$ and $\langle \ , \ \rangle_{\mathbb{C}}$.

Lemma 1.2.17 *Let* $(V, \langle \ , \ \rangle)$ *be an euclidian vector space with a compatible almost complex structure* I. *Under the canonical isomorphism* $(V, I) \cong (V^{1,0}, i)$ *one has* $\frac{1}{2}(\ , \ ) = \langle \ , \ \rangle_{\mathbb{C}}|_{V^{1,0}}$

Proof. The natural isomorphism was given by $v \mapsto \frac{1}{2}(v - iI(v))$. Now use the definitions of $(\ , \ )$ to conclude

$$
\begin{aligned}
&\langle (v - iI(v)), (v' - iI(v')) \rangle_{\mathbb{C}} \\
&= \langle v, v' \rangle + i \langle v, I(v') \rangle - i \langle I(v), v' \rangle + \langle I(v), I(v') \rangle \\
&= 2 \langle v, v' \rangle + 2i \langle v, I(v') \rangle = 2(v, v')
\end{aligned}
$$

$\square$

Often, it is useful to do calculations in coordinates. Let us see how the above products can be expressed explicitly once suitable basis have been chosen.

Let $z_1, \ldots, z_n$ be a $\mathbb{C}$-basis of $V^{1,0}$. Write $z_i = \frac{1}{2}(x_i - iI(x_i))$ with $x_i \in V$. Then $x_1, y_1 := I(x_1), \ldots, x_n, y_n := I(x_n)$ is a $\mathbb{R}$-basis of V and $x_1, \ldots, x_n$ is a

$\mathbb{C}$-basis of (V, I). The hermitian form $\langle\,,\,\rangle_{\mathbb{C}}$ on $V^{1,0}$ with respect to the basis z_i is given by an hermitian matrix, say $\frac{1}{2}(h_{ij})$. Concretely,

$$\left\langle \sum_{i=1}^{n} a_i z_i, \sum_{j=1}^{n} b_j z_j \right\rangle_{\mathbb{C}} = \frac{1}{2} \sum_{i,j=1}^{n} h_{ij} a_i \bar{b}_j.$$

Using the lemma, we obtain $(x_i, x_j) = h_{ij}$. Since $(\,,\,)$ is hermitian on (V, I), this yields $(x_i, y_j) = -ih_{ij}$ and $(y_i, y_j) = h_{ij}$.

By definition of $(\,,\,)$, one has $\omega = -\mathrm{Im}(\,,\,)$ and $\langle\,,\,\rangle = \mathrm{Re}(\,,\,)$. Hence, $\omega(x_i, x_j) = \omega(y_i, y_j) = -\mathrm{Im}(h_{ij})$, $\omega(x_i, y_j) = \mathrm{Re}(h_{ij})$, $\langle x_i, x_j \rangle = \langle y_i, y_j \rangle = \mathrm{Re}(h_{ij})$, and $\langle x_i, y_j \rangle = \mathrm{Im}(h_{ij})$. Thus,

$$\omega = -\sum_{i<j} \mathrm{Im}(h_{ij})(x^i \wedge x^j + y^i \wedge y^j) + \sum_{i,j=1}^{n} \mathrm{Re}(h_{ij}) x^i \wedge y^j.$$

Using $z^i \wedge \bar{z}^j = (x^i + iy^i) \wedge (x^j - iy^j) = x^i \wedge x^j - i(x^i \wedge y^j + x^j \wedge y^i) + y^i \wedge y^j$ this yields

$$\omega = \frac{i}{2} \sum_{i,j=1}^{n} h_{ij} z^i \wedge \bar{z}^j.$$

If $x_1, y_1, \ldots, x_n, y_n$ is an orthonormal basis of V with respect to $\langle\,,\,\rangle$, i.e. $\langle\,,\,\rangle = \sum_{i=1}^{n} x^i \otimes x^i + \sum_{i=1}^{n} y^i \otimes y^i$, then

$$\omega = \frac{i}{2} \sum_{i=1}^{n} z^i \wedge \bar{z}^i = \sum_{i=1}^{n} x^i \wedge y^i.$$

Note that there always exists an orthonormal basis as above. Indeed, pick $x_1 \neq 0$ arbitrary of norm one and define $y_1 = I(x_1)$, which is automatically orthogonal to x_1. Then continue with the orthogonal complement of $x_1\mathbb{R} \oplus y_1\mathbb{R}$.

Definition 1.2.18 Let $(V, \langle\,,\,\rangle)$ be an euclidian vector space and let I be a compatible almost complex structure. Furthermore, let ω be the associated fundamental form. Then the *Lefschetz operator* $L : \bigwedge^* V_{\mathbb{C}}^* \to \bigwedge^* V_{\mathbb{C}}^*$ is given by $\alpha \mapsto \omega \wedge \alpha$.

Remark 1.2.19 The following properties are easy to verify:
i) L is the $\mathbb{C}$-linear extension of the real operator $\bigwedge^* V^* \to \bigwedge^* V^*$, $\alpha \mapsto \omega \wedge \alpha$.
ii) The Lefschetz operator is of bidegree $(1,1)$, i.e.

$$L\left(\bigwedge^{p,q} V^*\right) \subset \bigwedge^{p+1,q+1} V^*.$$

Furthermore the Lefschetz operator induces bijections

$$L^k : \bigwedge^k V^* \xrightarrow{\;\simeq\;} \bigwedge^{2n-k} V^*$$

for all $k \leq n$, where $\dim_{\mathbb{R}} V = 2n$. An elementary proof can be given by choosing a basis, but it is slightly cumbersome. A more elegant but less elementary argument, using $\mathfrak{sl}(2)$-representation theory, will be given in Proposition 1.2.30.

The Lefschetz operator comes along with its dual Λ. In order to define and to describe Λ we need to recall the Hodge $*$-operator on a real vector space.

Let $(V, \langle \ , \ \rangle)$ be an oriented euclidian vector space of dimension d, then $\langle \ , \ \rangle$ defines scalar products on all the exterior powers $\bigwedge^k V$. Explicitly, if $e_1, \ldots, e_d \in V$ is an orthonormal basis of V, then $e_I \in \bigwedge^k V$ with $I = \{i_1 < \ldots < i_k\}$ is an orthonormal basis of $\bigwedge^k V$. Let $\mathrm{vol} \in \bigwedge^d V$ be the orientation of V of norm 1 given by $\mathrm{vol} = e_1 \wedge \ldots \wedge e_d$.

Then the *Hodge $*$-operator* is defined by

$$\alpha \wedge *\beta = \langle \alpha, \beta \rangle \cdot \mathrm{vol}$$

for $\alpha, \beta \in \bigwedge^* V$. This determines $*$, for the exterior product defines a non-degenerate pairing $\bigwedge^k V \times \bigwedge^{d-k} V \to \bigwedge^d V = \mathrm{vol} \cdot \mathbb{R}$. One easily sees that $* : \bigwedge^k V \to \bigwedge^{d-k} V$.

The most important properties of the Hodge $*$-operator are collected in the following proposition. Their proofs are all elementary.

Proposition 1.2.20 *Let $(V, \langle \ , \ \rangle)$ be an oriented euclidian vector space of dimension d. Let $e_1, \ldots, e_d$ be an orthonormal basis of V and let $\mathrm{vol} \in \bigwedge^d V$ be the orientation of norm one given by $e_1 \wedge \ldots \wedge e_d$. The Hodge $*$-operator associated to $(V, \langle \ , \ \rangle, \mathrm{vol})$ satisfies the following conditions:*
  i) If $\{i_1, \ldots, i_k, j_1, \ldots, j_{d-k}\} = \{1, \ldots d\}$ one has

$$*(e_{i_1} \wedge \ldots \wedge e_{i_k}) = \varepsilon \cdot e_{j_1} \wedge \ldots \wedge e_{j_{d-k}},$$

where $\varepsilon = \mathrm{sgn}(i_1, \ldots, i_k, j_1 \ldots j_{d-k})$. In particular, $*1 = \mathrm{vol}$.
  ii) The $*$-operator is self-adjoint up to sign: For $\alpha \in \bigwedge^k V$ one has

$$\langle \alpha, *\beta \rangle = (-1)^{k(d-k)} \langle *\alpha, \beta \rangle.$$

  iii) The $*$-operator is involutive up to sign:

$$(*|_{\bigwedge^k V})^2 = (-1)^{k(d-k)}.$$

  iv) The Hodge $*$-operator is an isometry on $(\bigwedge^* V, \langle \ , \ \rangle)$. $\qquad\square$

In our situation we will usually have $d = 2n$ and $*$ and $\langle \ , \ \rangle$ will be considered on the dual space $\bigwedge^* V^*$.

Let us now come back to the situation considered before. Associated to $(V, \langle \ , \ \rangle, I)$ we had introduced the Lefschetz operator $L : \bigwedge^k V^* \to \bigwedge^{k+2} V^*$.

Definition 1.2.21 The *dual Lefschetz operator* Λ is the operator $\Lambda : \bigwedge^* V^* \to \bigwedge^* V^*$ that is adjoint to L with respect to $\langle \, , \, \rangle$, i.e. $\Lambda\alpha$ is uniquely determined by the condition

$$\langle \Lambda\alpha, \beta \rangle = \langle \alpha, L\beta \rangle \ \text{ for all } \ \beta \in \bigwedge{}^* V^*.$$

The $\mathbb{C}$-linear extension $\bigwedge^* V_{\mathbb{C}}^* \to \bigwedge^* V_{\mathbb{C}}^*$ of the dual Lefschetz operator will also be denoted by Λ.

Remark 1.2.22 Recall that I induces a natural orientation on V (Corollary 1.2.3). Thus, the Hodge $*$-operator is well-defined. Using an orthonormal basis $x_1, y_1 = I(x_1), \ldots, x_n, y_n = I(x_n)$ as above, a straightforward calculation yields

$$n! \cdot \omega^n = \text{vol},$$

where ω is the associated fundamental form. See Exercise 1.2.9 for a far reaching generalization of this.

Lemma 1.2.23 *The dual Lefschetz operator Λ is of degree -2, i.e. $\Lambda(\bigwedge^k V^*) \subset \bigwedge^{k-2} V^*$. Moreover, one has $\Lambda = *^{-1} \circ L \circ *$.*

Proof. The first assertion follows from the fact that L is of degree two and that $\bigwedge^* V^* = \bigoplus \bigwedge^k V^*$ is orthogonal.

By definition of the Hodge $*$-operator one has $\langle \alpha, L\beta \rangle \cdot \text{vol} = \langle L\beta, \alpha \rangle \cdot \text{vol} = L\beta \wedge *\alpha = \omega \wedge \beta \wedge *\alpha = \beta \wedge (\omega \wedge *\alpha) = \langle \beta, *^{-1}(L(*\alpha)) \rangle \cdot \text{vol}.$ $\square$

Recall that $\langle \, , \, \rangle_{\mathbb{C}}$ had been defined as the hermitian extension to $V_{\mathbb{C}}^*$ of the scalar product $\langle \, , \, \rangle$ on V^*. It can further be extended to a positive definite hermitian form on $\bigwedge^* V_{\mathbb{C}}^*$. Equivalently, one could consider the extension of $\langle \, , \, \rangle$ on $\bigwedge^* V^*$ to an hermitian form on $\bigwedge^* V_{\mathbb{C}}^*$. In any case, there is a natural positive hermitian product on $\bigwedge^* V_{\mathbb{C}}^*$ which will also be called $\langle \, , \, \rangle_{\mathbb{C}}$.

The Hodge $*$-operator associated to $(V, \langle \, , \, \rangle, \text{vol})$ is extended $\mathbb{C}$-linearly to $* : \bigwedge^k V_{\mathbb{C}}^* \to \bigwedge^{2n-k} V_{\mathbb{C}}^*$. On $\bigwedge^* V_{\mathbb{C}}^*$ these two operators are now related by

$$\alpha \wedge *\bar{\beta} = \langle \alpha, \beta \rangle_{\mathbb{C}} \cdot \text{vol}.$$

Clearly, the Lefschetz operator L and its dual Λ on $\bigwedge^* V_{\mathbb{C}}^*$ are also formally adjoint to each other with respect to $\langle \, , \, \rangle_{\mathbb{C}}$. Moreover, $\Lambda = *^{-1} \circ L \circ *$ on $\bigwedge^* V_{\mathbb{C}}^*$.

Lemma 1.2.24 *Let $\langle \, , \, \rangle_{\mathbb{C}}$, Λ, and $*$ be as above. Then*

i) *The decomposition $\bigwedge^k V_{\mathbb{C}}^* = \bigoplus \bigwedge^{p,q} V^*$ is orthogonal with respect to $\langle \, , \, \rangle_{\mathbb{C}}$.*

ii) *The Hodge $*$-operator maps $\bigwedge^{p,q} V^*$ to $\bigwedge^{n-q,n-p} V^*$, where $n = \dim_{\mathbb{C}}(V, I)$.*

iii) *The dual Lefschetz operator Λ is of bidegree $(-1, -1)$, i.e. $\Lambda(\bigwedge^{p,q} V^*) \subset \bigwedge^{p-1,q-1} V^*$.*

Proof. The first assertion follows directly from Lemma 1.2.16. The third assertion follows from the first and the fact that Λ is the formal adjoint of L with respect to $\langle\ ,\ \rangle_{\mathbb{C}}$. For the second assertion use $\alpha \wedge *\bar\beta = \langle \alpha, \beta \rangle_{\mathbb{C}} \cdot$ vol and that $\gamma_1 \wedge \gamma_2 = 0$ for $\gamma_i \in \bigwedge^{p_i, q_i} V^*$ with $p_1 + p_2 + q_1 + q_2 = 2n$ but $(p_1 + p_2, q_1 + q_2) \neq (n, n)$. □

Definition 1.2.25 Let $H : \bigwedge^* V \to \bigwedge^* V$ be the *counting operator* defined by $H|_{\bigwedge^k V} = (k - n) \cdot \mathrm{id}$, where $\dim_{\mathbb{R}} V = 2n$. Equivalently,

$$H = \sum_{k=0}^{2n} (k - n) \cdot \Pi^k.$$

With H, L, Λ, Π, etc., we dispose of a large number of linear operators on $\bigwedge^* V^*$ and one might wonder whether they commute. In fact, they do not, but their commutators can be computed. This is done in the next proposition. We use the notation $[A, B] = A \circ B - B \circ A$.

Proposition 1.2.26 *Let $(V, \langle\ ,\ \rangle)$ be an euclidian vector space endowed with a compatible almost complex structure I. Consider the following linear operators on $\bigwedge^* V^*$: The associated Lefschetz operator L, its dual Λ, and the counting operator H. They satisfy:*

$$\text{i) } [H, L] = 2L, \quad \text{ii) } [H, \Lambda] = -2\Lambda, \quad \text{and} \quad \text{iii) } [L, \Lambda] = H.$$

Proof. Let $\alpha \in \bigwedge^k V^*$. Then $[H, L](\alpha) = (k + 2 - n)(\omega \wedge \alpha) - \omega \wedge ((k - n)\alpha) = 2\omega \wedge \alpha$. Analogously, $[H, \Lambda](\alpha) = (k - 2 - n)(\Lambda\alpha) - \Lambda((k - n)\alpha) = -2\Lambda\alpha$.

The third assertion is the most difficult one. We will prove it by induction on the dimension of V. Assume we have a decomposition $V = W_1 \oplus W_2$ which is compatible with the scalar product and the almost complex structure, i.e. $(V, \langle\ ,\ \rangle, I) = (W_1, \langle\ ,\ \rangle_1, I_1) \oplus (W_2, \langle\ ,\ \rangle_2, I_2)$. Then $\bigwedge^* V^* = \bigwedge^* W_1^* \otimes \bigwedge^* W_2^*$ and in particular $\bigwedge^2 V^* = \bigwedge^2 W_1^* \oplus \bigwedge^2 W_2^* \oplus W_1^* \otimes W_2^*$. Since $V = W_1 \oplus W_2$ is orthogonal, the fundamental form ω on V decomposes as $\omega_1 \oplus \omega_2$, where ω_i is the fundamental form on W_i (no component in $W_1^* \otimes W_2^*$). Hence the Lefschetz operator L on $\bigwedge^* V^*$ is the direct sum of the Lefschetz operators L_1 and L_2 acting on $\bigwedge^* W_1^*$ and $\bigwedge^* W_2^*$, respectively, i.e. $L = L_1 + L_2$ with L_1 and L_2 acting as $L_1 \otimes 1$ respectively $1 \otimes L_2$ on $\bigwedge^* W_1^* \otimes \bigwedge^* W_2^*$.

Let $\alpha, \beta \in \bigwedge^* V^*$ and suppose that both are split, i.e. $\alpha = \alpha_1 \otimes \alpha_2$, $\beta = \beta_1 \otimes \beta_2$, with $\alpha_i, \beta_i \in \bigwedge^* W_i^*$. Then $\langle \alpha, \beta \rangle = \langle \alpha_1, \beta_1 \rangle \cdot \langle \alpha_2, \beta_2 \rangle$. Therefore,

$$\begin{aligned}
\langle \alpha, L\beta \rangle &= \langle \alpha, L_1(\beta_1) \otimes \beta_2 \rangle + \langle \alpha, \beta_1 \otimes L_2(\beta_2) \rangle \\
&= \langle \alpha_1, L_1\beta_1 \rangle \langle \alpha_2, \beta_2 \rangle + \langle \alpha_1, \beta_1 \rangle \langle \alpha_2, L_2\beta_2 \rangle \\
&= \langle \Lambda_1\alpha_1, \beta_1 \rangle \langle \alpha_2, \beta_2 \rangle + \langle \alpha_1, \beta_1 \rangle \langle \Lambda_2\alpha_2, \beta_2 \rangle \\
&= \langle \Lambda_1(\alpha_1) \otimes \alpha_2, \beta_1 \otimes \beta_2 \rangle + \langle \alpha_1 \otimes \Lambda_2(\alpha_2), \beta_2 \rangle.
\end{aligned}$$

Hence, $\varLambda = \varLambda_1 + \varLambda_2$, where $\varLambda_i$ is the dual Lefschetz operator on $\bigwedge^* W_i^*$. This yields

$$
\begin{aligned}
[L, \varLambda](\alpha_1 \otimes \alpha_2) &= (L_1 + L_2)(\varLambda_1(\alpha_1) \otimes \alpha_2 + \alpha_1 \otimes \varLambda_2(\alpha_2)) \\
&\quad -(\varLambda_1 + \varLambda_2)(L_1(\alpha_1) \otimes \alpha_2 + \alpha_1 \otimes L_2(\alpha_2)) \\
&= [L_1, \varLambda_1](\alpha_1) \otimes \alpha_2 + \alpha_1 \otimes [L_2, \varLambda_2](\alpha_2).
\end{aligned}
$$

By induction hypothesis $[L_i, \varLambda_i] = H_i$ and, therefore,

$$
\begin{aligned}
[L, \alpha](\alpha_1 \otimes \alpha_2) &= H_1(\alpha_1) \otimes \alpha_2 + \alpha_1 \otimes H_2(\alpha_2) \\
&= (k_1 - n_1)(\alpha_1 \otimes \alpha_2) + (k_2 - n_2)(\alpha_1 \otimes \alpha_2) \\
&= (k_1 + k_2 - n_1 - n_2)(\alpha_1 \otimes \alpha_2),
\end{aligned}
$$

for $\alpha_i \in \bigwedge^{k_i} W_i^*$ and $n_i = \dim_{\mathbb{C}}(W_i, I_i)$.

It remains to prove the case $\dim_{\mathbb{C}}(V, I) = 1$. With respect to a basis x_1, y_1 of V one has

$$
\begin{aligned}
\textstyle\bigwedge^* V^* &= \textstyle\bigwedge^0 V^* \oplus \quad \textstyle\bigwedge^1 V^* \quad \oplus \textstyle\bigwedge^2 V^* \\
&= \mathbb{R} \quad \oplus (x^1 \mathbb{R} \oplus y^1 \mathbb{R}) \oplus \quad \omega \mathbb{R}
\end{aligned}
$$

Moreover, $L : \bigwedge^0 V^* \to \bigwedge^2 V^*$ and $\varLambda : \bigwedge^2 V^* \to \bigwedge^0 V^*$ are given by $1 \mapsto \omega$ and $\omega \mapsto 1$, respectively. Hence, $[L, \varLambda]|_{\bigwedge^0 V^*} = -\varLambda L|_{\bigwedge^0 V^*} = -1$, $[L, \varLambda]|_{\bigwedge^1 V^*} = 0$, and $[L, \varLambda]|_{\bigwedge^2 V^*} = 1$. □

Corollary 1.2.27 *Let* $(V, \langle \ , \ \rangle, I)$ *be an euclidian vector space with a compatible almost complex structure. The action of* L, $\varLambda$, *and* H *defines a natural* $\mathfrak{sl}(2)$-*representation on* $\bigwedge^* V^*$.

Proof. Recall, that $\mathfrak{sl}(2)$ is the three-dimensional (over $\mathbb{C}$ or over $\mathbb{R}$) Lie algebra of all 2×2-matrices of trace zero. A basis is given by $X = \left(\begin{smallmatrix} 0 & 1 \\ 0 & 0 \end{smallmatrix}\right)$, $Y = \left(\begin{smallmatrix} 0 & 0 \\ 1 & 0 \end{smallmatrix}\right)$, and $B = \left(\begin{smallmatrix} 1 & 0 \\ 0 & -1 \end{smallmatrix}\right)$. A quick calculation shows that they satisfy $[B, X] = 2X$, $[B, Y] = -2Y$, and $[X, Y] = B$. Thus mapping $X \mapsto L$, $Y \mapsto \varLambda$, and $B \mapsto H$ defines a Lie algebra homomorphism $\mathfrak{sl}(2) \to \mathrm{End}(\bigwedge^* V^*)$. The $\mathfrak{sl}(2, \mathbb{C})$-representation is obtained by tensorizing with $\mathbb{C}$. □

Assertion iii) of Proposition 1.2.26 can be generalized to

Corollary 1.2.28 $[L^i, \varLambda](\alpha) = i(k - n + i - 1)L^{i-1}(\alpha)$ *for all* $\alpha \in \bigwedge^k V^*$.

Proof. This is easily seen by induction on i as follows:

$$
\begin{aligned}
[L^i, \varLambda](\alpha) &= L^i \varLambda \alpha - \varLambda L^i \alpha \\
&= L(L^{i-1} \varLambda \alpha - \varLambda L^{i-1} \alpha) + L \varLambda L^{i-1} \alpha - \varLambda L L^{i-1} \alpha \\
&= L[L^{i-1}, \varLambda](\alpha) + [L, \varLambda](L^{i-1} \alpha) \\
&= (i - 1)(k - n + (i - 1) - 1)L^{i-1}(\alpha) + (2i - 2 + k - n)L^{i-1}(\alpha) \\
&= i(k - n + i - 1)L^{i-1}(\alpha).
\end{aligned}
$$

□

Definition 1.2.29 Let $(V, \langle \ , \ \rangle, I)$ and the induced operators L, Λ, and H be as before. An element $\alpha \in \bigwedge^k V^*$ is called *primitive* if $\Lambda \alpha = 0$. The linear subspace of all primitive elements $\alpha \in \bigwedge^k V^*$ is denoted by $P^k \subset \bigwedge^k V^*$.

Accordingly, an element $\alpha \in \bigwedge^k V_{\mathbb{C}}^*$ is called primitive if $\Lambda \alpha = 0$. Clearly, the subspace of those is just the complexification of P^k.

Proposition 1.2.30 *Let $(V, \langle \ , \ \rangle, I)$ be an euclidian vector space of dimension $2n$ with a compatible almost complex structure and let L and Λ be the associated Lefschetz operators.*
   i) *There exists a direct sum decomposition of the form:*

$$\bigwedge^k V^* = \bigoplus_{i \geq 0} L^i(P^{k-2i}). \tag{1.9}$$

This is the Lefschetz decomposition. *Moreover, (1.9) is orthogonal with respect to $\langle \ , \ \rangle$.*
   ii) *If $k > n$, then $P^k = 0$.*
   iii) *The map $L^{n-k} : P^k \to \bigwedge^{2n-k} V^*$ is injective for $k \leq n$.*
   iv) *The map $L^{n-k} : \bigwedge^k V^* \to \bigwedge^{2n-k} V^*$ is bijective for $k \leq n$.*
   v) *If $k \leq n$, then $P^k = \{\alpha \in \bigwedge^k V^* \mid L^{n-k+1}\alpha = 0\}$.*

The following two diagrams might to help memorize the above facts:

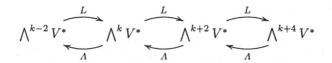

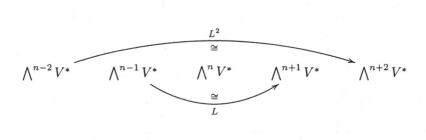

Proof. i) The easiest way to prove i) is to apply some small amount of representation theory. Since $\bigwedge^* V_{\mathbb{C}}^*$ is a finite-dimensional $\mathfrak{sl}(2)$-representation, it is a direct sum of irreducible ones. Any finite-dimensional $\mathfrak{sl}(2)$-representation admits a primitive vector v, i.e. $\Lambda v = 0$. Indeed, for any vector v the sequence $\Lambda^i v$ for $i = 0, 1, \ldots$ has to terminate by dimension reasons. (Use $H\Lambda^i v = (\deg(v) - 2i - n)\Lambda^i v$.) Using Corollary 1.2.28 one finds that for any primitive v the subspace $v, Lv, L^2 v, \ldots$ defines a subrepresentation. Thus, the irreducible $\mathfrak{sl}(2)$-representations are of this form. Altogether this proves the existence of the direct sum decomposition (1.9). The orthogonality with respect to $\langle \ , \ \rangle$ follows from Corollary 1.2.28.

ii) If $\alpha \in P^k$, $k > n$, and $0 < i$ minimal with $L^i\alpha = 0$, then by Corollary 1.2.28 one has $0 = [L^i, \Lambda](\alpha) = i(k - n + i - 1)L^{i-1}\alpha$. This yields $i = 0$, i.e. $\alpha = 0$.

iii) Let $0 \neq \alpha \in P^k$, $k \leq n$ and $0 < i$ minimal with $L^i\alpha = 0$. Then again by Corollary 1.2.28 one finds $0 = [L^i, \Lambda](\alpha) = i(k-n+i-1)L^{i-1}\alpha$ and, therefore, $k - n + i - 1 = 0$. In particular, $L^{n-k}(\alpha) \neq 0$. Moreover, $L^{n-k+1}\alpha = 0$, which will be used in the proof of v).

Assertion iv) follows from i), ii), and iii).

v) We have seen already that $P^k \subset \mathrm{Ker}(L^{n-k+1})$. Conversely, let $\alpha \in \bigwedge^k V^*$ with $L^{n-k+1}\alpha = 0$. Then $L^{n-k+2}\Lambda\alpha = L^{n-k+2}\Lambda\alpha - \Lambda L^{n-k+2}\alpha = (n - k + 2)L^{n-k+1}\alpha = 0$. But by iv) the map L^{n-k+2} is injective on $\bigwedge^{k-2} V^*$. Hence, $\Lambda\alpha = 0$. $\qquad\square$

Let us consider a few special cases. Obviously, $\bigwedge^0 V^* = P^0 = \mathbb{R}$ and $\bigwedge^1 V^* = P^1$. In degree two and four one has $\bigwedge^2 V^* = \omega\mathbb{R} \oplus P^2$ and $\bigwedge^4 V^* = \omega^2\mathbb{R} \oplus L(P^2) \oplus P^4$.

Roughly, the Lefschetz operators and its dual Λ induce a reflection of $\bigwedge^* V^*$ in the middle exterior product $\bigwedge^n V^*$. But there is another operator with this property, namely the Hodge $*$-operator. The interplay between these two is described in the following mysterious but extremely useful proposition.

Proposition 1.2.31 *For all $\alpha \in P^k$ one has*

$$*L^j\alpha = (-1)^{\frac{k(k+1)}{2}} \frac{j!}{(n - k - j)!} \cdot L^{n-k-j}\mathbf{I}(\alpha).$$

Proof. The proof will be given by induction. Suppose that $\dim_{\mathbb{C}}(V) = 1$. Choose an orthonormal basis $V = x_1\mathbb{R} \oplus y_1\mathbb{R}$ such that $I(x_1) = y_1$. Thus, $\omega = x^1 \wedge y^1$. Moreover, $\bigwedge^* V^* = \bigwedge^0 V^* \oplus \bigwedge^1 V^* \oplus \bigwedge^2 V^*$ and the primitive part of $\bigwedge^* V^*$ is $\bigwedge^0 V^* \oplus \bigwedge^1 V^*$. Thus, in order to prove the assertion in the one-dimensional case one has to compare $*1 = \omega$, $*\omega = 1$, $*x^1 = y^1$, and $*y^1 = -x^1$ with the corresponding expressions on the right hand side. Using $\mathbf{I}(x^1) = -y^1$ this is easily verified.

Next, let V be of arbitrary dimension and let $(V, \langle \ , \ \rangle, I) = (W_1, \langle \ , \ \rangle_1, I_1) \oplus (W_2, \langle \ , \ \rangle_2, I_2)$ be a direct sum decomposition. As has been used already in the proof of Proposition 1.2.26, one has $L = L_1 \otimes 1 + 1 \otimes L_2$ and $\Lambda = \Lambda_1 \otimes 1 + 1 \otimes \Lambda_2$ on $\bigwedge^* V^* = \bigwedge^* W_1^* \otimes \bigwedge^* W_2^*$. Moreover, for $\delta_i \in \bigwedge^{k_i} W_i^*$, $i = 1, 2$, the Hodge $*$-operator of $\delta_1 \otimes \delta_2$ is given by $*(\delta_1 \otimes \delta_2) = (-1)^{k_1 k_2}(*_1\delta_1) \otimes (*_2\delta_2)$.

Assuming the assertion for W_1 and W_2 one could in principle deduce the assertion for V. However, as the Lefschetz decomposition of $\bigwedge^* V^*$ is not the product of the Lefschetz decompositions of $\bigwedge^* W_1^*$ and $\bigwedge^* W_2^*$, the calculation is slightly cumbersome. It is actually more convenient to assume in addition that W_2 is complex one-dimensional. Of course, the induction argument is still valid. So, we let $W_2 = x_1\mathbb{R} \oplus y_1\mathbb{R}$ as in the one-dimensional case.

Any $\alpha \in \bigwedge^k V^*$ can thus be written as

$$\alpha = \beta_k + \beta'_{k-1} \otimes x^1 + \beta''_{k-1} \otimes y^1 + \beta_{k-2} \otimes \omega,$$

where $\beta_k \in \bigwedge^k W_1^*$, $\beta'_{k-1}, \beta''_{k-1} \in \bigwedge^{k-1} W_1^*$, and $\beta_{k-2} \in \bigwedge^{k-2} W_1^*$. Hence, $\Lambda\alpha = \Lambda_1\beta_k + (\Lambda_1\beta'_{k-1}) \otimes x^1 + (\Lambda_1\beta''_{k-1}) \otimes y^1 + (\Lambda_1\beta_{k-2}) \otimes \omega + \beta_{k-2}$. Thus, α is primitive if and only if $\beta'_{k-1}, \beta''_{k-1}, \beta_{k-2} \in \bigwedge^* W_1^*$ are primitive and $\Lambda_1\beta_k + \beta_{k-2} = 0$. The latter condition holds true if and only if the Lefschetz decomposition of β_k is of the form $\beta_k = \gamma_k + L_1\gamma_{k-2}$ and $\beta_{k-2} = (k - n - 1)\gamma_{k-2}$.

Next one computes $L^j\alpha$. Since W_2 is one-dimensional, one has $L^j = L_1^j \otimes 1 + jL_1^{j-1} \otimes L_2$ and, therefore,

$$L^j\alpha = L_1^j\gamma_k + L_1^{j+1}\gamma_{k-2} + j(L_1^{j-1}\gamma_k) \otimes \omega + j(L_1^j\gamma_{k-2}) \otimes \omega$$
$$+ (L_1^j\beta'_{k-1}) \otimes x^1 + (L_1^j\beta''_{k-1}) \otimes y^1 + (k - n - 1)(L_1^j\gamma_{k-2}) \otimes \omega.$$

In order to compute $*L^j\alpha$, one uses this equation and the induction hypothesis:

$$*_1 L_1^\ell\gamma_k = (-1)^{\frac{k(k+1)}{2}} \frac{\ell!}{(n-1-k-\ell)!} L_1^{n-1-k-\ell}\mathbf{I}_1(\gamma_k), \quad \ell = j-1, j$$

$$*_1 L_1^\ell\gamma_{k-2} = (-1)^{\frac{(k-2)(k-1)}{2}} \frac{\ell!}{(n+1-k-\ell)!} L_1^{n+1-k-\ell}\mathbf{I}_1(\gamma_{k-2}), \quad \ell = j, j+1$$

$$*_1 L_1^j\beta_{k-1}^{\prime(\prime)} = (-1)^{\frac{k(k-1)}{2}} \frac{j!}{(n-k-j)!} L_1^{n-k-j}\mathbf{I}_1(\beta_{k-1}^{\prime(\prime)}).$$

This yields

$$(-1)^{\frac{k(k+1)}{2}} \frac{(n-k-j)!}{j!} * L^j\alpha$$

$$= (n-k-j)(L_1^{n-k-j}\mathbf{I}_1(\gamma_k)) \otimes \omega - (j+1)(L_1^{n-k-j}\mathbf{I}_1(\gamma_{k-2})) \otimes \omega$$
$$- (L_1^{n-k-j}\mathbf{I}_1(\beta'_{k-1})) \otimes *_2(x^1) - (L_1^{n-k-j}\mathbf{I}_1(\beta''_{k-1})) \otimes *_2(y^1)$$
$$+ L_1^{n-k-j}\mathbf{I}_1(\gamma_k) - L_1^{n+1-k-j}\mathbf{I}_1(\gamma_{k-2})$$

On the other hand,

$$L^{n-k-j}\mathbf{I}(\alpha) = L_1^{n-k-j}\mathbf{I}_1(\gamma_k) + (n-k-j)(L_1^{n-k-j-1}\mathbf{I}_1(\gamma_k)) \otimes \omega$$
$$+ (L_1^{n-k-j+1}\mathbf{I}_1(\gamma_{k-2}) + (n-k-j)(L_1^{n-k-j}\mathbf{I}_1(\gamma_{k-2})) \otimes \omega$$
$$+ (L_1^{n-k-j}\mathbf{I}_1(\beta'_{k-1})) \otimes (-y^1) + (L_1^{n-k-j}\mathbf{I}_1(\beta''_{k-1})) \otimes x^1$$
$$+ (k - n - 1)(L_1^{n-k-j}\mathbf{I}_1(\gamma_{k-2})) \otimes \omega$$

Comparing both expressions yields the result. $\qquad\square$

Observe that the above proposition shows once again that L^{n-k} is bijective on $\bigwedge^k V^*$ for $k \leq n$ (cf. iv), Proposition 1.2.30).

Example 1.2.32 Here are a few instructive special cases. Let $j = k = 0$ and $\alpha = 1$, then we obtain $*1 = \frac{1}{n!}L^n1 = \frac{\omega^n}{n!}$. Thus, vol $= \frac{\omega^n}{n!}$ as was claimed before (Remark 1.2.22).

For $k = 0$, $\alpha = 1$, and $j = 1$, the proposition yields $*\omega = \frac{1}{(n-1)!}\omega^{n-1}$.

If α is a primitive $(1,1)$-form, i.e. $\alpha \in P^2 \cap \bigwedge^{1,1} V^*$, then $*\alpha = \frac{-1}{(n-2)!}\omega^{n-2} \wedge \alpha$.

Remark 1.2.33 Since L, Λ, and H are of pure type $(1,1)$, $(-1,-1)$ and $(0,0)$, respectively, the Lefschetz decomposition is compatible with the bidegree decomposition. Thus, $P_{\mathbb{C}}^k = \bigoplus_{p+q=k} P^{p,q}$, where $P^{p,q} = P_{\mathbb{C}}^k \cap \bigwedge^{p,q} V^*$. Since Λ and L are real, one also has $\overline{P^{p,q}} = P^{q,p}$.

Example 1.2.34 In particular, $\bigwedge^0 V_{\mathbb{C}}^* = P^{0,0} = P_{\mathbb{C}}^0 = \mathbb{C}$, $\bigwedge^1 V_{\mathbb{C}}^* = P^{1,0} \oplus P^{0,1}$, and

$$\bigwedge^2 V_{\mathbb{C}}^* = \bigwedge^{2,0} V^* \oplus \bigwedge^{1,1} V^* \oplus \bigwedge^{0,2} V^*$$
$$= P^{2,0} \oplus (P^{1,1} \oplus \omega\mathbb{C}) \oplus P^{0,2}.$$

Definition 1.2.35 Let $(V, \langle\ ,\ \rangle, I)$ be as before and let ω be the associated fundamental form. The *Hodge–Riemann pairing* is the bilinear form

$$Q : \bigwedge^k V^* \times \bigwedge^k V^* \longrightarrow \mathbb{R}, \quad (\alpha, \beta) \longmapsto (-1)^{\frac{k(k-1)}{2}} \alpha \wedge \beta \wedge \omega^{n-k},$$

where $\bigwedge^{2n} V^*$ is identified with $\mathbb{R}$ via the volume form vol.

By definition $Q = 0$ on $\bigwedge^k V^*$ for $k > n$. We will also denote by Q the $\mathbb{C}$-linear extension of the Hodge–Riemann pairing to $\bigwedge^* V_{\mathbb{C}}^*$.

Corollary 1.2.36 (Hodge–Riemann bilinear relation) *Let $(V, \langle\ ,\ \rangle, I)$ be an euclidian vector space endowed with a compatible almost complex structure. Then the associated Hodge–Riemann pairing Q satisfies:*

$$Q(\bigwedge^{p,q} V^*, \bigwedge^{p',q'} V^*) = 0$$

for $(p, q) \neq (q', p')$ and

$$i^{p-q} Q(\alpha, \bar{\alpha}) = (n - (p+q))! \cdot \langle \alpha, \alpha \rangle_{\mathbb{C}} > 0$$

for $0 \neq \alpha \in P^{p,q}$ with $p + q \leq n$.

Proof. Only the second assertion needs a proof. By definition

$$Q(\alpha, \bar{\alpha}) \cdot \text{vol} = (-1)^{\frac{k(k-1)}{2}} \alpha \wedge \bar{\alpha} \wedge \omega^{n-k}$$
$$= (-1)^{\frac{k(k-1)}{2}} \alpha \wedge L^{n-k}\bar{\alpha}$$
$$= (-1)^{\frac{k(k-1)}{2}} \langle \alpha, \beta \rangle_{\mathbb{C}} \cdot \text{vol},$$

where $k = p + q$ and $\beta \in \bigwedge^k V^*$ such that $*\bar{\beta} = L^{n-k}\bar{\alpha}$. Hence, $*^2\bar{\beta} = (-1)^k\bar{\beta}$ and, on the other hand,

$$*^2\bar{\beta} = *L^{n-k}\bar{\alpha} = (-1)^{\frac{k(k+1)}{2}}(n-k)!i^{q-p}\bar{\alpha}$$

by Proposition 1.2.31. Thus, $\beta = (-1)^{k+\frac{k(k+1)}{2}}(n-k)! \cdot i^{p-q}\alpha$ and, therefore,

$$Q(\alpha, \bar{\alpha}) = (-1)^{k+\frac{k(k+1)}{2}+\frac{k(k-1)}{2}}(n-k)! \cdot i^{q-p} \cdot \langle \alpha, \alpha \rangle_{\mathbb{C}}.$$

This yields $i^{p-q}Q(\alpha, \bar{\alpha}) = (n-k)! \cdot \langle \alpha, \alpha \rangle_{\mathbb{C}} > 0$ for $0 \neq \alpha \in P^{p,q}$. $\square$

Example 1.2.37 Suppose $n \geq 2$ and consider the decomposition $(\bigwedge^{1,1} V^*)_{\mathbb{R}} = \omega\mathbb{R} \oplus P^{1,1}_{\mathbb{R}}$, where $(\ )_{\mathbb{R}}$ denotes the intersection with $\bigwedge^2 V^*$. Then, the decomposition is Q-orthogonal, because $(\alpha \wedge \omega) \wedge \omega^{n-2} = \alpha \wedge \omega^{n-1} = 0$ for $\alpha \in P^2$. Moreover, Q is a positive definite symmetric bilinear form on $\omega\mathbb{R}$ and a negative definite symmetric bilinear from on $P^{1,1}_{\mathbb{R}}$. This is what will lead to the Hodge index theorem in Section 3.3.

Exercises

1.2.1 Let $(V, \langle\ ,\ \rangle)$ be a four-dimensional euclidian vector space. Show that the set of all compatible almost complex structures consist of two copies of S^2.

1.2.2 Show that the two decompositions $\bigwedge^k V^* = \bigoplus_{0 \leq i} L^i P^{k-2i}$ and $L^i P^{k-2i} = \bigoplus_{p+q=k-2i} L^i P^{p,q}$ are orthogonal with respect to the Hodge–Riemann pairing.

1.2.3 Prove the following identities: $*\Pi^{p,q} = \Pi^{n-q,n-p}*$ and $[L, \mathbf{I}] = [\Lambda, \mathbf{I}] = 0$.

1.2.4 Is the product of two primitive forms again primitive?

1.2.5 Let $(V, \langle\ ,\ \rangle)$ be an euclidian vector space and let I, J, and K be compatible almost complex structures where $K = I \circ J = -J \circ I$. Show that V becomes in a natural way a vector space over the quaternions. The associated fundamental forms are denoted by ω_I, ω_J, and ω_K. Show that $\omega_J + i\omega_K$ with respect to I is a form of type $(2, 0)$. How many natural almost complex structures do you see in this context?

1.2.6 Let $\omega \in \bigwedge^2 V^*$ be non-degenerate, i.e. the induced homomorphism $\tilde{\omega} : V \to V^*$ is bijective. Study the relation between the two isomorphisms $L^{n-k} : \bigwedge^k V^* \to \bigwedge^{2n-k} V^*$ and $\bigwedge^k V^* \cong \bigwedge^{2n-k} V \cong \bigwedge^{2n-k} V^*$, where the latter is given by $\bigwedge^{2n-k} \tilde{\omega}$. Here, $2n = \dim_{\mathbb{R}}(V)$.

1.2.7 Let V be a vector space endowed with a scalar product and a compatible almost complex structure. What is the signature of the pairing $(\alpha, \beta) \mapsto \frac{\alpha \wedge \beta \wedge \omega^{n-2}}{\text{vol}}$ on $\bigwedge^2 V^*$?

1.2.8 Let $\alpha \in P^k$ and $s \geq r$. Prove the following formula $\Lambda^s L^r \alpha = r(r-1)\ldots(r - s + 1)(n - k - r + 1)\ldots(n - k - r + s)L^{r-s}\alpha$.

1.2.9 (Wirtinger inequality) Let $(V, \langle\ ,\ \rangle)$ be an euclidian vector space endowed with a compatible almost complex structure I and the associated fundamental form ω. Let $W \subset V$ be an oriented subspace of dimension $2m$. The induced scalar product on W together with the chosen orientation define a natural volume form $\mathrm{vol}_W \in \bigwedge^{2m} W^*$. Show that
$$\omega^m|_W \leq m! \cdot \mathrm{vol}_W$$
and that equality holds if and only if $W \subset V$ is a complex subspace, i.e. $I(W) = W$, and the orientation is the one induced by the almost complex structure. (The inequality is meant with respect to the isomorphism $\bigwedge^{2m} W^* \cong \mathbb{R}$, $\mathrm{vol}_W \mapsto 1$. Hint: Use that there exists an oriented orthonormal base $e_1, \ldots, e_{2m}$ such that $\omega|_W = \sum_{i=1}^m \lambda_i e^{2i} \wedge e^{2i-1}$ and the Cauchy–Schwarz inequality.)

1.2.10   Choose an orthonormal basis $x_1, y_1 = I(x_1), \ldots, x_n, y_n = I(x_n)$ of an euclidian vector space V endowed with a compatible almost complex structure I. Show that the dual Lefschetz operator applied to a two-form α is explicitly given by $\Lambda\alpha = \sum \alpha(x_i, y_i)$.

Comments: The preceding section was essentially a chapter in linear algebra. I believe that the original source for this is [114]. Most of it, though not always with complete proofs, can be found in [35, 59, 116]. In [116] the point of view of $\mathfrak{sl}(2)$-representations is emphasized.

1.3 Differential Forms

A real manifold M is studied by means of its tangent bundle TM, the collection of all tangent spaces $T_x M$ for $x \in M$, and its k-form bundle $\bigwedge^k (TM)^*$. In this section we will apply the linear algebra developed previously to the form bundles of an open subset $M = U \subset \mathbb{C}^n$. The bidegree decomposition induces a decomposition of the exterior differential d which is well suited for the study of holomorphic functions on U. We conclude by a local characterization of so called Kähler metrics which will be of central interest in the global setting.

Let $U \subset \mathbb{C}^n$ be an open subset. Thus, U can in particular be considered as a $2n$-dimensional real manifold. For $x \in U$ we have its real tangent space $T_x U$ at the point x which is of real dimension $2n$. A canonical basis of $T_x U$ is given by the tangent vectors

$$\frac{\partial}{\partial x_1}, \dots, \frac{\partial}{\partial x_n}, \frac{\partial}{\partial y_1}, \dots, \frac{\partial}{\partial y_n},$$

where $z_1 = x_1 + iy_1, \dots, z_n = x_n + iy_n$ are the standard coordinates on $\mathbb{C}^n$. Moreover, the vectors $\frac{\partial}{\partial x_1}, \dots, \frac{\partial}{\partial y_n}$ are global trivializing sections of TU.

Each tangent space $T_x U$ admits a natural almost complex structure defined by

$$I : T_x U \longrightarrow T_x U, \quad \frac{\partial}{\partial x_i} \longmapsto \frac{\partial}{\partial y_i}, \quad \frac{\partial}{\partial y_i} \longmapsto -\frac{\partial}{\partial x_i},$$

which is compatible with the global trivialization. We shall regard I as a vector bundle endomorphism of the real vector bundle TU over U.

The dual basis of $(T_x U)^*$ is denoted by $dx_1, \dots, dx_n, dy_1, \dots, dy_n$. Recall that the induced almost complex structure on $T_x U$ in terms of this dual basis is described by $I(dx_i) = -dy_i$, $I(dy_i) = dx_i$ (cf. page 28).

The general theory developed in the previous section applies to this almost complex structure and yields the following

Proposition 1.3.1 *The complexified tangent bundle $T_\mathbb{C} U := TU \otimes \mathbb{C}$ decomposes as a direct sum of complex vector bundles*

$$T_\mathbb{C} U = T^{1,0} U \oplus T^{0,1} U,$$

such that the complex linear extension of I satisfies

$$I|_{T^{1,0}U} = i \cdot \mathrm{id} \quad \text{and} \quad I|_{T^{0,1}U} = -i \cdot \mathrm{id}.$$

The vector bundles $T^{1,0}U$ and $T^{0,1}U$ are trivialized by the sections $\frac{\partial}{\partial z_i} := \frac{1}{2}(\frac{\partial}{\partial x_i} - i\frac{\partial}{\partial y_i})$ and $\frac{\partial}{\partial \bar{z}_i} := \frac{1}{2}(\frac{\partial}{\partial x_i} + i\frac{\partial}{\partial y_i})$, $i = 1, \dots, n$, respectively. $\square$

The complexified cotangent bundle $T_\mathbb{C}^* U := T^* U \otimes \mathbb{C}$ admits an analogous decomposition $T_\mathbb{C}^* U = (T^* U)^{1,0} \oplus (T^* U)^{0,1}$ and $(T^* U)^{1,0}$ and $(T^* U)^{0,1}$ are trivialized by the dual basis $dz_i := dx_i + idy_i$ and $d\bar{z}_i := dx_i - idy_i$, $i = 1, \dots, n$, respectively.

Note that these decompositions are compatible with restriction to smaller open subsets $U' \subset U$. As we have seen and used already in Section 1.1, one has

Proposition 1.3.2 *Let* $f : U \to V$ *be a holomorphic map between open subsets* $U \subset \mathbb{C}^m$ *and* $V \subset \mathbb{C}^n$. *The* $\mathbb{C}$-*linear extension of the differential* $df : T_x U \to T_{f(x)} V$ *respects the above decomposition, i.e.* $df(T_x^{1,0} U) \subset T_{f(x)}^{1,0} V$ *and* $df(T_x^{0,1} U) \subset T_{f(x)}^{0,1} V$. $\hfill\square$

In a similar fashion, we can use the results of the previous section in order to decompose the bundles of k-forms.

Definition 1.3.3 Let $U \subset \mathbb{C}^n$ be an open subset. Over U one defines the complex vector bundles

$$\bigwedge\nolimits^{p,q} U := \bigwedge\nolimits^{p}((T^*U)^{1,0}) \otimes \bigwedge\nolimits^{q}((T^*U)^{0,1})).$$

By $\mathcal{A}_{\mathbb{C}}^k(U)$ and $\mathcal{A}^{p,q}(U)$ we denote the spaces of sections of $\bigwedge_{\mathbb{C}}^k U := \bigwedge^k T_{\mathbb{C}}^* U$ and $\bigwedge^{p,q} U$, respectively.

Proposition 1.2.8 immediately yields

Corollary 1.3.4 *There are natural decompositions* $\bigwedge_{\mathbb{C}}^k U = \bigoplus_{p+q=k} \bigwedge^{p,q} U$ *and* $\mathcal{A}_{\mathbb{C}}^k(U) = \bigoplus_{p+q=k} \mathcal{A}^{p,q}(U)$. $\hfill\square$

The restriction map $\mathcal{A}_{\mathbb{C}}^k(U) \to \mathcal{A}_{\mathbb{C}}^k(U')$ for an open subset $U' \subset U$ respects this decomposition. As before, the projection operators $\bigwedge_{\mathbb{C}}^k U \to \bigwedge^{p,q} U$ and $\mathcal{A}_{\mathbb{C}}^k(U) \to \mathcal{A}^{p,q}(U)$ will be denoted by $\Pi^{p,q}$.

Definition 1.3.5 Let $d : \mathcal{A}_{\mathbb{C}}^k(U) \to \mathcal{A}_{\mathbb{C}}^{k+1}(U)$ be the complex linear extension of the usual exterior differential. Then

$$\partial : \mathcal{A}^{p,q}(U) \longrightarrow \mathcal{A}^{p+1,q}(U) \quad \text{and} \quad \bar{\partial} : \mathcal{A}^{p,q}(U) \longrightarrow \mathcal{A}^{p,q+1}(U)$$

are defined as $\partial := \Pi^{p+1,q} \circ d$ and $\bar{\partial} := \Pi^{p,q+1} \circ d$.

For any local function f one has

$$d(f) = \sum_i \frac{\partial f}{\partial x_i} dx_i + \sum_i \frac{\partial f}{\partial y_i} dy_i = \sum_i \frac{\partial f}{\partial z_i} dz_i + \sum_i \frac{\partial f}{\partial \bar{z}_i} d\bar{z}_i. \qquad (1.10)$$

Thus, f is holomorphic if and only if $\bar{\partial} f = 0$. Moreover, using (1.10) the operators ∂ and $\bar{\partial}$ can be expressed explicitly as follows:

$$\partial(f dz_{i_1} \wedge \ldots \wedge dz_{i_p} \wedge d\bar{z}_{j_1} \ldots \wedge d\bar{z}_{j_q}) = \sum_{k=1}^{n} \frac{\partial f}{\partial z_k} dz_k \wedge dz_{i_1} \wedge \ldots \wedge dz_{i_p} \wedge d\bar{z}_{j_1} \ldots \wedge d\bar{z}_{j_q}$$

and

$$\bar\partial(f dz_{i_1}\wedge\ldots\wedge dz_{i_p}\wedge d\bar z_{j_1}\ldots\wedge d\bar z_{j_q}) = \sum_{\ell=1}^{n} \frac{\partial f}{\partial \bar z_\ell} d\bar z_\ell \wedge dz_{i_1}\wedge\ldots\wedge dz_{i_p}\wedge d\bar z_{j_1}\ldots\wedge d\bar z_{j_q}.$$

Lemma 1.3.6 *For the differential operators ∂ and $\bar\partial$ one has:*
 i) $d = \partial + \bar\partial$.
 ii) $\partial^2 = \bar\partial^2 = 0$ *and* $\partial\bar\partial = -\bar\partial\partial$.
 iii) *They satisfy the Leibniz rule, i.e.*

$$\partial(\alpha\wedge\beta) = \partial(\alpha)\wedge\beta + (-1)^{p+q}\alpha\wedge\partial(\beta)$$
$$\bar\partial(\alpha\wedge\beta) = \bar\partial(\alpha)\wedge\beta + (-1)^{p+q}\alpha\wedge\bar\partial(\beta)$$

for $\alpha \in \mathcal{A}^{p,q}(U)$ and $\beta \in \mathcal{A}^{r,s}(U)$.

Proof. i) follows from the local description of ∂ and $\bar\partial$ given above and ii) is deduced from $d^2 = 0$.

To see iii) we recall that the exterior differential satisfies

$$d(\alpha\wedge\beta) = d(\alpha)\wedge\beta + (-1)^{p+q}\alpha\wedge d(\beta).$$

Taking the $(p+r+1, q+s)$-parts on both sides one obtains the Leibniz rule for ∂. Similarly, taking $(p+r, q+s+1)$-parts proves the assertion for $\bar\partial$. ☐

Since $\bar\partial$ and ∂ share the usual properties of the exterior differential d and reflect the holomorphicity of functions, it seems natural to build up a holomorphic analogue of the de Rham complex. As we work here exclusively in the local context, only the local aspects will be discussed. Of course, locally the de Rham complex is exact (see Appendix A) due to the standard Poincaré lemma. We will show that this still holds true for $\bar\partial$ (and ∂).

Proposition 1.3.7 ($\bar\partial$-Poincaré lemma in one variable) *Consider an open neighbourhood of the closure of a bounded one-dimensional disc $B_\varepsilon \subset \overline{B}_\varepsilon \subset U \subset \mathbb{C}$. For $\alpha = f d\bar z \in \mathcal{A}^{0,1}(U)$ the function*

$$g(z) := \frac{1}{2\pi i}\int_{B_\varepsilon} \frac{f(w)}{w-z} dw \wedge d\bar w$$

on B_ε satisfies $\alpha = \bar\partial g$.

Proof. Note first, that for $w = x+iy$ one has $dw\wedge d\bar w = (dx+idy)\wedge(dx-idy) = -2idx \wedge dy$. The existence of g as well as the assertion $\alpha = \bar\partial g$ will be shown by splitting g into two parts. This splitting will depend on a chosen point $z_0 \in B_\varepsilon$ or rather on a neighbourhood of such a point.

Let $z_0 \in B := B_\varepsilon$ and let $\psi : B \to \mathbb{R}$ be a differentiable function with compact supp$(\psi) \subset B$ and such that $\psi|_V \equiv 1$ for some open neighbourhood of $z_0 \in V \subset B$. If $f_1 := \psi \cdot f$ and $f_2 := (1 - \psi) \cdot f$, then $f = f_1 + f_2$. In order to see that the above integral is well-defined we consider first the following integrals

$$g_i(z) := \frac{1}{2\pi i} \int_B \frac{f_i(w)}{w - z} dw \wedge d\bar{w}, \ i = 1, 2.$$

Since $f_2|_V \equiv 0$, the second one is obviously well defined for $z \in V$. The first integral can be rewritten as

$$g_1(z) = \frac{1}{2\pi i} \int_B \frac{f_1(w)}{w - z} dw \wedge d\bar{w}$$

$$= \frac{1}{2\pi i} \int_{\mathbb{C}} \frac{f_1(w)}{w - z} dw \wedge d\bar{w}, \text{ since } \operatorname{supp}(f_1) \subset B \text{ is compact}$$

$$= \frac{1}{2\pi i} \int_{\mathbb{C}} \frac{f_1(u + z)}{u} du \wedge d\bar{u}, \text{ for } u := w - z$$

$$= \frac{1}{\pi} \int_{\mathbb{C}} f_1 \left(z + re^{i\varphi} \right) e^{-i\varphi} d\varphi \wedge dr, \text{ for } u = re^{i\varphi}.$$

The last integral is clearly well-defined. Since the integral defining g splits into the two integrals just considered, we see that the function g in the assertion is well-defined on V and thus everywhere on B.

In order to compute $\bar{\partial} g$, we use the same splitting of $g = g_1 + g_2$ as before. Let us first consider $\bar{\partial} g_2$. Since $(w - z)^{-1}$ is holomorphic as a function of z for w in the complement of V, one finds

$$\frac{\partial g_2}{\partial \bar{z}}(z) = \frac{1}{2\pi i} \int_B f_2(w) \frac{\partial (w - z)^{-1}}{\partial \bar{z}} dw \wedge d\bar{w} = 0$$

for all $z \in V$.

Using the above expression for g_1 we get

$$\frac{\partial g_1}{\partial \bar{z}}(z) = \frac{1}{\pi} \int_{\mathbb{C}} \frac{\partial f_1(z + re^{i\varphi})}{\partial \bar{z}} e^{-i\varphi} d\varphi \wedge dr$$

$$= \frac{1}{\pi} \int_{\mathbb{C}} \left(\frac{\partial f_1}{\partial \bar{w}} \frac{\partial (\bar{z} + re^{-i\varphi})}{\partial \bar{z}} + \frac{\partial f_1}{\partial w} \frac{\partial (z + re^{-i\varphi})}{\partial \bar{z}} \right) e^{-i\varphi} d\varphi \wedge dr$$

$$= \frac{1}{\pi} \int_{\mathbb{C}} \frac{\partial f_1}{\partial \bar{w}} (z + re^{i\varphi}) e^{-i\varphi} d\varphi \wedge dr$$

$$= \frac{1}{2\pi i} \int_B \frac{\partial f_1}{\partial \bar{w}} (w) \frac{dw \wedge d\bar{w}}{w - z}.$$

Thus, for $z \in V$ one has

$$\frac{\partial g}{\partial \bar{z}}(z) = \frac{\partial g_1}{\partial \bar{z}} + \frac{\partial g_2}{\partial \bar{z}} = \frac{\partial g_1}{\partial \bar{z}}$$

$$= \frac{1}{2\pi i} \int_B \frac{\partial f_1}{\partial \bar{w}}(w) \frac{dw \wedge d\bar{w}}{w - z} \stackrel{(*)}{=} f_1(z) = f(z).$$

Here, (*) is a consequence of Stokes' theorem:

$$\frac{1}{2\pi i} \int_B \frac{\partial f_1}{\partial \bar{w}}(w) \frac{dw \wedge d\bar{w}}{w - z} = \frac{1}{2\pi i} \lim_{\delta \to 0} \int_{B \setminus B_\delta(z)} \frac{\partial f_1}{\partial \bar{w}}(w) \frac{dw \wedge d\bar{w}}{w - z}$$

$$= \frac{-1}{2\pi i} \lim_{\delta \to 0} \int_{B \setminus B_\delta(z)} d\left(\frac{f_1(w)}{w - z} dw\right), \quad \begin{array}{l} \text{since } (w - z)^{-1} \text{ is} \\ \text{holomorphic on } B \setminus B_\delta(z) \end{array}$$

$$= \frac{1}{2\pi i} \lim_{\delta \to 0} \int_{\partial B_\delta(z)} \frac{f_1(w)}{w - z} dw, \quad \text{since } \operatorname{supp}(f_1) \subset B$$

$$= \frac{1}{2\pi} \lim_{\delta \to 0} \int_0^{2\pi} f_1(z + \delta e^{i\varphi}) d\varphi = f_1(z).$$

$\square$

The following proposition and its corollary are known as the Grothendieck–Poincaré lemma. The first proof of it is due to Grothendieck and was presented by Serre in the Séminaire Cartan in 1958.

Proposition 1.3.8 ($\bar{\partial}$-Poincaré lemma in several variables) *Let U be an open neighbourhood of the closure of a bounded polydisc $B_\varepsilon \subset \overline{B}_\varepsilon \subset U \subset \mathbb{C}^n$. If $\alpha \in \mathcal{A}^{p,q}(U)$ is $\bar{\partial}$-closed and $q > 0$, then there exists a form $\beta \in \mathcal{A}^{p,q-1}(B_\varepsilon)$ with $\alpha = \bar{\partial}\beta$ on B_ε.*

Proof. We first reduce the assertion to the case that $p = 0$. To this end, we write the form $\alpha \in \mathcal{A}^{p,q}(U)$ as

$$\alpha = \sum_{I,J} f_{IJ} dz_I \wedge \bar{z}_J = \sum_I dz_I \wedge \alpha_I,$$

where $|I| = p$, $|J| = q$, and $\alpha_I = \sum_J f_{IJ} d\bar{z}_J \in \mathcal{A}^{0,q}(U)$. Then $\bar{\partial}\alpha = \sum_{i,I,J} \frac{\partial f_{IJ}}{\partial \bar{z}_i} d\bar{z}_i \wedge dz_I \wedge d\bar{z}_J = 0$ if and only if $\bar{\partial}\alpha_I = 0$ for all I. Moreover, $\alpha = \bar{\partial}(\sum_{K,L} g_{KL} dz_K \wedge d\bar{z}_L)$ if and only if $\alpha_I = \bar{\partial}(\sum_L g_{IL} d\bar{z}_L)$ for all I.

Next, let $\alpha \in \mathcal{A}^{0,q}(U)$ be $\bar{\partial}$-closed and write $\alpha = \sum f_I d\bar{z}_I$. Choose k minimal such that no $d\bar{z}_i$ occurs in this sum for $i > k$. Thus, we can write $\alpha = \alpha_1 \wedge d\bar{z}_k + \alpha_2$, with α_2 free of $d\bar{z}_i$ for $i \geq k$. By assumption, $0 = \bar{\partial}\alpha = (\bar{\partial}\alpha_1) \wedge d\bar{z}_k + \bar{\partial}\alpha_2$. If we set $\bar{\partial}_i := (\partial/\partial \bar{z}_i) d\bar{z}_i$, then this implies $\bar{\partial}_i \alpha_1 = \bar{\partial}_i \alpha_2 = 0$ for $i > k$. Therefore, the functions f_I are holomorphic in $z_{k+1}, \ldots, z_n$.

By the one-dimensional Poincaré lemma 1.3.7 the function

$$g_I(z) = \frac{1}{2\pi i} \int_{B_{\varepsilon_k}} \frac{f_I(z_1, \ldots, z_{k-1}, w, z_{k+1}, \ldots,)}{w - z_k} dw \wedge d\bar{w}$$

satisfies $\frac{\partial g_I}{\partial \bar{z}_k} = f_I$ on $B_{\varepsilon_k} \subset \mathbb{C}$. Moreover, the function g_I is holomorphic in $z_{k+1}, \ldots, z_n$ and differentiable in the other variables.

Set $\gamma := (-1)^q \sum_{k \in I} g_I d\bar{z}_{I \setminus \{k\}}$. Then $\bar{\partial}_i \gamma(z) = 0$ for $i > k$ and $\bar{\partial}_k \gamma(z) = -\alpha_1$. Hence, $\alpha + \bar{\partial}\gamma$ is still $\bar{\partial}$-closed, but it does not involve any $d\bar{z}_i$ for $i \geq k$ anymore. Then one concludes by induction. $\qquad \square$

In the following, B shall denote a polydisc B_ε which can be unbounded, i.e. $\varepsilon_i = \infty$ is allowed.

Corollary 1.3.9 ($\bar{\partial}$-Poincaré lemma on the open disc) *If $\alpha \in \mathcal{A}^{p,q}(B)$ is $\bar{\partial}$-closed and $q > 0$, then there exists $\beta \in \mathcal{A}^{p,q-1}(B)$ with $\alpha = \bar{\partial}\beta$.*

Proof. Choose strictly monoton increasing sequences $\varepsilon_i(m)$ with $\varepsilon_i(m) \to \varepsilon_i$ for $m \to \infty$. Let B_m be the polydisc $\{z \mid |z_i| < \varepsilon_i(m)\}$. Thus, $B_1 \subset B_2 \subset \ldots \subset \bigcup B_m = B$.

Let us first show that for any m there exists a form $\beta_m \in \mathcal{A}^{p,q-1}(B)$ with $\bar{\partial}\beta_m = \alpha$ on B_m. Due to Proposition 1.3.8 we find $\beta'_m \in \mathcal{A}^{p,q-1}(B_{m+1})$ with $\bar{\partial}\beta'_m = \alpha$ on B_{m+1}. Choose a function ψ on B_ε with $\text{supp}(\psi) \subset B_{m+1}$ and $\psi|_{B_m} \equiv 1$. Then define $\beta_m := \psi \cdot \beta'_m$, which is a form on B_{m+1} that extends smoothly to a form on B. Clearly, $\bar{\partial}\beta_m = \bar{\partial}\beta'_m = \alpha$ on B_m.

Next, we claim that for $q > 1$ we can choose a sequence (β_m) as above with the additional property that $\beta_m = \beta_{m+1}$ on B_{m-1}. Assume we have constructed $\beta_1, \ldots, \beta_{m-1}$ already. Choose any $\tilde{\beta}_{m+1} \in \mathcal{A}^{p,q-1}(B)$ with $\bar{\partial}\tilde{\beta}_{m+1} = \alpha$ on B_{m+1}. Hence, $\bar{\partial}(\beta_m - \tilde{\beta}_{m+1}) = 0$ on B_m. By induction hypothesis we may assume that there exists $\gamma \in \mathcal{A}^{p,q-2}(B_m)$ with $\beta_m - \tilde{\beta}_{m+1} = \bar{\partial}\gamma$. Choose a function ψ on B such that $\text{supp}(\psi) \subset B_m$ and $\psi|_{B_{m-1}} \equiv 1$ and let $\beta_{m+1} := \tilde{\beta}_{m+1} + \bar{\partial}(\psi \cdot \gamma)$. Then $\beta_{m+1}|_{B_{m-1}} = \tilde{\beta}_{m+1}|_{B_{m-1}} + \bar{\partial}(\gamma)|_{B_{m-1}} = \beta_m|_{B_{m-1}}$ and $\bar{\partial}\beta_{m+1} = \bar{\partial}\tilde{\beta}_{m+1} = \alpha$ on B_{m+1}. Clearly, the sequence (β_m) obtained in this way converges to a form $\beta \in \mathcal{A}^{p,q-1}(B)$ with $\bar{\partial}\beta = \alpha$ on B.

For $q = 1$ we proceed as follows. This time one constructs a sequence (β_m) of functions on B with $\bar{\partial}\beta_m = \alpha$ on B_m and $|\beta_{m+1} - \beta_m| < 2^{-m}$ on B_{m-1}. This would yield a locally uniformly convergent sequence the limit of which would provide the desired function on B. Assume $\beta_1, \ldots, \beta_m$ have been constructed. Choose $\tilde{\beta}_{m+1} \in \mathcal{A}^{p,0}(B)$ with $\bar{\partial}\tilde{\beta}_{m+1} = \alpha$ on B_{m+1}. Then the function $\beta_m - \tilde{\beta}_{m+1}$ on B_m is holomorphic and can therefore be expanded in a power series. On the smaller disc $B_{m-1} \subset B_m$ it can be approximated by a polynomial P such that $|\beta_m - \tilde{\beta}_{m+1} - P| < 2^{-m}$. The polynomial P defines a holomorphic function on B and we set $\beta_{m+1} := \tilde{\beta}_{m+1} + P$. Then $\bar{\partial}\beta_{m+1} = \bar{\partial}\tilde{\beta}_{m+1} = \alpha$ on B_{m+1} and $|\beta_{m+1} - \beta_m| < 2^{-m}$ on B_{m-1}. $\qquad \square$

So far, only the consequences of the existence of a natural (almost) complex structure on each $T_x U$ have been discussed. Following the presentation in Section 1.2 we shall conclude by combining this with certain metric aspects of the manifold U.

Let $U \subset \mathbb{C}^n$ be an open subset and consider a Riemannian metric g on U. For what follows we may always assume that U is a polydisc. The metric g is *compatible* with the natural (almost) complex structure on U if for any $x \in U$ the induced scalar product g_x on $T_x U$ is compatible with the induced almost complex structure I, i.e. $g_x(v, w) = g_x(I(v), I(w))$ for all $v, w \in T_x U$. Recall, that by Definition 1.2.13 one has in this situation a natural $(1,1)$-form $\omega \in \mathcal{A}^{1,1}(U) \cap \mathcal{A}^2(U)$ defined by

$$\omega := g(I(\ ),(\ )),$$

which is called the *fundamental form* of g. Moreover, $h := g - i\omega$ defines a positive hermitian form on the complex vector spaces $(T_x U, g_x)$ for any $x \in U$ (cf. Lemma 1.2.15).

Example 1.3.10 Let g be the constant standard metric such that

$$\frac{\partial}{\partial x_1}, \ldots, \frac{\partial}{\partial x_n}, \frac{\partial}{\partial y_1}, \ldots, \frac{\partial}{\partial y_n}$$

is an orthonormal basis for any $T_x U$. Clearly, complex structure and g are compatible. The form ω in this is case is (see page 30)

$$\omega = \frac{i}{2} \sum_{i=1}^{n} dz_i \wedge d\bar{z}_i.$$

An arbitrary metric g on U, if compatible with the almost complex structure, is uniquely determined by the matrix $h_{ij}(z) := h(\frac{\partial}{\partial x_i}, \frac{\partial}{\partial x_j})$. The fundamental form can then be written as

$$\omega = \frac{i}{2} \sum_{i,j=1}^{n} h_{ij} dz_i \wedge d\bar{z}_j.$$

Even if g is not the standard metric, one might try to change the complex coordinates $z_1, \ldots, z_n$ such that it becomes the standard metric with respect to the new coordinates. Of course, this cannot always be achieved, but a reasonable class of metrics is the following one.

Definition 1.3.11 The metric g *osculates in the origin to order two* to the standard metric if $(h_{ij}) = \mathrm{id} + O(|z|^2)$.

Explicitly, the condition means $\frac{\partial h_{ij}}{\partial z_k}(0) = \frac{\partial h_{ij}}{\partial \bar{z}_k}(0) = 0$ for all i, j, k. In other words, the power series expansion of (h_{ij}) differs from the constant matrix id by terms of order at least two, thus terms of the form $a_{ijk}z_k + a'_{ijk}\bar{z}_k$ do not occur.

Osculating metrics will provide the local models of Kähler metrics which will be extensively studied in the later chapters. Here is the crucial fact:

Proposition 1.3.12 *Let g be a compatible metric on U and let ω be the associated fundamental form. Then, $d\omega = 0$ if and only if for any point $x \in U$ there exist a neighbourhood U' of $0 \in \mathbb{C}^n$ and a local biholomorphic map $f : U' \cong f(U') \subset U$ with $f(0) = x$ and such that f^*g osculates in the origin to order two to the standard metric.*

Proof. First note that for any local biholomorphic map f the pull-back $f^*\omega$ is the associated fundamental form to f^*g. In particular, ω is closed on $f(U')$ if and only if $f^*\omega$ is closed. Thus, in order to show that $d\omega = 0$ one can assume that the metric g osculates to order two to the standard metric and then one verifies that $d\omega$ vanishes in the origin. But the latter follows immediately from $\frac{\partial h_{ij}}{\partial z_k}(0) = \frac{\partial h_{ij}}{\partial \bar{z}_k}(0) = 0$.

For the other direction let us assume that $d\omega = 0$. We fix a point $x \in U$. After translating we may assume that $x = 0$. By a linear coordinate change we may furthermore assume that $(h_{ij})(0) = \mathrm{id}$. Thus

$$h_{ij} = \delta_{ij} + \sum_k a_{ijk}z_k + \sum_k a'_{ijk}\bar{z}_k + O(|z|^2).$$

Thus, $a_{ijk} = \frac{\partial h_{ij}}{\partial z_k}(0)$ and $a'_{ijk} = \frac{\partial h_{ij}}{\partial \bar{z}_k}(0)$. The assumption $d\omega(0) = 0$ implies $a_{ijk} = a_{kji}$ and $a'_{ijk} = a'_{ikj}$. Furthermore, since ω is real, $h_{ij} = \bar{h}_{ji}$ and thus $a'_{ijk} = \bar{a}_{jik}$. New holomorphic coordinates in a neighbourhood of the origin can now be defined by

$$w_j := z_j + \frac{1}{2}\sum_{i,k=1}^n a_{ijk}z_iz_k.$$

Then, $dw_j = dz_j + \frac{1}{2}\sum_{i,k=1}^n a_{ijk}(dz_i)z_k + \frac{1}{2}\sum_{i,k=1}^n a_{ijk}z_i(dz_k) = dz_j + \sum_{i,k=1}^n a_{ijk}z_kdz_i$ and similarly $d\bar{w}_j = d\bar{z}_j + \sum_{i,k=1}^n a'_{jik}\bar{z}_kd\bar{z}_i$. Therefore, up to terms of order at least two, one finds

$$\frac{i}{2}\sum_{j=1}^n dw_j \wedge d\bar{w}_j$$

$$= \frac{i}{2}\sum_{j=1}^n \left(dz_j \wedge d\bar{z}_j + (\sum_{i,k=1}^n a_{ijk}z_kdz_i) \wedge d\bar{z}_j + dz_j \wedge (\sum_{i,k=1}^n a'_{jik}\bar{z}_kd\bar{z}_i) \right)$$

$$= \frac{i}{2}\left(\sum_{j=1}^n dz_j \wedge d\bar{z}_j + \sum_{i,j=1}^n (\sum_{k=1}^n a_{ijk}z_k)dz_i \wedge d\bar{z}_j + \sum_{i,j}(\sum_{k=1}^n a'_{jik}\bar{z}_k)dz_j \wedge d\bar{z}_i \right)$$

$$= \omega$$

$\square$

Example 1.3.13 Any compatible metric on $U \subset \mathbb{C}$ satisfies the above condition. Clearly, the three-form $d\omega$ vanishes for dimension reasons.

Exercises

1.3.1   Let $f : U \to V$ be a holomorphic map. Show that the natural pull-back $f^* : \mathcal{A}^k(V) \to \mathcal{A}^k(U)$ induces maps $\mathcal{A}^{p,q}(V) \to \mathcal{A}^{p,q}(U)$.

1.3.2   Show that $\overline{\partial \alpha} = \bar{\partial} \bar{\alpha}$. In particular, this implies that a real (p,p)-form $\alpha \in \mathcal{A}^{p,p}(U) \cap \mathcal{A}^{2p}(U)$ is ∂-closed (exact) if and only if α is $\bar{\partial}$-closed (exact). Formulate the ∂-versions of the three Poincaré lemmas.

1.3.3   Let $B \subset \mathbb{C}^n$ be a polydisc and let $\alpha \in \mathcal{A}^{p,q}(B)$ be a d-closed form with $p, q \geq 1$. Show that there exists a form $\gamma \in \mathcal{A}^{p-1,q-1}(B)$ such that $\partial\bar{\partial}\gamma = \alpha$. (This is a local version of the $\partial\bar{\partial}$-lemma for compact Kähler manifolds Corollary 3.2.10.)

1.3.4   Show that for a polydisc $B \subset \mathbb{C}^n$ the sequence

$$\mathcal{A}^{p-1,q-1}(B) \xrightarrow{\partial\bar{\partial}} \mathcal{A}^{p,q}(B) \xrightarrow{d} \mathcal{A}_{\mathbb{C}}^{p+q+1}(B)$$

is exact. (For more general open subsets or complex manifolds the sequence is no longer exact. This gives rise to the so called Bott–Chern cohomology which shall be introduced in Exercise 2.6.7.)

1.3.5   Show that $\omega = \frac{i}{2\pi}\partial\bar{\partial}\log(|z|^2 + 1) \in \mathcal{A}^{1,1}(\mathbb{C})$ is the fundamental form of a compatible metric g that osculates to order two in any point. (This is the local shape of the Fubini–Study Kähler form on $\mathbb{P}^1$, cf. Section 3.1.)

1.3.6   Analogously to Exercise 1.3.5, study the form $\omega = \frac{i}{2\pi}\partial\bar{\partial}\log(1 - |z|^2)$ on $B_1 \subset \mathbb{C}$. (This is the local example of a negatively curved Kähler structure. See Section 3.1.)

1.3.7   Let $\omega = \frac{i}{2\pi}\sum dz_i \wedge d\bar{z}_i$ be the standard fundamental form on $\mathbb{C}^n$. Show that one can write $\omega = \frac{i}{2\pi}\partial\bar{\partial}\varphi$ for some positive function φ and determine φ. The function φ is called the *Kähler potential*.

1.3.8   Let $\omega \in \mathcal{A}^{1,1}(B)$ be the fundamental form associated to a compatible metric on a polydisc $B \subset \mathbb{C}^n$ which osculates in every point $z \in B$ to order two. Show that $\omega = \frac{i}{2\pi}\partial\bar{\partial}\varphi$ for some real function $\varphi \in \mathcal{A}^0(B)$.

1.3.9   Let g be a compatible metric on $U \subset \mathbb{C}^n$ that osculates to order two in any point $z \in U$. For which real function f has the conformally equivalent metric $e^f \cdot g$ the same property?

2

Complex Manifolds

Complex manifolds are topological spaces that are locally modeled on open polydiscs $B \subset \mathbb{C}^n$ with holomorphic transition functions. They are close relatives of differentiable manifolds, but yet very different in many aspects. All similarities and differences between differentiable and holomorphic functions find their global counterpart in the theory of complex manifolds. In some (imprecise) sense, complex manifolds are more rigid than differentiable manifolds. Just to name two examples, a compact complex manifold does not admit any non-constant holomorphic function and cannot holomorphically be embedded into $\mathbb{C}^n$. On the other hand, complex manifolds can often be described in very explicit terms, e.g. by polynomials.

This chapter intends to clarify the relation between differentiable and complex manifolds. Two equivalent definitions of complex manifolds are given and compared in Sections 2.1 and 2.6. Section 2.1 also introduces a number of interesting (usually compact) examples of complex manifolds and presents Siegel's theorem 2.1.9, which bounds the size of the field of meromorphic functions on a compact complex manifold.

As one is used to from differential geometry, any complex manifold possesses a tangent bundle, by means of which the geometry of the manifold can effectively be studied. In the complex setting, the tangent bundle is holomorphic, and the general concept of holomorphic vector bundles is discussed and compared to its real counterpart. Section 2.2 also presents a number of basic constructions for holomorphic vector bundles.

The particular case of holomorphic line bundles is treated in Section 2.3. Many important examples and constructions can be studied in terms of holomorphic line bundles. The idea behind this is to try to describe the geometry of a complex manifold by polynomials. This approach leads to a picture that relates divisors, e.g. hypersurfaces, of a given complex manifold X, and holomorphic line bundles on X.

The projective space is a particular example of a compact complex manifold. Section 2.4 is devoted to it. The projective space has the advantage to be tangible by linear algebra methods, but also contains many intriguing complex submanifolds. In fact, complex algebraic geometry just studies those complex manifolds that can be embedded into some projective space. Maybe the role of the projective space in complex geometry can be compared to the role played by spheres in differential geometry.

The short Section 2.5 introduces a kind of complex surgery, called blow-up, that changes a complex manifold along a given submanifold. This is a fundamental construction and a useful tool, as we will see in later sections.

We conclude by Section 2.6, which discusses almost complex structures and the Newlander–Niernberg integrability criterion 2.6.19 (the only result that is quoted in this chapter). The reader will also be introduced to a cohomology theory adapted to complex manifolds, the Dolbeault cohomology.

2.1 Complex Manifolds: Definition and Examples

This section introduces the holomorphic analogue of differential manifolds, complex manifolds, and provides a list of fundamental examples of complex manifolds. Holomorphic functions on compact complex manifolds are automatically constant (Proposition 2.1.5), but non-constant meromorphic functions often exist. Siegel's theorem 2.1.9 shows however that the field of all meromorphic functions, the function field, is not too big.

We assume that the reader is familiar with the notion of a differentiable manifold. Appendix A collects some of the most basic results.

Definition 2.1.1 A *holomorphic atlas* on a differentiable manifold is an atlas $\{(U_i, \varphi_i)\}$ of the form $\varphi_i : U_i \simeq \varphi_i(U_i) \subset \mathbb{C}^n$, such that the transition functions $\varphi_{ij} := \varphi_i \circ \varphi_j^{-1} : \varphi_j(U_i \cap U_j) \to \varphi_i(U_i \cap U_j)$ are holomorphic. The pair (U_i, φ_i) is called a *holomorphic chart*. Two holomorphic atlases $\{(U_i, \varphi_i)\}, \{(U'_j, \varphi'_j)\}$ are called equivalent if all maps $\varphi_i \circ \varphi_j'^{-1} : \varphi'_j(U_i \cap U'_j) \to \varphi_i(U_i \cap U'_j)$ are holomorphic.

Definition 2.1.2 A *complex manifold* X of dimension n is a (real) differentiable manifold of dimension $2n$ endowed with an equivalence class of holomorphic atlases.

Note that this definition is very similar to the one of a differentiable manifold itself (see Definition A.0.1). A differentiable manifold is a topological manifold endowed with an equivalence class of differentiable atlases. In fact, we could define a complex manifold as a topological manifold with a holomorphic atlas. In order to distinguish clearly between the complex manifold X and the underlying real manifold, we sometimes write M for the latter. Thus, X is M endowed with a holomorphic atlas.

A complex manifold is called connected, compact, simply connected, etc., if the underlying differentiable (or, topological) manifold has this property. A complex manifold of dimension one (two, three, ...) is called a complex *curve* (a complex *surface*, a complex *threefold*,..., respectively). Clearly, any open subset of a complex manifold is in a natural way a complex manifold itself.

Definition 2.1.3 A *holomorphic function* on a complex manifold X is a function $f : X \to \mathbb{C}$, such that $f \circ \varphi_i^{-1} : \varphi_i(U_i) \to \mathbb{C}$ is holomorphic for any chart (U_i, φ_i) of a holomorphic atlas in the equivalence class defining X.

Definition 2.1.4 Let X be a complex manifold. By $\mathcal{O}_X$ we denote the *sheaf of holomorphic functions* on X, i.e. for any open subsets $U \subset X$ one has

$$\mathcal{O}_X(U) = \Gamma(U, \mathcal{O}_X) = \{f : U \to \mathbb{C} \mid f \text{ is holomorphic}\}.$$

It is obvious from the definition that via a holomorphic chart (U, φ) with $x \in U$ and $\varphi(x) = 0 \in \mathbb{C}^n$ the stalk $\mathcal{O}_{X,x}$ is isomorphic to $\mathcal{O}_{\mathbb{C}^n,0}$. Let $Q(\mathcal{O}_{X,x})$ denote the quotient field of $\mathcal{O}_{X,x}$. If $U \subset X$ is an open connected subset, then holomorphic functions $0 \neq h, g : U \to \mathbb{C}$ define elements $g, h \in \mathcal{O}_{X,x}$ and $\frac{g}{h} \in Q(\mathcal{O}_{X,x})$ for all $x \in U$.

The fundamental difference between complex and differentiable manifolds becomes manifest already in the following proposition, which is in fact an easy consequence of the maximum principle.

Proposition 2.1.5 *Let X be a compact connected complex manifold. Then $\Gamma(X, \mathcal{O}_X) = \mathbb{C}$, i.e. any global holomorphic function on X is constant.*

Proof. Since X is compact, any holomorphic function $f : X \to \mathbb{C}$, which is in particular continuous, attains its maximum at some point $x \in X$. If (U_i, φ_i) is a holomorphic chart with $x \in U_i$, then $f \circ \varphi_i^{-1}$ is locally constant due to the maximum principle on $\varphi_i(U_i) \subset \mathbb{C}^n$ (see page 5). Since X is connected, this shows that f must be constant. Thus, $\mathbb{C} = \Gamma(X, \mathcal{O}_X)$. $\square$

Using Hartogs' theorem 1.1.4, the compactness assumption on X can be weakened. We only note the following special case.

Corollary 2.1.6 *Let X be a complex manifold of dimension at least two and let $x \in X$. Then $\Gamma(X, \mathcal{O}_X) = \Gamma(X \setminus \{x\}, \mathcal{O}_X)$. If X is in addition compact and connected, then $\Gamma(X \setminus \{x\}, \mathcal{O}_X) = \mathbb{C}$.* $\square$

Also note that one of the main technical tools in real analysis, the partition of unity, is of limited use in complex geometry.

Definition 2.1.7 Let X and Y be two complex manifolds. A continuous map $f : X \to Y$ is a *holomorphic map* (or, a *morphism*) if for any holomorphic charts (U, φ) and (U', φ') of X and Y, respectively, the map $\varphi' \circ f \circ \varphi^{-1} : \varphi(f^{-1}(U') \cap U) \to \varphi'(U')$ is holomorphic. Two complex manifolds X and Y are called *isomorphic* (or, *biholomorphic*) if there exists a holomorphic homeomorphism $f : X \to Y$.

Note that the inverse f^{-1} of a holomorphic homeomorphism is holomorphic by Proposition 1.1.13.

Definition 2.1.8 A *meromorphic* function on a complex manifold X is a map

$$f : X \longrightarrow \bigsqcup_{x \in X} Q(\mathcal{O}_{X,x})$$

which associates to any $x \in X$ an element $f_x \in Q(\mathcal{O}_{X,x})$ such that for any $x_0 \in X$ there exists an open neighbourhood $x_0 \in U \subset X$ and two holomorphic functions $g, h : U \to \mathbb{C}$ with $f_x = \frac{g}{h}$ for all $x \in U$. The *sheaf of meromorphic functions* is denoted by $\mathcal{K}_X$. Furthermore, its space of global sections is denoted $K(X) := \Gamma(X, \mathcal{K}_X)$.

It is straightforward to see that $\mathcal{K}_X(U)$ is a field if U is connected. If X is connected, then $K(X)$ is called the *function field* of X.

Clearly, there is a natural inclusion $\mathcal{O}_X \subset \mathcal{K}_X$. But how many global sections has $\mathcal{K}_X$? This actually depends very much on X and its geometry as we will see later (e.g. page 88). For the moment we only give an upper bound for $K(X)$ expressed in terms of its transcendence degree.

Proposition 2.1.9 (Siegel) *Let X be a compact connected complex manifold of dimension n. Then*

$$\mathrm{trdeg}_\mathbb{C} K(X) \leq n \ .$$

Proof. Recall that $\mathrm{trdeg}_\mathbb{C} K \leq n$ for a field extension $\mathbb{C} \subset K$ if and only if for any $(n + 1)$ elements $f_1, \ldots, f_{n+1} \in K$ there exists a non-trivial polynomial $F \in \mathbb{C}[x_1 \ldots x_{n+1}]$ such that $F(f_1, \ldots, f_{n+1}) = 0$. The set $P_m \subset \mathbb{C}[x_1, \ldots, x_{n+1}]$ of all polynomials of degree $\leq m$ is a linear subspace. It is easy to see that $\dim(P_m) = \binom{m+n+1}{m}$.

Let $f_1, \ldots, f_{n+1} \in K(X)$. For any point $x \in X$ there exist an open neighbourhood $x \in U_x \subset X$ and holomorphic functions $g_{i,x}, h_{i,x} : U_x \to \mathbb{C}$ with $f|_{U_x} = \frac{g_{i,x}}{h_{i,x}}$. Moreover, we may assume that $g_{i,x}$ and $h_{i,x}$ are relatively prime in $\mathcal{O}_{X,y}$ for all $y \in U_x$ (use Proposition 1.1.35).

Using holomorphic charts around each point x one finds open neighbourhoods $x \in W_x \subset V_x \subset \overline{V}_x \subset U_x$ such that $V_x \cong \{z \mid |z_i| < 1\}$ and such that under this isomorphism W_x is identified with the disc $\{z \mid \|z\| < 1/2\}$.

Since X is compact, there exist finitely many points $x_1, \ldots, x_N \in X$ such that

$$X = \bigcup_{k=1}^{N} W_{x_k}.$$

To shorten notation we will write $U_k, V_k, W_k, g_{i,k}, h_{i,k}$ instead of $U_{x_k}, V_{x_k}, W_{x_k}, g_{i,x_k}, h_{i,x_k}$.

Since $\frac{g_{i,k}}{h_{i,k}} = \frac{g_{i,\ell}}{h_{i,\ell}}$ on $U_k \cap U_\ell$ and since $g_{i,\ell}$ and $h_{i,\ell}$ are relatively prime everywhere on $U_k \cap U_\ell$, one can write $h_{i,k} = h_{i,\ell} \cdot \varphi_{i,k\ell}$, where $\varphi_{i,k\ell} = \frac{g_{i,k}}{g_{i,\ell}} : U_k \cap U_\ell \to \mathbb{C}^*$ is a holomorphic function. Thus, $\varphi_{k\ell} := \prod_{i=1}^{n+1} \varphi_{i,k\ell}$ is a holomorphic and bounded function on $V_k \cap V_\ell$. Let $C := \max\{|\varphi_{k\ell}(x)| \mid k, \ell = 1, \ldots, N, \ x \in \overline{V}_k \cap \overline{V}_\ell\}$. From $\varphi_{k\ell} \cdot \varphi_{\ell k} = 1$ one deduces $C \geq 1$.

Let $F \in \mathbb{C}[x_1, \ldots x_{n+1}]$ be a polynomial of degree m. Then

$$F\left(\frac{g_{1,k}}{h_{1,k}}, \ldots, \frac{g_{n+1,k}}{h_{n+1,k}}\right) = \frac{G_k}{\left(\prod\limits_{i=1}^{n+1} h_{i,k}\right)^m},$$

with $G_k : U_k \to \mathbb{C}$ holomorphic. Furthermore, $G_k = \varphi_{\ell k}^m \cdot G_\ell$.

Let $m' > 0$. We claim that there exists a polynomial $F \in \mathbb{C}[x_1, \ldots, x_{n+1}]$ such that G_k vanishes of order m' in x_k for all k. Indeed, G_k vanishes of order m' in x_k if for all differential operators $D = \frac{\partial^a}{\partial z_1^{a_1} \ldots \partial z_n^{a_n}}$ with $a = \sum\limits_{j=1}^{n} a_j < m'$ one has $DG_k(x_k) = 0$.

These differential operators D span a space of dimension $\binom{m'-1+n}{m'-1}$. Thus, the condition that G_k vanishes of order m' in x_k poses at most $\binom{m'-1+n}{m'-1}$ linear conditions on the coefficients of F. For all points $x_1, \ldots, x_N$ together this accounts for at most $N \cdot \binom{m'-1+n}{m'-1}$ linear conditions. Thus, if

$$\binom{m+n+1}{m} > N \cdot \binom{m'-1+n}{m'-1},$$

then a polynomial F with the above property of degree $\leq m$ can be found. We will indeed use that a polynomial F of degree m can be found as soon as this inequality is satisfied.

By the Schwarz lemma 1.1.36 this yields

$$|G_k(x)| \leq \left(\frac{1}{2}\right)^{m'} \cdot C'$$

for $x \in W_k$ and $C' := \max\{|G_k(x)| \mid k = 1, \ldots, N, \ x \in \overline{V}_k\}$.

Now it is enough to show that $C' = 0$ for an appropriate choice of m' (and m).

Let $x \in X$ and $1 \leq k \leq N$ such that $C' = |G_k(x)|$. Then there exists $1 \leq \ell \leq N$ with $x \in W_\ell$. Hence

$$C' = |G_k(x)| = |G_\ell(x)| \cdot |\varphi_{\ell k}^m(x)| \leq \frac{C'}{2^{m'}} \cdot C^m.$$

For $\frac{C^m}{2^{m'}} < 1$ this yields a contradiction except for $C' = 0$. Let us show that we can always achieve $\frac{C^m}{2^{m'}} < 1$. We set $C = 2^\lambda$ with $\lambda \geq 0$, then $\frac{C^m}{2^{m'}} = 2^{\lambda m - m'}$. It is not hard to see that $\lambda m < m'$ and $\binom{m+1+n}{m} > N \cdot \binom{m'-1+n}{m'-1}$ can be satisfied at the same time. In fact, $\binom{m+1+n}{m}$ and $\binom{m'-1+n}{m'-1}$ are polynomials of degree $n+1$ in m and of degree n in m', respectively. $\qquad \square$

Definition 2.1.10 The *algebraic dimension* of a compact connected complex manifold X is $a(X) := \operatorname{trdeg}_{\mathbb{C}} K(X)$.

Thus, the function field $K(X)$ contains a purely transcendental extension of $\mathbb{C}$ of degree $a(X) \leq \dim(X)$. Any other meromorphic function on X will be algebraic over this extension. In fact, due to a theorem of Remmert [97] one knows that the function field is a finite extension of this purely transcendental extension. In the examples we will see that the algebraic dimension can take arbitrary values between zero and the dimension of the manifold.

The rest of this section is devoted to the construction of various examples of complex manifolds. Each of them has specific properties which we will come back to in the later chapters. Many of the results presented in this course will be applied to the examples of the following list. If not done explicitly, the reader is strongly encouraged to do so on its own.

At first glance, there seem to be many different methods for constructing non-trivial complex manifolds and, indeed, the classification of complex surfaces is already tremendously complicated and a detailed list of complex threefolds seems out of reach for the time being. On the other hand, it is not always easy to find complex (sub-)manifolds of a specific type. In fact, it often seems that there is a serious lack of techniques to uncover the richness and the beauty of the complex landscape.

Affine space. In fact, this is just the algebraic name for the most basic complex manifold provided by the n-dimensional complex space $\mathbb{C}^n$. Of course, any complex vector space is also a complex manifold. The open subsets of $\mathbb{C}^n$ serve as the local models for arbitrary complex manifolds.

Note that there is an essential difference between differentiable and complex manifolds. A differentiable manifold can always be covered by open subsets diffeomorphic to $\mathbb{R}^n$. In contrast, a general complex manifold cannot be covered by open subsets biholomorphic to $\mathbb{C}^n$. This phenomenon is due to the fact that $\mathbb{C}$ is not biholomorphic to a bounded open disc (Liouville's theorem, see page 4).

Projective space. The complex projective space $\mathbb{P}^n := \mathbb{P}^n_{\mathbb{C}}$ is the most important compact complex manifold. By definition, $\mathbb{P}^n$ is the set of lines in $\mathbb{C}^{n+1}$ or, equivalently,

$$\mathbb{P}^n = (\mathbb{C}^{n+1} \setminus \{0\})/\mathbb{C}^*,$$

where $\mathbb{C}^*$ acts by multiplication on $\mathbb{C}^{n+1}$. The points of $\mathbb{P}^n$ are written as $(z_0 : z_1 : \ldots : z_n)$. Here, the notation intends to indicate that for $\lambda \in \mathbb{C}^*$ the two points $(\lambda z_0 : \lambda z_1 : \ldots : \lambda z_n)$ and $(z_0 : z_1 : \ldots : z_n)$ define the same point in $\mathbb{P}^n$. Only the origin $(0 : 0 : \ldots : 0)$ does not define a point in $\mathbb{P}^n$.

The *standard open covering* of $\mathbb{P}^n$ is given by the $n + 1$ open subsets

$$U_i := \{(z_0 : \ldots : z_n) \mid z_i \neq 0\} \subset \mathbb{P}^n.$$

If $\mathbb{P}^n$ is endowed with the quotient topology via

$$\pi : \mathbb{C}^{n+1} \setminus \{0\} \longrightarrow (\mathbb{C}^{n+1} \setminus \{0\})/\mathbb{C}^* = \mathbb{P}^n,$$

then the U_i's are indeed open.

Consider the bijective maps

$$\varphi_i : U_i \longrightarrow \mathbb{C}^n, \quad (z_0 : \ldots : z_n) \longmapsto \left(\frac{z_0}{z_i}, \ldots, \frac{z_{i-1}}{z_i}, \frac{z_{i+1}}{z_i}, \ldots, \frac{z_n}{z_i} \right).$$

For the transition maps $\varphi_{ij} := \varphi_i \circ \varphi_j^{-1} : \varphi_j(U_i \cap U_j) \to \varphi_i(U_i \cap U_j)$ one has

$$\varphi_{ij}(w_1, \ldots, w_n) = \left(\frac{w_1}{w_i}, \ldots, \frac{w_{i-1}}{w_i}, \frac{w_{i+1}}{w_i}, \ldots, \frac{w_{j-1}}{w_i}, \frac{1}{w_i}, \frac{w_j}{w_i}, \ldots, \frac{w_n}{w_i} \right).$$

Note that $\varphi_j(U_i \cap U_j) = \mathbb{C}^n \setminus Z(w_i)$. These maps are obviously bijective and holomorphic.

There is a more elegant way to describe the transition functions. Namely, we may identify $\varphi_i(U_i)$ with the affine subspace $\{(z_0, \ldots, z_n) \mid z_i = 1\} \subset \mathbb{C}^{n+1}$. Then $\varphi_j(U_i \cap U_j) = \{(z_0, \ldots, z_n) \mid z_j = 1, z_i \neq 0\}$ and $\varphi_{ij}(z_0, \ldots z_n) = z_i^{-1} \cdot (z_0, \ldots, z_n)$.

If V is a complex vector space, then $\mathbb{P}(V)$ denotes the projective space associated to V, i.e. $\mathbb{P}(V) := (V \setminus \{0\})/\mathbb{C}^*$. Of course, after choosing a basis of V the complex manifold $\mathbb{P}(V)$ becomes isomorphic to $\mathbb{P}^n$ with $n = \dim_{\mathbb{C}} V - 1$.

Warning: Sometimes, $\mathbb{P}(V)$ is used to denote the space of all hyperplanes in V. In our notation this space would be $\mathbb{P}(V^*)$.

Complex tori. Let X be the quotient $\mathbb{C}^n/\mathbb{Z}^{2n}$, where $\mathbb{Z}^{2n} \subset \mathbb{R}^{2n} = \mathbb{C}^n$ is the natural inclusion. Then X can be endowed with the quotient topology of $\pi : \mathbb{C}^n \to \mathbb{C}^n/\mathbb{Z}^{2n} = X$. If $U \subset \mathbb{C}^n$ is a small open subset such that $(U + (a_1 + ib_1, \ldots, a_n + ib_n)) \cap U = \varnothing$ for all $0 \neq (a_1, b_1, \ldots, a_n, b_n) \in \mathbb{Z}^{2n}$, then $U \to \pi(U)$ is bijective. Covering X by those provides a holomorphic atlas of X. The transition functions are just translations by vectors in $\mathbb{Z}^{2n}$. Explicitly, if $z \in \mathbb{C}^n$, then the polydisc $U = B_\epsilon(z)$ with $\epsilon = (1/2, \ldots, 1/2)$ has the above property.

More generally, if V is a complex vector space and $\Gamma \subset V$ is a free abelian, discrete subgroup of order $2n$, i.e. Γ is freely generated by an $\mathbb{R}$-basis of V, then $X = V/\Gamma$ is a complex manifold. From a differential geometric point of view these manifolds are basic, but not very exciting; they are diffeomorphic to $(S^1)^{2n}$. However, from the complex view point the situation is very different. If you pick two lattices $\Gamma_1, \Gamma_2 \subset \mathbb{C}^n$ randomly, then $\mathbb{C}^n/\Gamma_1$ and $\mathbb{C}^n/\Gamma_2$ will not be isomorphic as complex manifolds. Of course, if $f : \mathbb{C}^n \cong \mathbb{C}^n$ is a $\mathbb{C}$-linear bijective map with $f(\Gamma_1) = \Gamma_2$, then $\mathbb{C}^n/\Gamma_1 \cong \mathbb{C}^n/\Gamma_2$.

Let us consider the one-dimensional case a bit more in detail. A lattice $\Gamma \subset \mathbb{C}$ is of the form $\Gamma = z_1\mathbb{Z} + z_2\mathbb{Z}$. By a coordinate change we can achieve that $\Gamma = \mathbb{Z} + \tau\mathbb{Z}$ with $\tau \in \mathbb{H}$. More precisely, multiplication with $\pm 1/z_1$ defines a bijection $f : \mathbb{C} \to \mathbb{C}$ such that $f(\Gamma)$ is of the required form. There is a natural action

$$\mathrm{Sl}(2,\mathbb{R}) \times \mathbb{H} \longrightarrow \mathbb{H}, \quad (A = \left(\begin{smallmatrix} a & b \\ c & d \end{smallmatrix}\right), (\tau)) \longmapsto \frac{a\tau+b}{c\tau+d}$$

and $\mathbb{C}/(\mathbb{Z} + \tau\mathbb{Z}) \cong \mathbb{C}/(\mathbb{Z} + A(\tau)\mathbb{Z})$ for $A \in \mathrm{Sl}(2,\mathbb{Z})$. Indeed, $\mathbb{Z} + \tau\mathbb{Z} = (c\tau + d)\mathbb{Z} + (a\tau+b)\mathbb{Z}$ for $A \in \mathrm{Sl}(2,\mathbb{Z})$ and multiplication with $1/(c\tau+d)$ defines a $\mathbb{C}$-linear isomorphism $f : \mathbb{C} \to \mathbb{C}$ with $f(\mathbb{Z}+\tau\mathbb{Z}) = \mathbb{Z}+A(\tau)\mathbb{Z}$. In fact, $\tau_1, \tau_2 \in \mathbb{H}$ define isomorphic tori if and only if $\tau_1 = A(\tau_2)$ for some $A \in \mathrm{Sl}(2,\mathbb{Z})$, but this is slightly more difficult to prove (see page 136). A one-dimensional complex torus $\mathbb{C}/\Gamma$ is also called an *elliptic curve*, which refers to the fact that there is an obvious group structure on $\mathbb{C}/\Gamma$.

Also note in passing that for $n \geq 2$ and a very general lattice $\Gamma \subset \mathbb{C}^n$ the function field $K(\mathbb{C}^n/\Gamma)$ is trivial, i.e. $K(\mathbb{C}^n/\Gamma) = \mathbb{C}$ (cf. Exercise 3.3.6). However, for special lattices $\Gamma \subset \mathbb{C}^n$ and for all lattices $\Gamma \subset \mathbb{C}$ one knows that $\mathrm{trdeg}_{\mathbb{C}} K(\mathbb{C}^n/\Gamma) = n$. This will be discussed in subsequent sections (see Exercise 5.3.5).

Affine hypersurfaces. Let $f : \mathbb{C}^n \to \mathbb{C}$ be a holomorphic function such that $0 \in \mathbb{C}$ is a regular value. Let

$$X := f^{-1}(0) = Z(f) \subset \mathbb{C}^n.$$

By the implicit function theorem 1.1.11 there exists an open cover $X = \bigcup U_i$, open subsets $V_i \subset \mathbb{C}^{n-1}$ and holomorphic maps $g_i : V_i \to \mathbb{C}^n$ inducing bijective maps $g_i : V_i \to U_i$. The transition maps $\varphi_{ij} = g_i^{-1} \circ g_j : g_j^{-1}(U_i) \to g_i^{-1}(U_j)$ are holomorphic (use Corollary 1.1.12). Thus, X is a complex manifold of dimension $n - 1$. Of course, the same arguments also apply when f is only given on an open subset $U \subset \mathbb{C}^n$.

Projective hypersurfaces. Let f be a homogeneous polynomial in $n + 1$ variables $z_0, \ldots, z_n$. Assume that $0 \in \mathbb{C}$ is a regular value for the induced holomorphic map $f : \mathbb{C}^{n+1} \setminus \{0\} \to \mathbb{C}$. By the previous example we know that $f^{-1}(0) = Z(f)$ is a complex manifold. The subset

$$X := V(f) := f^{-1}(0)/\mathbb{C}^* \subset \mathbb{P}^n$$

is the set of all points $(z_0 : \ldots : z_n) \in \mathbb{P}^n$ with $f(z_0, \ldots, z_n) = 0$. Note that the value $f(z_0, \ldots, z_n)$ in general depends on the chosen representative of $(z_0 : \ldots : z_n)$, but the zero set $V(f)$ is well-defined as f is homogeneous. We claim that X is a complex manifold of dimension $n - 1$. In fact, X is covered by the open subset $X \cap U_i$, where the U_i are the standard charts of $\mathbb{P}^n$. Using the standard isomorphisms $U_i \cong \mathbb{C}^n$, the set $X \cap U_i$ is identified with the fibre over $0 \in \mathbb{C}$ of the map $f_i : (w_1 \ldots w_n) \mapsto f(w_1, \ldots, w_{i-1}, 1, w_i, \ldots, w_n)$. Check that 0 is a regular value of this map. Thus, by the implicit function theorem one can find charts for X such that the transition maps are holomorphic.

Complete intersections. The two last examples can be generalized in a straightforward way as follows. Instead of looking at one holomorphic function (resp. homogeneous polynomial of degree d) one can consider k holomorphic functions (resp. homogeneous polynomials of degree $d_1, \ldots, d_k$). If 0 is a regular value of $(f_1, \ldots, f_k) : \mathbb{C}^n \to \mathbb{C}^k$ (resp. $(f_1, \ldots, f_k) : \mathbb{C}^{n+1} \setminus \{0\} \to \mathbb{C}^k$), then

$$X := Z(f_1) \cap \ldots \cap Z(f_k) \subset \mathbb{C}^n \quad \text{resp.} \quad X = V(f_1) \cap \cdots \cap V(f_k) \subset \mathbb{P}^n$$

is a complex manifold of dimension $n - k$.

Complex Lie groups. Let G be a group and a complex manifold at the same time. Then G is called a complex Lie group if $G \times G \to G$, $(x, y) \mapsto x \cdot y^{-1}$ is a holomorphic map. Examples of complex Lie groups are provided by $\mathrm{Gl}(n, \mathbb{C})$, $\mathrm{Sl}(n, \mathbb{C})$, $\mathrm{Sp}(n, \mathbb{C})$. They are certainly not abelian for $n > 1$. A compact example is provided by a complex torus $X = \mathbb{C}^n / \Gamma$, where the group operations are induced by the natural ones on $\mathbb{C}^n$. Clearly, X is an abelian complex Lie group. Also $\mathbb{C}^n$ is an abelian Lie group, but non-compact. One can prove that any connected compact complex Lie group is abelian (Exercise 2.6.9) and in fact a torus (cf. Section 3.3).

Note that certain classical groups like $\mathrm{U}(n)$ or $\mathrm{O}(n)$ are often not complex, but just ordinary real Lie groups. E.g. $\mathrm{U}(1) \cong S^1$ which if of odd real dimension.

Quotients. Before giving more examples we will discuss a general technique, that allows one to obtain new and interesting examples by taking quotients of the action of a complex Lie group on a complex manifold. The complex torus, that has been discussed before, is a particular example of this.

Let X be a topological space and let G be a group that acts on X, i.e. there exists an action $G \times X \to X$ such that for any $g \in G$ the induced map $g : X \to X$ is continuous.

The quotient space (or orbit space) X/G can be endowed with a topology such that the projection map $\pi : X \to X/G$ is continuous by saying that $V \subset X/G$ is open if and only if $\pi^{-1}(V) \subset X$ is open. In general, however, the topology of the quotient X/G might be bad, e.g. non-Hausdorff, although X is a nice manifold. (Already the natural action of $\mathbb{C}^*$ on $\mathbb{C}$ provides such an example.) Additional conditions need to be imposed on the action in order to ensure that X/G is a reasonable space.

Definition 2.1.11 The *action is free* if for all $1 \neq g \in G$ and all $x \in X$ one has $g \cdot x \neq x$.

The action is *proper* if the map $G \times X \to X \times X$, $(g, x) \mapsto (g \cdot x, x)$ is proper.

If X is a complex manifold and G a complex Lie group acting on X, then we will tacitly assume that the map $G \times X \to X$ is holomorphic. In particular, for any $g \in G$ the induced map $g : X \to X$ is biholomorphic.

Examples 2.1.12 i) If G is discrete and the action on X is free and proper, then the action satisfies the following two conditions. (The action is also called *properly discontinuous* in this case. See [17, III,4] or [110] for detailed proofs.)
– Any point $x \in X$ admits a neighbourhood $x \in U_x \subset X$ such that $g(U_x) \cap U_x = \emptyset$ for any $1 \neq g \in G$.
– If $x, y \in X$ such that $x \notin G \cdot y$, then there exist neighbourhoods $x \in U_x \subset X$ and $y \in U_y \subset X$ with $U_x \cap g(U_y) = \emptyset$ for any $g \in G$.

The first condition ensures that the quotient X/G admits holomorphic charts and the second that the quotient is Hausdorff. Indeed, one finds an open covering $X = \bigcup U_i$ by charts (U_i, φ_i) such that $g(U_i) \cap U_i = \emptyset$ for any $1 \neq g \in G$. Hence, $U_i \to \pi(U_i)$ is bijective and $\pi(U_i)$ is open. Thus, holomorphic charts for the quotient X/G are defined by $\pi(U_i) \xrightarrow{\pi^{-1}} U_i \xrightarrow{\varphi_i} \varphi_i(U_i) \subset \mathbb{C}^n$.

ii) The natural $\mathbb{C}^*$-action on $\mathbb{C}^n \setminus \{0\}$ is free and proper.

Generalizing the arguments of the first example to the case of proper free actions yields the following proposition. A proof for differentiable manifolds can be found in [40, Thm. 1.11.4].

Proposition 2.1.13 *Let $G \times X \to X$ be the proper and free action of a complex Lie group G on a complex manifold X. Then the quotient X/G is a complex manifold in a natural way and the quotient map $\pi : X \to X/G$ is holomorphic.* □

Examples 2.1.14 i) A discrete lattice $\Gamma \subset \mathbb{C}^n$ certainly acts freely and discretely by translations. Thus, the fact that the torus $X = \mathbb{C}^n/\Gamma$ is a complex manifold could also be seen as a consequence of the above proposition.

ii) The projective space $\mathbb{P}^n$ is the quotient of the natural $\mathbb{C}^*$-action on $\mathbb{C}^{n+1} \setminus \{0\}$, which is proper and free.

Remark 2.1.15 If a quotient as above exists then one often uses the following universality property of the quotient map $\pi : X \to X/G$: Any G-invariant holomorphic map $f : X \to Y$ factorizes via the quotient map π and a holomorphic map $\tilde{f} : X/G \to Y$.

If G does not act freely on a complex manifold X, then the quotient X/G may or may not be a manifold. E.g. let $X = \mathbb{C}/\Gamma$ be a one-dimensional complex torus and let $G = \mathbb{Z}/2\mathbb{Z}$ act by $z \mapsto -z$. This action has four fix points. If $\Gamma = \tau_1 \mathbb{Z} + \tau_2 \mathbb{Z}$ these are the points $0, \tau_1/2, \tau_2/2$, and $(\tau_1 + \tau_2)/2$. However, the quotient X/G is a complex manifold (see Exercise 2.1.8).

Ball quotients. Consider the unit disc $D^n = \{z \mid \|z\| < 1\} \subset \mathbb{C}^n$, which can also be viewed as an open subset of the standard open subset $U_0 \subset \mathbb{P}^n$. Often, one introduces on $\mathbb{C}^{n+1}$ the (non-positive) hermitian product $\langle\ ,\ \rangle$ given by the diagonal matrix $\mathrm{diag}(1, -1, \ldots, -1)$. Then D^n is the open subset of $\mathbb{P}^n$ of points z with $\langle z, z \rangle > 0$, which is clearly invariant under the action of $\mathrm{SU}(\langle\ ,\ \rangle) = \mathrm{SU}(1, n)$.

A *ball quotient* is a quotient of D by any discrete group $\Gamma \subset \mathrm{SU}(1, n)$ acting freely on D. Usually one also assumes the quotient D/Γ to be compact.

As an explicit example one might look at the one-dimensional case $D^1 \cong \mathbb{H}$ and the genus g surface group Γ_g generated by $2g$ generators $a_1, b_1, \ldots, a_g, b_g$ with the only relation $\prod [a_i, b_i] = 1$. This group can be mapped into $\mathrm{PGl}(2, \mathbb{R})$ in many ways. Here, the action of $\mathrm{PGl}(2, \mathbb{R})$ on $\mathbb{H}$ is similarly to the action of $\mathrm{Sl}(2, \mathbb{Z})$ discussed earlier.

Finite quotients of product of curves. Let C be a curve and let $E = \mathbb{C}/\Gamma$ be an elliptic curve. Let G be a finite group that acts on C. If in addition G is subgroup of E, and therefore acts by translations on E, then there is a natural free G-action on the product $C \times E$. Therefore, the quotient $X = (C \times E)/G$ is a complex surface and there exists a surjective holomorphic map $f : X \to C/G$. If G acts freely in $z \in C$, then the fibre $f^{-1}(z)$ is isomorphic to the elliptic curve E. In general, $f^{-1}(z)$ is isomorphic to the quotient of E by the stabilizer $\mathrm{Stab}(z) := \{g \mid g \cdot z = z\}$.

Hopf manifolds. Let $\mathbb{Z}$ act on $\mathbb{C}^n \setminus \{0\}$ by $(z_1, \ldots, z_n) \mapsto (\lambda^k z_1, \ldots, \lambda^k z_n)$ for $k \in \mathbb{Z}$. For $0 < \lambda < 1$ the action is free and discrete. The quotient complex manifold $X = (\mathbb{C}^n \setminus \{0\})/\mathbb{Z}$ is diffeomorphic to $S^1 \times S^{2n-1}$. For $n = 1$ this manifold is isomorphic to a complex torus $\mathbb{C}/\Gamma$. The lattice Γ can be determined explicitly (cf. Exercise 2.1.5). There are generalizations of this construction notably for surfaces by letting act different scalars λ_1 and λ_2. See [8] and Exercise 2.1.6.

Iwasawa manifold. Let G be the complex Lie group that consists of all matrices of the form $\left(\begin{smallmatrix} 1 & z_1 & z_2 \\ 0 & 1 & z_3 \\ 0 & 0 & 1 \end{smallmatrix} \right) \in \mathrm{Gl}(3, \mathbb{C})$. Clearly, G is biholomorphic (as a complex manifold not as a complex Lie group) to $\mathbb{C}^3$. The group G (and every subgroup of it) acts on G by multiplication in $\mathrm{Gl}(3, \mathbb{C})$.

Consider the subgroup $\Gamma := G \cap \mathrm{Gl}(3, \mathbb{Z} + i\mathbb{Z})$. Then $(w_1, w_2, w_3) \in \Gamma$ acts on G by $(z_1, z_2, z_3) \mapsto (z_1 + w_1, z_2 + w_1 z_3 + w_2, z_3 + w_3)$. Obviously, this action is properly discontinuous. Thus, the quotient $X := G/\Gamma$ is a complex manifold of dimension three. It is easy to see that the first and third coordinates define a holomorphic map $f : X \to \mathbb{C}/(\mathbb{Z} + i\mathbb{Z}) \times \mathbb{C}/(\mathbb{Z} + i\mathbb{Z})$. The fibres are isomorphic to $\mathbb{C}/(\mathbb{Z} + i\mathbb{Z})$.

Grassmannian manifolds. Let V be a complex vector space of dimension $n + 1$. As a generalization of the projective space $\mathbb{P}(V)$, which is naturally identified with the set of all lines in V, one defines the Grassmannian $\mathrm{Gr}_k(V)$ for $k \leq n + 1$ as the set of all k-dimensional subspaces of V, i.e.

$$\mathrm{Gr}_k(V) := \{W \subset V \mid \dim(W) = k\}.$$

In particular, $\mathrm{Gr}_1(V) = \mathbb{P}(V)$ and $\mathrm{Gr}_n(V) = \mathbb{P}(V^*)$.

In order to show that $\mathrm{Gr}_k(V)$ is a complex manifold, we may assume that $V = \mathbb{C}^{n+1}$. Any $W \in \mathrm{Gr}_k(V)$ is generated by the rows of a $(k, n+1)$-matrix A of rank k. Let us denote the set of these matrices by $M_{k,n+1}$, which is an open subset of the set of all $(k, n+1)$-matrices. The latter space is a complex manifold canonically isomorphic to $\mathbb{C}^{k \cdot (n+1)}$. Thus, we obtain a natural surjection $\pi : M_{k,n+1} \to \mathrm{Gr}_k(\mathbb{C}^{n+1})$, which is the quotient by the natural action of $\mathrm{Gl}(k, \mathbb{C})$ on $M_{k,n+1}$.

Let us fix an ordering $\{B_1, \ldots, B_m\}$ of all (k, k)-minors of matrices $A \in M_{k,n+1}$. Define an open covering $\mathrm{Gr}_k(\mathbb{C}^{n+1}) = \bigcup_{i=1}^{m} U_i$ where U_i is the open subset $\{\pi(A) \mid \det(B_i) \neq 0\}$. Note that if $\pi(A) = \pi(A')$, then $\det(B_i) \neq 0$ if and only if $\det(B_i') \neq 0$, i.e. the open subsets U_i are well-defined. After permuting the columns of $A \in \pi^{-1}(U_i)$ we may assume that A is of the form (B_i, C_i), where C_i is a $(k, n+1-k)$-matrix. Then the map $\varphi_i : U_i \to \mathbb{C}^{k \cdot (n+1-k)}$, $\pi(A) \mapsto B_i^{-1} C_i$ is well-defined. We leave it to the reader to verify that $\{(U_i, \varphi_i)\}$ defines a holomorphic atlas of $\mathrm{Gr}_k(\mathbb{C}^{n+1})$, such that any $\mathbb{C}$-linear isomorphism of $\mathbb{C}^{n+1}$ induces a biholomorphic map of $\mathrm{Gr}_k(\mathbb{C}^{n+1})$.

The argument also gives the dimension of the Grassmannian manifolds: $\dim_{\mathbb{C}} \mathrm{Gr}_k(V) = k \cdot (n+1-k)$.

Flag manifolds. The previous example can further be generalized to so called flag manifolds. Let again V be a complex vector space of dimension $n+1$ and fix $0 \leq k_1 \leq k_2 \leq \ldots \leq k_\ell \leq n+1$. Then $\mathrm{Flag}(V, k_1, \ldots, k_\ell)$ is the manifold of all flags $W_1 \subset W_2 \subset \ldots \subset W_\ell \subset V$ with $\dim(W_i) = k_i$. E.g. $\mathrm{Flag}(V, k) = \mathrm{Gr}_k(V)$. Furthermore, $\mathrm{Flag}(V, 1, n)$ is the incidence variety $\{(\ell, H) \mid \ell \subset H\} \subset \mathbb{P}(V) \times \mathbb{P}(V^*)$ of all pairs (ℓ, H) consisting of a line $\ell \subset V$ contained in a hyperplane $H \subset V$. The fibre of the projections $\mathrm{Flag}(V, 1, n) \to \mathbb{P}(V)$, $(\ell, H) \mapsto \ell$ and $\mathrm{Flag}(V, 1, n) \to \mathbb{P}(V^*)$, $(\ell, H) \mapsto H$ over ℓ respectively H are canonically isomorphic to $\mathbb{P}(V/\ell)$ and $\mathbb{P}(H)$, respectively.

In some of the above examples the complex manifold was given as a submanifold of another one. We conclude this section with the precise definition of complex submanifolds and, more generally, analytic subvarieties of a complex manifold.

Definition 2.1.16 Let X be a complex manifold of complex dimension n and let $Y \subset X$ be a differentiable submanifold of real dimension $2k$. Then Y is a *complex submanifold* if there exists a holomorphic atlas $\{(U_i, \varphi_i)\}$ of X such that $\varphi_i : U_i \cap Y \cong \varphi_i(U_i) \cap \mathbb{C}^k$.

Here, $\mathbb{C}^k$ is embedded into $\mathbb{C}^n$ via $(z_1, \ldots, z_k) \mapsto (z_1, \ldots, z_k, 0, \ldots, 0)$. The *codimension* of Y in X is by definition $\dim(X) - \dim(X) = n - k$.

Definition 2.1.17 A complex manifold X is *projective* if X is isomorphic to a closed complex submanifold of some projective space $\mathbb{P}^N$.

Projective hypersurfaces and, more generally, projective complete intersections which have been discussed earlier provide examples of projective manifolds. Some complex tori, but not all (see Exercise 3.3.6), are projective.

Any $\mathcal{O}_Y$-sheaf $\mathcal{F}$ on a complex submanifold $Y \subset X$ can be considered as an $\mathcal{O}_X$-sheaf on X supported on Y. More precisely, one identifies $\mathcal{F}$ with its direct image $i_*\mathcal{F}$ under the inclusion $i : Y \subset X$. The restriction of holomorphic functions yields a natural surjection $\mathcal{O}_X \to \mathcal{O}_Y$. This gives rise to the *structure sheaf sequence* of $Y \subset X$:

$$0 \longrightarrow \mathcal{I}_Y \longrightarrow \mathcal{O}_X \longrightarrow \mathcal{O}_Y \longrightarrow 0,$$

where $\mathcal{I}_Y$ is the *ideal sheaf* of all holomorphic functions vanishing on Y.

It frequently happens that submanifolds become singular. Most of what can be said about complex submanifolds holds true also for singular subvarieties as defined below, they are just not manifolds themselves. The following is the global version of the notion of an analytic subset given in Section 1.1.

Definition 2.1.18 Let X be a complex manifold. An *analytic subvariety* of X is a closed subset $Y \subset X$ such that for any point $x \in X$ there exists an open neighbourhood $x \in U \subset X$ such that $Y \cap U$ is the zero set of finitely many holomorphic functions $f_1, \ldots, f_k \in \mathcal{O}(U)$. A point $x \in Y$ is a *smooth* or *regular* point of Y if the functions $f_1, \ldots, f_k$ can be chosen such that $\varphi(x) \in \varphi(U)$ is a regular point of the holomorphic map $f := (f_1 \circ \varphi^{-1}, \ldots, f_k \circ \varphi^{-1}) : \varphi(U) \to \mathbb{C}^k$, i.e. its Jacobian has rank k. Here, (U, φ) is a local chart around x. A point $x \in Y$ is *singular* if it is not regular.

Using Corollary 1.1.12 we see that an analytic subvariety in a neighbourhood of a regular point is nothing but a complex submanifold. Analogously to Exercise 1.1.14 one shows that the set of regular points $Y_{\text{reg}} = Y \setminus Y_{\text{sing}}$ is a non-empty complex submanifold of X (cf. Exercise 2.6.4). An analytic subvariety Y is *irreducible* if it cannot be written as the union $Y = Y_1 \cup Y_2$ of two proper analytic subvarieties $Y_i \subset Y$. The *dimension* of an irreducible analytic subvariety $Y \subset X$ is by definition $\dim(Y) = \dim(Y_{\text{reg}})$. A *hypersurface* is an analytic subvariety of codimension one.

As for submanifolds, one has for any subvariety $Y \subset X$ a short exact sequence

$$0 \longrightarrow \mathcal{I}_Y \longrightarrow \mathcal{O}_X \longrightarrow \mathcal{O}_Y \longrightarrow 0.$$

For singular $Y \subset X$ one defines $\mathcal{O}_Y$ by this sequence. This amounts to giving Y the *induced reduced structure*.

Example 2.1.19 Let V be a vector space and $W_1 \subsetneq W_2 \subsetneq \ldots \subsetneq W_\ell \subsetneq V$ be a flag of subspaces. The associated *Schubert variety* is defined as

$$\Omega(W_\bullet) := \{W \in \text{Gr}_k(V) \mid \dim(W \cap W_i) \geq i, \ i = 1, \ldots, \ell\}.$$

E.g. if $\ell = k = 1$, then $\Omega(W_\bullet) = \mathbb{P}(W) \subset \mathbb{P}(V)$. But this is rather exceptional as in most cases $\Omega(W_\bullet)$ will not be a submanifold of the Grassmannian

$\mathrm{Gr}_k(V)$, it will be singular over those $W \in \mathrm{Gr}_k(V)$ for which $\dim(W \cap W_i) > i$ for at least on i.

Exercises

2.1.1 Show that $\mathbb{P}^n$ is a compact complex manifold. Describe a diffeomorphism of $\mathbb{P}^1$ with the two-dimensional sphere S^2. Conclude that $\mathbb{P}^1$ is simply connected.

2.1.2 Show that $\mathbb{C}^n$ does not have any compact submanifolds of positive dimension. (This is in contrast to the real situation, where any manifold, compact or not, can be realized as a submanifold of some $\mathbb{R}^N$.)

2.1.3 Determine the algebraic dimension of the following manifolds: $\mathbb{P}^1$, $\mathbb{P}^n$, and $\mathbb{C}/(\mathbb{Z} + i\mathbb{Z})$. For the latter, you might need to recall some basic facts on the Weierstrass $\wp$-function (e.g. [98]). How big is the function field of $\mathbb{C}$?

2.1.4 Show that any holomorphic map from $\mathbb{P}^1$ into a complex torus is constant. What about maps from $\mathbb{P}^n$ into a complex torus?

2.1.5 Consider the Hopf curve $X = (\mathbb{C} \setminus \{0\})/\mathbb{Z}$, where $k \in \mathbb{Z}$ acts by $z \mapsto \lambda^k z$, for $\lambda \in \mathbb{R}_{>0}$. Show that X is isomorphic to an elliptic curve $\mathbb{C}/\Gamma$ and determine Γ explicitly.

2.1.6 Generalize the construction of the Hopf manifolds by considering the action of $\mathbb{Z}$ given by $(z_1, \ldots, z_n) \mapsto (\lambda_1^k z_1, \ldots, \lambda_n^k z_n)$, where $0 < \lambda_i < 1$. Show that the quotient $(\mathbb{C}^n \setminus \{0\})/\mathbb{Z}$ is again diffeomorphic to $S^1 \times S^{2n-1}$.

2.1.7 Show that any Hopf surface contains elliptic curves.

2.1.8 Describe the quotient of the torus $\mathbb{C}/\mathbb{Z} + \tau\mathbb{Z}$ by the involution $z \mapsto -z$ locally and globally. Using the Weierstrass function $\wp$ again, one can show that the quotient is isomorphic to $\mathbb{P}^1$. What happens in higher dimensions?

2.1.9 Let E_1, E_2 be two elliptic curves and let $G = \mathbb{Z}/2\mathbb{Z}$ act by translation on E_1 and by the involution $z \mapsto -z$ on E_2. Study the quotient $(E_1 + E_2)/G$.

2.1.10 Describe connected complex manifolds X and Y together with a holomorphic map $X \to Y$ such that every complex torus of dimension one is isomorphic to one of the fibres.

2.1.11 Show that
$$(W \subset V) \mapsto (\bigwedge^k W \subset \bigwedge^k V)$$
defines a (holomorphic) embedding $\mathrm{Gr}_k(V) \hookrightarrow \mathbb{P}(\bigwedge^k V)$. (This is called the *Plücker embedding* $\mathrm{Gr}_k(V)$. In particular, the Grassmannians are all projective.)

2.1.12   Let ρ be a fifth root of unity. The group $G = \langle \rho \rangle \cong \mathbb{Z}/5\mathbb{Z}$ acts on $\mathbb{P}^3$ by

$$(z_0 : z_1 : z_2 : z_3) \mapsto (z_0 : \rho z_1 : \rho^2 z_2 : \rho^3 z_3).$$

Describe all fix points of this action. Show that the surface Y defined by $\sum_{i=0}^{3} z_i^5 = 0$ is G-invariant and that the induced action is fix point free. (The quotient $X = Y/G$ is called *Godeaux surface* and historically was the first compact complex surface with $H^i(X, \mathcal{O}) = 0$, $i = 1, 2$, that is not rational.)

2.1.13   Let $G = \langle \rho \rangle$ as before and let $\tilde{G}$ be the following subgroup of G^5:

$$\tilde{G} = \left\{ (\xi_0, \ldots, \xi_4) \mid \xi_i \in G, \ \prod_{i=0}^{4} \xi_i = 1 \right\}.$$

We let act $\tilde{G}$ on $\mathbb{P}^4$ by $(z_0 : z_1 : z_2 : z_3 : z_4) \mapsto (\xi_0 z_0 : \xi_1 z_1 : \xi_2 z_2 : \xi_3 z_3 : \xi_4 z_4)$. Describe the subgroup H that acts trivially. Show that the hypersurface

$$X = V\left(\sum_{i=0}^{4} z_i^5 - 5t \prod_{i=0}^{4} z_i \right) \subset \mathbb{P}^4,$$

with $t \in \mathbb{C}$, is invariant under $\tilde{G}$. Study the action of $\tilde{G}/H$ on X, in particular the points with non-trivial stabilizer. (The quotient $X/(\tilde{G}/H)$ is not a manifold, but its singularities are rather mild.)

Comments: - We are mainly interested in complex manifolds. However, sometimes one cannot avoid to talk also about singular analytic varieties. A thorough treatment of those is provided by the theory of complex spaces which parallels the theory of schemes in the algebraic setting. Avoiding the general theory makes certain definitions and arguments less natural. E.g. we only introduced reduced subvarieties $Y \subset X$ but sometimes it is important to distinguish between the zero set defined by $f(z) = z$ and by $f(z) = z^2$. For the formal definition of a complex space see Section 6.2.

- There are many more interesting examples of compact complex manifolds in the literature. In [8] most of the available techniques to construct compact surfaces are described in detail. We have not touched upon certain general construction methods provided by toric geometry (cf. [48]). They are very combinatorial in spirit and often allow very explicit calculations, e.g. of Hodge numbers.

- Quotients by more general group actions are much harder to study. Even the concept of a quotient is modified in order to produce geometrically relevant objects. In the realm of algebraic geometry this is studied by means of Geometric Invariant Theory (see [77, 92]).

- For the general theory of complex tori we recommend [13, 28, 76].

2.2 Holomorphic Vector Bundles

In order to understand the geometry of a complex manifold X one has to study in particular:

- Holomorphic maps from other complex manifolds into X. In particular, complex submanifolds of X.
- Holomorphic maps from X to other (easier) complex manifolds.
- Holomorphic vector bundles on X.

All three things are closely related. E.g. hypersurfaces in X and holomorphic line bundles on X always come together (see the discussion in Section 2.3). In this section we introduce holomorphic vector bundles and explain how to perform simple operations with them. A few basic examples, like tangent and cotangent bundles, are discussed in more detail.

Definition 2.2.1 Let X be a complex manifold. A *holomorphic vector bundle* of rank r on X is a complex manifold E together with a holomorphic map

$$\pi : E \longrightarrow X$$

and the structure of an r-dimensional complex vector space on any *fibre* $E(x) := \pi^{-1}(x)$ satisfying the following condition: There exists an open covering $X = \bigcup U_i$ and biholomorphic maps $\psi_i : \pi^{-1}(U_i) \cong U_i \times \mathbb{C}^r$ commuting with the projections to U_i such that the induced map $\pi^{-1}(x) \cong \mathbb{C}^r$ is $\mathbb{C}$-linear.
A *holomorphic line bundle* is a holomorphic vector bundle of rank one.

Note that in the above definition the induced transition functions

$$\psi_{ij}(x) := (\psi_i \circ \psi_j^{-1})(x,\ ) : \mathbb{C}^r \longrightarrow \mathbb{C}^r$$

are $\mathbb{C}$-linear for all $x \in U_i \cap U_j$.

Let $\pi_E : E \to X$ and $\pi_F : F \to X$ be two holomorphic vector bundles. A *vector bundle homomorphism* from E to F is a holomorphic map $\varphi : E \to F$ with $\pi_E = \pi_F \circ \varphi$ such that the induced map $\varphi(x) : E(x) \to F(x)$ is linear with $\mathrm{rk}(\varphi(x))$ independent of $x \in X$. Two vector bundles E and F are *isomorphic* if there exists a bijective vector bundle homomorphism $\varphi : E \to F$.

Remarks 2.2.2 i) Any complex manifold is in particular a differentiable manifold and any holomorphic vector bundle is in particular a differentiable vector bundle (see Appendix A). A holomorphic vector bundle should not be confused with a complex vector bundle. The latter is just a differentiable vector bundle whose fibers are complex vector spaces and the transition maps are complex linear. An important and often highly non-trivial question is in how many ways a complex vector bundle can be seen as a holomorphic vector bundle. Sometimes, a complex vector bundle does not admit any holomorphic

structure, but it might also happen that there are many of them. Both phenomena can be observed already in the case of line bundles. See Section 3.3
and in particular Exercises 3.3.7 and 3.3.8.

ii) Analogously to differentiable real or complex vector bundles, a holomorphic rank r vector bundle $\pi : E \to X$ is determined by the holomorphic
cocycle $\{(U_i, \psi_{ij} : U_i \cap U_j \to \mathrm{Gl}(r, \mathbb{C}))\}$.

As in the differentiable situation one has the following

Meta-theorem 2.2.3 *Any canonical construction in linear algebra gives rise
to a geometric version for holomorphic vector bundles.*

In the spirit of this meta-theorem we collect in the following a list of some
of the main constructions. The reader is advised to work out the cocycle
description for them.

Examples 2.2.4 Let E and F be holomorphic vector bundles over a complex
manifold X.

i) The *direct sum* $E \oplus F$ is the holomorphic vector bundle over X whose
fibre $(E \oplus F)(x)$ for any $x \in X$ is canonically isomorphic to $E(x) \oplus F(x)$ (as
complex vector spaces).

ii) The *tensor product* $E \otimes F$ is the holomorphic vector bundle over X
whose fibre $(E \otimes F)(x)$ for any $x \in X$ is canonically isomorphic to $E(x) \otimes F(x)$
(again as complex vector spaces).

iii) The i-th *exterior power* $\bigwedge^i E$ and the i-th *symmetric power* $S^i E$ are the
holomorphic vector bundles over X whose fibres for any $x \in X$ are canonically
isomorphic to $\bigwedge^i E(x)$ and $S^i(E(x))$, respectively.

iv) The *dual bundle* (E^*) is the holomorphic vector bundle over X whose
fibre $(E^*)(x)$ over $x \in X$ is naturally isomorphic to the dual vector space
$E(x)^*$. Note that $(E^*)^*$ is canonically isomorphic to E.

v) The *determinant line bundle* of a holomorphic vector bundle $E \to X$ of
rank r is the holomorphic line bundle $\det(E) := \bigwedge^r E$.

vi) The map $s : X \to E$ that maps any $x \in X$ to $0 \in E(x)$ is a holomorphic
section of $\pi_E : E \to X$, the *zero section*. On the complement $E \setminus s(X)$ one
has a natural $\mathbb{C}^*$-action. The quotient

$$\mathbb{P}(E) := (E \setminus s(X))/\mathbb{C}^*$$

is a complex manifold that admits a holomorphic projection $\pi : \mathbb{P}(E) \to X$
such that $\pi^{-1}(x)$ is isomorphic to $\mathbb{P}(E(x))$. One calls $\mathbb{P}(E)$ the *projective
bundle associated to E* or, simply, the *projectivization of E*.

vii) Let $\varphi : E \to F$ be a vector bundle homomorphism. There exist holo-
morphic vector bundles $\mathrm{Ker}(\varphi)$ and $\mathrm{Coker}(\varphi)$ over X, such that the fibers
over $x \in X$ are canonically isomorphic to $\mathrm{Ker}(\varphi(x) : E(x) \to F(x))$ and
$\mathrm{Coker}(\varphi(x) : E(x) \to F(x))$, respectively.

viii) For a *short exact sequence* of holomorphic vector bundles

$$0 \longrightarrow E \longrightarrow F \longrightarrow G \longrightarrow 0$$

one requires that $f : E \to F$ is a vector bundle homomorphism with $\mathrm{Ker}(f) = 0$ and $\mathrm{Coker}(f) = G$. In this case there exists a canonical isomorphism

$$\det(F) \cong \det(E) \otimes \det(G).$$

Recall that in the category of differentiable vector bundles every short exact sequence splits (see Appendix A). This is no longer true in the holomorphic setting, e.g. the Euler sequence in Section 2.4 does not split.

ix) Let E and F be two holomorphic vector bundles given by cocycles $\{\psi_{ij}\}$ and $\{\psi'_{ij}\}$ with respect to the same open covering. If for all $x \in U_i \cap U_j$ the matrix $\psi'_{ij}(x)$ has the form $\psi'_{ij}(x) = \begin{pmatrix} \psi_{ij}(x) & * \\ 0 & \phi_{ij}(x) \end{pmatrix}$, then E is a holomorphic subbundle of F, i.e. there exists a canonical injection $E \subset F$.

Conversely, if E is a holomorphic subbundle of F then cocycles of this form can be found. Moreover, in this situation the cokernel $G = F/E$ is described by the cocycle $\{\phi_{ij}\}$.

Definition 2.2.5 Let $f : Y \to X$ be a holomorphic map between complex manifolds and let E be a holomorphic vector bundle on X given by a cocycle $\{(U_i, \psi_{ij})\}$.

Then the *pull-back* f^*E of E is the holomorphic vector bundle over Y that is given by the cocycle $\{(f^{-1}(U_i), \psi_{ij} \circ f)\}$. For any $y \in Y$ there is a canonical isomorphism $(f^*E)(y) \cong E(f(y))$.

If Y is a complex submanifold of X and $i : Y \to X$ is the inclusion, then $E|_Y := i^*E$ is the *restriction* of E to Y.

It turns out that over $\mathbb{P}^n$ there exists essentially only one holomorphic line bundle, dubbed the *tautological line bundle* $\mathcal{O}(1)$. Its dual can be described as follows:

Proposition 2.2.6 *The set $\mathcal{O}(-1) \subset \mathbb{P}^n \times \mathbb{C}^{n+1}$ that consists of all pairs $(\ell, z) \in \mathbb{P}^n \times \mathbb{C}^{n+1}$ with $z \in \ell$ forms in a natural way a holomorphic line bundle over $\mathbb{P}^n$.*

Proof. The projection $\pi : \mathcal{O}(-1) \to \mathbb{P}^n$ is given by projecting to the first factor. Let $\mathbb{P}^n = \bigcup_{i=0}^{n} U_i$ be the standard open covering (see page 56). A canonical trivialization of $\mathcal{O}(-1)$ over U_i is given by $\psi_i : \pi^{-1}(U_i) \cong U_i \times \mathbb{C}$, $(\ell, z) \to (\ell, z_i)$. The transition maps $\psi_{ij}(\ell) : \mathbb{C} \to \mathbb{C}$ are given by $w \mapsto \frac{z_i}{z_j} \cdot w$, where $\ell = (z_0 : \ldots : z_n)$. The maps ψ_i provide at the same time holomorphic charts for $\mathcal{O}(-1)$. $\square$

Note that the fibre $\mathcal{O}(-1) \to \mathbb{P}^n$ over $\ell \in \mathbb{P}^n$ is naturally isomorphic to ℓ.

Definition 2.2.7 The line bundle $\mathcal{O}(1)$ is the dual $\mathcal{O}(-1)^*$ of $\mathcal{O}(-1)$. For $k > 0$ let $\mathcal{O}(k)$ be the line bundle $\mathcal{O}(1)^{\otimes k} = \mathcal{O}(1) \otimes \ldots \otimes \mathcal{O}(1)$ (k-times). Analogously, for $k < 0$ one defines $\mathcal{O}(k) := \mathcal{O}(-k)^*$. If E is any vector bundle on $\mathbb{P}^n$ one writes $E(k)$ instead of $E \otimes \mathcal{O}(k)$.

Remark 2.2.8 The fibre of $\pi : \mathcal{O}(1) \to \mathbb{P}^n$ over $\ell \in \mathbb{P}^n$ is canonically isomorphic to ℓ^*. The global linear coordinates $z_0, \ldots, z_n$ on $\mathbb{C}^{n+1}$ define natural sections of $\mathcal{O}(1)$.

We leave it to the reader to write down a coordinate free version of the above definitions, i.e. work with $\mathbb{P}(V)$ instead of $\mathbb{P}^n$. In particular, V^* is a subspace of the space of all global sections of $\mathcal{O}(1)$. However, for the proof of Proposition 2.2.6 it seems preferable to introduce a basis.

If we let $\mathcal{O}(0)$ be the trivial line bundle $X \times \mathbb{C} \to X$, then we have found a set of line bundles on $\mathbb{P}^n$ which forms a group isomorphic to the abelian group $\mathbb{Z}$. This is no coincidence, as is shown by the next proposition. In fact, for $\mathbb{P}^n$ one can show that any (holomorphic) line bundle is of the form $\mathcal{O}(k)$ (cf. Exercise 3.2.11).

Proposition 2.2.9 *The tensor product and the dual endow the set of all isomorphism classes of holomorphic line bundles on a complex manifold X with the structure of an abelian group. This group is the* Picard group $\mathrm{Pic}(X)$ *of X.*

Proof. By definition the product of two line bundles L_1, L_2 on X is the tensor product $L_1 \otimes L_2$ and the inverse of L is the dual L^*. The only thing that needs a proof is that $L \otimes L^*$ is isomorphic to the trivial line bundle. This is best seen by using the cocycle description of L and the induced one for L^*. $\square$

Corollary 2.2.10 *There is a natural isomorphism* $\mathrm{Pic}(X) \cong H^1(X, \mathcal{O}_X^*)$.

Proof. The description of line bundles in terms of their cocycles provides us with an isomorphism $\mathrm{Pic}(X) \cong \check{H}^1(X, \mathcal{O}_X^*)$. By general arguments, there always exists a natural homomorphism $\check{H}^p(X, \mathcal{F}) \to H^p(X, \mathcal{F})$ which is bijective for any reasonable topological space, e.g. for manifolds. In fact, for $p = 1$ it is always bijective (cf. Proposition B.0.43). $\square$

Recall that a holomorphic vector bundle E on a complex manifold X can be pulled-back under a holomorphic map $f : Y \to X$ to a holomorphic vector bundle f^*E on Y. For line bundles this operation yields a group homomorphism due to the following corollary, the verification of which is left to the reader.

Corollary 2.2.11 *Let $f : Y \to X$ be a holomorphic map. Then the pull-back under f defines a group homomorphism*

$$f^* : \mathrm{Pic}(X) \longrightarrow \mathrm{Pic}(Y).$$

In the description of the Picard group given in the previous corollary this map is induced by the sheaf homomorphism $f^*\mathcal{O}_X^* \to \mathcal{O}_Y^*$.   □

Even the cohomological description of the Picard group of a complex manifold X as $H^1(X, \mathcal{O}_X^*)$ doesn't reveal immediately how many (if at all) nontrivial line bundles exist on X. The most efficient way to study the Picard group is via the exponential sequence.

Definition 2.2.12 The *exponential sequence* on a complex manifold X is the short exact sequence

$$0 \longrightarrow \mathbb{Z} \longrightarrow \mathcal{O}_X \longrightarrow \mathcal{O}_X^* \longrightarrow 0.$$

Here, $\mathbb{Z}$ is the locally constant sheaf and $\mathbb{Z} \hookrightarrow \mathcal{O}_X$ is the natural inclusion. The map $\mathcal{O}_X \to \mathcal{O}_X^*$ is given by the exponential $f \mapsto \exp(2\pi i \cdot f)$.

That the sequence is exact in $\mathbb{Z}$ and $\mathcal{O}_X$ is obvious. The surjectivity of $\mathcal{O}_X \to \mathcal{O}_X^*$ is due to the existence of the logarithm, i.e. the locally defined inverse of $\exp : \mathbb{C} \to \mathbb{C}^*$.

As to any short exact sequence of sheaves, one can associate to the exponential sequence its long exact cohomology sequence

$$H^1(X, \mathbb{Z}) \longrightarrow H^1(X, \mathcal{O}_X) \longrightarrow H^1(X, \mathcal{O}_X^*) \longrightarrow H^2(X, \mathbb{Z})$$

Note that for X compact, the map $H^1(X, \mathbb{Z}) \to H^1(X, \mathcal{O}_X)$ is injective.

Thus, $\mathrm{Pic}(X)$ can in principle be computed by the cohomology groups $H^i(X, \mathbb{Z})$ and $H^i(X, \mathcal{O}_X)$, $i = 1, 2$, and the induced maps between them. Very roughly, $\mathrm{Pic}(X)$ has two parts. A discrete part, measured by its image in $H^2(X, \mathbb{Z})$, and a continuous part coming from the (possibly trivial) vector space $H^1(X, \mathcal{O}_X)$. In order to study the discrete part one introduces the first Chern class.

Definition 2.2.13 The *first Chern class* $c_1(L)$ of a holomorphic line bundle $L \in \mathrm{Pic}(X)$ on X is the image of L under the boundary map

$$c_1 : \mathrm{Pic}(X) \cong H^1(X, \mathcal{O}_X^*) \longrightarrow H^2(X, \mathbb{Z}).$$

The first Chern class will be used throughout this book and we will give alternative descriptions of it in the later chapters.

Higher rank vector bundles, even on $\mathbb{P}^n$, are not easily constructed. Of course, each of the constructions in Example 2.2.4 yields new bundles of arbitrary rank, but these are not genuinely new ones. However, there is one natural holomorphic vector bundle on any complex manifold which often is not trivial: the holomorphic tangent bundle. The holomorphic tangent bundle and its relatives are of uttermost importance in the study of complex manifolds.

Let $X = \bigcup U_i$ be an open covering by charts $\varphi_i : U_i \cong \varphi_i(U_i) \subset \mathbb{C}^n$. By definition the *Jacobian* of the transition maps $\varphi_{ij} := \varphi_i \circ \varphi_j^{-1} : \varphi_j(U_i \cap U_j) \cong \varphi_i(U_i \cap U_j)$ is the matrix $J(\varphi_{ij})(\varphi_j(z)) := \left(\frac{\partial \varphi_{ij}^k}{\partial z_\ell}(\varphi_j(z)) \right)_{k,\ell}$.

Definition 2.2.14 The *holomorphic tangent bundle* of a complex manifold X of dimension n is the holomorphic vector bundle $\mathcal{T}_X$ on X of rank n which is given by the transition functions $\psi_{ij}(z) = J(\varphi_{ij})(\varphi_j(z))$.

The *holomorphic cotangent bundle* Ω_X is the dual of $\mathcal{T}_X$. The bundle of holomorphic p-forms is $\Omega_X^p := \bigwedge^p \Omega_X$ for $0 \leq p \leq n$ and $K_X := \det(\Omega_X) = \Omega_X^n$ is called the *canonical bundle* of X.

The definition is independent of the open covering and the coordinate maps φ_i. For different choices the resulting vector bundles are canonical isomorphic. Thus, $\mathcal{T}_X$, Ω_X, and K_X are basic invariants of the complex manifold X.

Lemma 2.2.15 *Let $Y \subset X$ be a complex submanifold. Then there is a canonical injection $\mathcal{T}_Y \subset \mathcal{T}_X|_Y$.*

Proof. We apply ix) of Example 2.2.4 to our situation as follows. Let $\{(U_i, \varphi_i)\}$ be a holomorphic atlas such that $\varphi_i(Y \cap U_i) = \{z \mid z_{m+1} = \ldots = z_n = 0\} \cap \varphi_i(U_i)$. Then the transition functions φ_{ij} map $\mathbb{C}^m$ to $\mathbb{C}^m$. Therefore, the restriction ψ_{ij}' of the cocycle $J(\varphi_{ij}) \circ \varphi_j$, which defines $\mathcal{T}_X$, to the submanifold Y has the form required by ix) of Example 2.2.4. Here, $\psi_{ij} = J(\varphi_{ij}|_Y) \circ \varphi_j|_Y$. $\square$

Definition 2.2.16 Let $Y \subset X$ be a complex submanifold. The *normal bundle* of Y in X is the holomorphic vector bundle $\mathcal{N}_{Y/X}$ on Y is the cokernel of the natural injection $\mathcal{T}_Y \subset \mathcal{T}_X|_Y$. Thus, there exists a short exact sequence of holomorphic vector bundles, the *normal bundle sequence*:

$$0 \longrightarrow \mathcal{T}_Y \longrightarrow \mathcal{T}_X|_Y \longrightarrow \mathcal{N}_{Y/X} \longrightarrow 0.$$

Proposition 2.2.17 (Adjunction formula) *Let Y be a submanifold of a complex manifold X. Then the canonical bundle K_Y of Y is naturally isomorphic to the line bundle $K_X|_Y \otimes \det(\mathcal{N}_{Y/X})$.*

Proof. Let $\{\psi_{ij}'\}$ be a cocycle defining $\mathcal{T}_X|_Y$ as in the proof of Lemma 2.2.15. Then ψ_{ij}' has the form $\begin{pmatrix} \psi_{ij} & * \\ 0 & \phi_{ij} \end{pmatrix}$. Here, $\{\psi_{ij}\}$ is the cocycle that defines $\mathcal{T}_Y$. Moreover, by definition $\mathcal{N}_{Y/X}$ is determined by the cocycle $\{\phi_{ij}\}$ (cf. ix) of Example 2.2.4). On the other hand, $K_X|_Y$ is given by $\{\det(\psi_{ij}')^{-1}\} = \{\det(\psi_{ij})^{-1} \cdot \det(\phi_{ij})^{-1}\}$ and, therefore, $K_Y \cong K_X|_Y \otimes \det(\mathcal{N}_{Y/X})$.

One could equally well just evoke viii) of Example 2.2.4. $\square$

The vector bundles Ω_X^p can be used to define numerical invariants of X which are very similar to Betti numbers of compact differentiable manifolds. In order to do so, we shall first explain how to define the cohomology of a holomorphic vector bundle in general.

Definition 2.2.18 Let $\pi : E \to X$ be a holomorphic vector bundle. Its *sheaf of sections*, also called E, is given by

$$U \longmapsto \{s : U \to \pi^{-1}(U) \text{ holomorphic} \mid \pi \circ s = \mathrm{id}_U\}.$$

It is straightforward to verify that the so defined presheaf is in fact a sheaf. Also note that the sheaf of sections of a holomorphic vector bundle is in a natural way a sheaf of $\mathcal{O}_X$-modules.

In particular, $\mathcal{O}_X$ is the sheaf of sections of the trivial line bundle $X \times \mathbb{C} \to X$. The use of the same symbol E for a holomorphic vector bundle and its sheaf of sections might look confusing, but these two objects are essentially equivalent:

Proposition 2.2.19 *Associating to a holomorphic vector bundle its sheaf of sections defines a canonical bijection between the set of holomorphic vector bundles of rank r and the set of locally free $\mathcal{O}_X$-modules of rank r.*

Proof. First recall that a locally free $\mathcal{O}_X$-module of rank r is a sheaf $\mathcal{F}$ of $\mathcal{O}_X$-modules on X which is locally isomorphic to $\mathcal{O}_X^{\oplus r}$. Clearly, the sheaf of sections of a holomorphic vector bundle $\pi : E \to X$ of rank r is locally free of rank r, for E is locally isomorphic to $X \times \mathbb{C}^r$.

Conversely, if we have chosen trivializations $\psi_i : \mathcal{F}|_{U_i} \cong \mathcal{O}_{U_i}^{\oplus r}$, then the transition maps $\psi_{ij} := (\psi_i \circ \psi_j^{-1})|_{U_i \cap U_j} : \mathcal{O}_{U_i \cap U_j}^{\oplus r} \cong \mathcal{O}_{U_i \cap U_j}^{\oplus r}$ are given by multiplication with a matrix of holomorphic functions on $U_i \cap U_j$. Therefore, $\{(U_i, \psi_{ij})\}$ can be used as a cocycle defining a holomorphic vector bundle. One easily checks that the two constructions are inverse to each other. $\square$

Example 2.2.20 Consider the holomorphic tangent bundle $\mathcal{T}_X$. The associated sheaf is sometimes denoted Θ_X. Furthermore, Θ_X admits an alternative description as the sheaf $\mathrm{Der}(\mathcal{O}_X)$ of derivations. More precisely, $\mathrm{Der}(\mathcal{O}_X)$ is the sheaf that associates to an open subset $U \subset X$ the set of all $\mathbb{C}$-linear maps $D : \mathcal{O}_X(U) \to \mathcal{O}_X(U)$ satisfying the Leibniz rule $D(f \cdot g) = f \cdot D(g) + D(f) \cdot g$. As we will see later in Section 2.6, local sections of $\mathcal{T}_X$ are of the form $\sum a_i \frac{\partial}{\partial z_i}$, where the $a_i(z)$ are holomorphic, and these can indeed be viewed as derivations of the above type.

There is, however, subtlety in the above bijection

$$\{\text{holomorphic vector bundles}\} \longleftrightarrow \{\text{locally free } \mathcal{O}_X-\text{modules}\}.$$

Warning: Homomorphisms between holomorphic vector bundles are, by definition (see page 66), of constant rank, whereas on the right hand side one usually allows arbitrary $\mathcal{O}_X$-module homomorphisms. More precisely, a $\mathcal{O}_X$-module homomorphism is induced by a vector bundle homomorphism if and only if the induced maps between the fibres are of constant rank. Furthermore, the $\mathcal{O}_X$-module associated to the cokernel of a vector bundle homomorphism is naturally isomorphic to the sheaf cokernel of the induced $\mathcal{O}_X$-module homomorphism.

Definition 2.2.21 If E is a holomorphic vector bundle on a complex manifold X, then $H^q(X, E)$ denotes the q-th *cohomology* of its sheaf of sections.

In particular, $H^0(X, E)$ is the space $\Gamma(X, E)$ of global holomorphic sections of the vector bundle projection $\pi : E \to X$.

Remark 2.2.22 If $f : Y \to X$ is a holomorphic map and E is a holomorphic vector bundle on X, then there exists a natural homomorphism $H^0(X, E) \to H^0(Y, f^*E)$ given by $s \mapsto (y \mapsto s(f(x)))$, where we use the natural isomorphism $(f^*E)(y) \cong E(f(y))$.

Definition 2.2.23 The *Hodge numbers* of a compact complex manifold X are the numbers $h^{p,q}(X) := \dim H^q(X, \Omega_X^p)$.

In Section 3.2 we will see that the Hodge numbers are indeed finite. They will later be related to the cohomology of X considered as a differentiable manifold and in particular to the Betti numbers of X.

The canonical bundle can be used to define another numerical invariant of compact complex manifolds, the Kodaira dimension. In order to define it, we have to note that for two holomorphic vector bundles E and F on a manifold X, there exists a natural map

$$H^0(X, E) \otimes H^0(X, F) \longrightarrow H^0(X, E \otimes F).$$

This yields the following

Lemma 2.2.24 *For every line bundle L on a complex manifold X the space*

$$R(X, L) := \bigoplus_{m \geq 0} H^0(X, L^{\otimes m})$$

has a natural ring structure. By definition $L^{\otimes 0} = \mathcal{O}_X$. $\square$

Definition 2.2.25 The *canonical ring* of a complex manifold X is the ring

$$R(X) := R(X, K_X) = \bigoplus_{m \geq 0} H^0(X, K_X^{\otimes m}).$$

If X is connected, then $R(X)$ is an integral domain. Thus, one can form the quotient field $Q(R(X))$ of $R(X)$.

Definition 2.2.26 The *Kodaira dimension* of a connected complex manifold X is defined as

$$\mathrm{kod}(X) := \begin{cases} -\infty & \text{if } R(X) = \mathbb{C} \\ \mathrm{trdeg}_{\mathbb{C}} Q(X) - 1 & \text{else.} \end{cases}$$

Analogously, one may define the Kodaira dimension kod(X, L) with respect to any line bundle $L \in \text{Pic}(X)$, but the case of the canonical line bundle is the most natural one. Note that in general $R(X, L)$ is not finitely generated. However, the canonical ring $R(X) = R(X, K_X)$ is expected to be finitely generated (abundance conjecture), at least if X is projective.

In order to relate the Kodaira dimension to the algebraic dimension (see Definition 2.1.10) we have to explain how to produce meromorphic functions from sections of line bundles. Let $s_1, s_2 \in H^0(X, L)$ be two non-zero sections of a holomorphic line bundle L on a connected complex manifold X. Then we associate to this pair a meromorphic function $s_1/s_2 \in K(X)$ as follows. Let $X = \bigcup U_i$ be an open covering, such that L can be trivialized over the open sets U_i, i.e. there exist isomorphisms $\psi_i : L|_{U_i} \cong \mathcal{O}_{U_i}$. Hence, $\psi_i(s_1|_{U_i})$ and $\psi_i(s_2|_{U_i})$ are holomorphic functions on U_i. The meromorphic function s_1/s_2 on U_i is by definition $\psi_i(s_1|_{U_i})/\psi_i(s_2|_{U_i})$. One easily verifies that the definition does not depend on $\{(U_i, \psi_i)\}$. In particular, any pair of non-zero sections $s_1, s_2 \in H^0(X, K_X^{\otimes n})$ yields a meromorphic function on X. For a more general version of this method to construct meromorphic functions see page 84.

Definition 2.2.27 Let $R = \bigoplus_{m \geq 0} R_m$ be a graded integral domain. Then $Q_0(R)$ is the subfield of the quotient field $Q(R)$ that consists of elements of the form f/g with $0 \neq g, f \in R_m$ for some m.

Proposition 2.2.28 *If X is a connected complex manifold, then there exists a natural inclusion $Q_0(R(X)) \subset K(X)$. Moreover, for $R(X) \neq \mathbb{C}$ one has*

$$\text{kod}(X) = \text{trdeg}_{\mathbb{C}} Q(R(X)) - 1 = \text{trdeg}_{\mathbb{C}} Q_0(R(X)) \leq \text{trdeg}_{\mathbb{C}} K(X) = a(X).$$

Proof. We only need to show that $\text{trdeg}_{\mathbb{C}} Q(R(X)) - 1 = \text{trdeg}_{\mathbb{C}} Q_0(R(X))$. Clearly, if $f_0, \ldots, f_k \in Q(R(X))$ are algebraically independent elements of degree $d_1, \ldots, d_k$, then $f_0^{e_0}/f_k^{e_k}, \ldots, f_{k-1}^{e_{k-1}}/f_k^{e_k} \in Q_0(R(X))$ are algebraically independent for $e_i = \prod_{j \neq i} d_j$. Hence, $\text{trdeg}_{\mathbb{C}} Q(X) - 1 \leq \text{trdeg}_{\mathbb{C}} Q_0(R(X))$. On the other hand, if $g_1, \ldots, g_k \in Q_0(R(X))$ are algebraically independent and g has positive degree, then also $g, g_1, \ldots, g_k \in Q(R(X))$ are algebraically independent. Thus, $\text{trdeg}_{\mathbb{C}} Q_0(R(X)) + 1 \leq \text{trdeg}_{\mathbb{C}} Q(R(X))$. $\square$

It should be clear from the proof that the assertion holds true also for the Kodaira dimension with respect to any line bundle L.

Exercises

2.2.1   Let E and F be vector bundles determined by the cocycles $\{(U_i, \psi_{ij} : U_i \cap U_j \to \text{Gl}(r, \mathbb{C})\}$ respectively $\{(U_i, \psi'_{ij} : U_i \cap U_j \to \text{Gl}(r', \mathbb{C}))\}$. Verify the cocycle description for the constructions in Example 2.2.4. The most important ones are:

i) The direct sum $E \oplus F$ corresponds to $\psi_{ij} \oplus \psi'_{ij} : U_i \cap U_j \to \mathrm{Gl}(r + r', \mathbb{C})$.
ii) The tensor product $E \otimes F$ corresponds to $\psi_{ij} \otimes \psi'_{ij} : U_i \cap U_j \to \mathrm{Gl}(r \cdot r', \mathbb{C})$.
iv) The dual bundle E^* corresponds to $(\psi_{ij}^t)^{-1}$.
v) The determinant bundle $\det(E)$ corresponds to $\det(\psi_{ij})$.

2.2.2 Show that any short exact sequence of holomorphic vector bundles $0 \to L \to E \to F \to 0$, where L is a line bundle, induces short exact sequences of the form
$$0 \longrightarrow L \otimes \bigwedge^{i-1} F \longrightarrow \bigwedge^i E \longrightarrow \bigwedge^i F \longrightarrow 0.$$

2.2.3 Show that for any holomorpic vector bundle E of rank r there exists a non-degenerate pairing
$$\bigwedge^k E \times \bigwedge^{r-k} E \longrightarrow \det(E).$$

Deduce from this the existence of a natural isomorphism of holomorphic vector bundles $\bigwedge^k E \cong \bigwedge^{r-k} E^* \otimes \det(E)$.

2.2.4 Show that any homomorphism $f : E \to F$ of holomorphic vector bundles E and F induces natural homomorphisms $f \otimes \mathrm{id}_G : E \otimes G \to F \otimes G$ for any holomorphic vector bundle G. If f is injective, then so is $f \otimes \mathrm{id}_G$.

2.2.5 Let L be a holomorphic line bundle on a compact complex manifold X. Show that L is trivial if and only if L and its dual L^* admit non-trivial global sections. (Hint: Use the non-trivial sections to construct a non trivial section of $\mathcal{O} \cong L \otimes L^*$.)

2.2.6 Let $L \in \mathrm{Pic}(X)$ and $Y \subset X$ a submanifold of codimension at least two. Show that the restriction $H^0(X, L) \to H^0(X \setminus Y, L)$ is bijective. This generalizes Corollary 2.1.6. (Use Exercise 1.1.20.)

2.2.7 Let L_1 and L_2 be two holomorphic line bundles on a complex manifold X. Suppose that $Y \subset X$ is a submanifold of codimension at least two such that L_1 and L_2 are isomorphic on $X \setminus Y$. Prove that $L_1 \cong L_2$.

2.2.8 Show that any non-trivial homogeneous polynomial $0 \neq s \in \mathbb{C}[z_0, \ldots, z_n]_k$ of degree k can be considered as a non-trivial section of $\mathcal{O}(k)$ on $\mathbb{P}^n$. (In fact, all sections are of this form, cf. Proposition 2.4.1.)

2.2.9 Show that $\mathcal{O}(-1) \setminus s(\mathbb{P}^n)$ is naturally identified with $\mathbb{C}^{n+1} \setminus \{0\}$, where $s : \mathbb{P}^n \to \mathcal{O}(-1)$ is the zero-section (see Example 2.2.4). Use this to construct a submersion $S^{2n+1} \to \mathbb{P}^n$ with fibre S^1. (For $n = 1$ this yields the *Hopf fibration* $S^3 \to S^2$.)

2.2.10 Let $\{(U_i, \varphi_i)\}$ be an atlas of the complex manifold X. Use the cocycle description of the holomorphic tangent bundle T_X to show that for $x \in U_i \subset X$ the fibre $T_X(x)$ can be identified with $T_{\varphi_i(x)}\varphi_i(U_i) \cong T^{1,0}_{\varphi_i(x)}\varphi_i(U_i)$. In particular, the vectors $\partial/\partial z_i$ can be viewed as a basis of $T_X(x)$. (This will be discussed in more detail in Section 2.6.)

2.2.11   Let $Y \subset X$ be a submanifold of dimension k of a complex manifold X of dimension n. Assume that $\{(U_i, \varphi_i)\}$ is a compatible atlas in the sense of Definition 2.1.16. Show that for $x \in Y \cap U_i$ the tangent vectors $\partial/\partial z_{k+1}, \ldots, \partial/\partial z_n \in T^{1,0}_{\varphi_i(x)}\varphi_i(U_i)$ form in a natural way a basis of $\mathcal{N}_{Y/X}(x)$. (Use the cocycle description of the normal bundle.)

2.2.12   Let $Y \subset X$ be a submanifold locally in $U \subset X$ defined by holomorphic functions $f_1, \ldots f_{n-k}$ (i.e. 0 is a regular value of $(f_1, \ldots, f_{n-k}) : U \to \mathbb{C}^{n-k}$ and Y is the pre-image of it.). Show that $f_1, \ldots, f_{n-k}$ naturally induce a basis of $\mathcal{N}_{Y/X}(x)^*$ for any $x \in U \cap Y$. (Use the map $\partial/\partial z_i \mapsto \partial f_j/\partial z_i$, for $i = k+1, \ldots, n$.)
   Go on and prove the existence of a natural isomorphism $\mathcal{N}^*_{Y/X} \cong \mathcal{I}_Y/\mathcal{I}_Y^2$.

2.2.13   Show that the holomorphic tangent bundle of a complex torus $X = \mathbb{C}^n/\Gamma$ is trivial, i.e. isomorphic to the trivial vector bundle $\mathcal{O}^{\oplus n}$. Compute $\mathrm{kod}(X)$.

2.2.14   Show that any submanifold of a complex torus has non-negative Kodaira dimension.

2.2.15   A *parallelizable* complex manifold is a manifold whose holomorphic tangent bundle is trivial. Thus, complex tori are parallelizable. Show that the Iwasawa manifold (cf. page 61) is parallelizable. Compute the Kodaira dimension of a compact parallelizable manifold.

2.2.16   What can you say at this point about the algebraic dimension of a projective manifold? Later we will see that projective manifolds are *Moishezon*, i.e. $a(X) = \dim(X)$. See Exercise 5.2.11.

Comments: - In this section we have relied on the reader's ability to manipulate (real or complex) differentiable vector bundles or, at least, the tangent bundle of a differentiable manifold. A more detailed understanding of the relation between differentiable and holomorphic bundles will be gained by working through the subsequent chapters. We will however not really touch upon the question how many different holomorphic structure a differentiable complex vector bundle may have. This question would lead us to moduli spaces of holomorphic vector bundles and is beyond the scope of this book. We have to refer to [78].
   - The examples introduced here, $\mathcal{O}(-1)$ and tangent and cotangent bundle, are the most interesting holomorphic vector bundles. For more details on line bundles on $\mathbb{P}^n$ see Section 2.4. A different view on $\mathcal{T}_X$ and Ω_X will be presented in Section 2.6.
   - The Kodaira dimension is a very coarse invariant, but it allows to put any compact manifold of dimension n in one of $n+1$ different classes. Each of them has a special feature and is studied by different means. For a thorough discussion of the Kodaira dimension in the algebraic context we recommend Ueno's book [111]. In particular, an alternative definition of the Kodaira dimension in terms of the growth of the pluri-genera $P_m(X) = \dim H^0(X, K_X^m)$ can be found there. The problem whether $R(X)$ is finitely generated is central in the classification theory of algebraic varieties (see e.g. [29]).

2.3 Divisors and Line Bundles

It turns out that hypersurfaces are always given as the zero locus of a global holomorphic section of a holomorphic line bundle. This will allow us to study the codimension one geometry of a compact complex manifold by means of holomorphic line bundles, their space of global sections and, eventually, their cohomology.

We start out by recalling the following

Definition 2.3.1 An analytic *hypersurface* of X is an analytic subvariety $Y \subset X$ of codimension one, i.e. $\dim(Y) = \dim(X) - 1$.

A hypersurface $Y \subset X$ is locally given as the zero set of a non-trivial holomorphic function. Indeed, locally $Y \subset X$ induces germs of codimension one and any such germ is the zero set of a single holomorphic function (cf. Remark 1.1.32).

Every hypersurface Y is the union $\bigcup Y_i$ of its irreducible components. Usually, we shall assume that Y has only finitely many irreducible components, e.g. this holds true if X is compact. In general, the union $\bigcup Y_i$ is only locally finite. Also note that an irreducible hypersurface $Y \subset X$ might define reducible germs (cf. Exercise 1.1.11).

Definition 2.3.2 A *divisor* D on X is a formal linear combination

$$D = \sum a_i[Y_i]$$

with $Y_i \subset X$ irreducible hypersurfaces and $a_i \in \mathbb{Z}$. The *divisor group* $\mathrm{Div}(X)$ is the set of all divisors endowed with the natural group structure.

In the above definition we want to assume that the sum is locally finite, i.e. for any $x \in X$ there exists an open neighbourhood U such that there exist only finitely many coefficients $a_i \neq 0$ with $Y_i \cap U \neq \emptyset$. If X is compact, this reduces to finite sums.

Remark 2.3.3 Every hypersurface defines a divisor $\sum[Y_i] \in \mathrm{Div}(X)$, where Y_i are the irreducible components of Y. Conversely, to any divisor $\sum a_i[Y_i]$ with $a_i \neq 0$ for all i one can associate the hypersurface $\bigcup Y_i$, but this construction is clearly not very natural, as the coefficients a_i do not enter the definition.

Definition 2.3.4 A divisor $D = \sum a_i[Y_i]$ is called *effective* if $a_i \geq 0$ for all i. In this case, one writes $D \geq 0$.

The divisor associated to a hypersurface is an example of an effective divisor.

Let $Y \subset X$ be a hypersurface and let $x \in Y$. Suppose that Y defines an irreducible germ in x. Hence, this germ is the zero set of an irreducible $g \in \mathcal{O}_{X,x}$.

Definition 2.3.5 Let f be a meromorphic function in a neighbourhood of $x \in Y$. Then the *order* $\operatorname{ord}_{Y,x}(f)$ of f in x with respect to Y is given by the equality $f = g^{\operatorname{ord(f)}} \cdot h$ with $h \in \mathcal{O}^*_{X,x}$.

Remarks 2.3.6 i) The definition of $\operatorname{ord}_{Y,x}(f)$ does not depend on the defining equation g, as long as we choose g irreducible (see Corollary 1.1.19). In fact, two irreducible functions $g, g' \in \mathcal{O}_{X,x}$ with $Z(g) = Z(g')$ only differ by a unit in $\mathcal{O}_{X,x}$.

ii) More globally, one defines the *order* $\operatorname{ord}_Y(f)$ of a meromorphic function $f \in K(X)$ along an irreducible hypersurface $Y \subset X$ as $\operatorname{ord}_Y(f) = \operatorname{ord}_{Y,x}(f)$ for $x \in Y$ such that Y is irreducible in x. Such a point $x \in Y$ always exist, e.g. if we may choose a regular point $x \in Y_{\operatorname{reg}}$ (cf. Exercise 1.1.14). Moreover, the definition does not depend on x. Here we use Proposition 1.1.35 and the fact that Y_{reg} is connected if Y is irreducible.

iii) Note that the order satisfies the equation

$$\operatorname{ord}_Y(f_1 f_2) = \operatorname{ord}_Y(f_1) + \operatorname{ord}_Y(f_2).$$

Definition 2.3.7 A meromorphic function $f \in K(X)$ has *zeros (poles) of order* $d \geq 0$ along an irreducible hypersurface $Y \subset X$ if

$$\operatorname{ord}_Y(f) = d \quad \text{(respectively, } \operatorname{ord}_Y(f) = -d\text{)}.$$

This information for all hypersurfaces is encoded by the divisor (f) associated to any meromorphic function $f \in K(X)$:

Definition 2.3.8 Let $f \in K(X)$. Then the *divisor associated to f* is

$$(f) := \sum \operatorname{ord}_Y(f)[Y] \in \operatorname{Div}(X),$$

where the sum is taken over all irreducible hypersurfaces $Y \subset X$. A divisor $D \in \operatorname{Div}(X)$ of this form is called *principal*.

The divisor (f) can be written as the difference of two effective divisors $(f) = Z(f) - P(f)$, where

$$Z(f) = \sum_{\operatorname{ord}_Y(f) > 0} \operatorname{ord}_Y(f)[Y] \quad \text{and} \quad P(f) = - \sum_{\operatorname{ord}_Y(f) < 0} \operatorname{ord}_Y(f)[Y]$$

is the *zero divisor* respectively the *pole divisor* of f.

Next, we shall generalize this construction, such that we cannot only associate divisors to globally meromorphic functions, but also to locally given ones. In order to do this we consider nowhere vanishing holomorphic functions as invertible meromorphic functions. This way we obtain an inclusion of sheaves $\mathcal{O}^*_X \subset \mathcal{K}^*_X$.

Proposition 2.3.9 *There exists a natural isomorphism*

$$H^0(X, \mathcal{K}_X^*/\mathcal{O}_X^*) \cong \operatorname{Div}(X).$$

Proof. The isomorphism is induced by associating to a meromorphic function its divisor. This is made precise as follows: An element $f \in H^0(X, \mathcal{K}_X^*/\mathcal{O}_X^*)$ is given by non-trivial meromorphic functions $f_i \in \mathcal{K}_X^*(U_i)$ on open subsets $U_i \subset X$ covering X such that $f_i \cdot f_j^{-1}$ is a holomorphic function without zeros. Thus, for any irreducible hypersurface $Y \subset X$ with $Y \cap U_i \cap U_j \neq \emptyset$ one has $\operatorname{ord}_Y(f_i) = \operatorname{ord}_Y(f_j)$. Hence, $\operatorname{ord}_Y(f)$ is well-defined for any irreducible hypersurface $Y \subset X$. Then one associated to $f \in H^0(X, \mathcal{K}_X^*/\mathcal{O}_X^*)$ the divisor $(f) = \sum \operatorname{ord}_Y(f)[Y] \in \operatorname{Div}(X)$.

Using the additivity of the order, one sees that the induced map

$$H^0(X, \mathcal{K}_X^*/\mathcal{O}_X^*) \longrightarrow \operatorname{Div}(X)$$

is a group homomorphism.

In order to show that it is bijective, we define the inverse map as follows. If $D = \sum a_i[Y_i] \in \operatorname{Div}(X)$ is given, then there exists an open cover $X = \bigcup U_i$ such that $Y_i \cap U_j$ is defined by some $g_{ij} \in \mathcal{O}(U_j)$ which is unique up to elements in $\mathcal{O}^*(U_j)$. Let $f_j := \prod_i g_{ij}^{a_i} \in \mathcal{K}_X^*(U_j)$. (The product can assumed to be finite.) Since g_{ij} and g_{ik} on $U_j \cap U_k$ define the same irreducible hypersurface, they only differ by an element in $\mathcal{O}^*(U_j \cap U_k)$. Thus, the functions f_j glue to an element $f \in H^0(X, \mathcal{K}_X^*/\mathcal{O}_X^*)$. Clearly, the two maps are inverse to each other. $\qquad\square$

In algebraic geometry (see e.g. [66]) elements in $H^0(X, \mathcal{K}_X^*/\mathcal{O}_X^*)$ are called *Cartier divisors* in contrast to the elements in $\operatorname{Div}(X)$, which are *Weil divisor*. The above equality holds in the algebraic setting under a weak smoothness assumption on X.

Corollary 2.3.10 *There exists a natural group homomorphism*

$$\operatorname{Div}(X) \longrightarrow \operatorname{Pic}(X), \quad D \longmapsto \mathcal{O}(D),$$

where the definition of $\mathcal{O}(D)$ will be given in the proof.

Proof. If $D = \sum a_i[Y_i] \in \operatorname{Div}(X)$ corresponds to $f \in H^0(X, \mathcal{K}_X^*/\mathcal{O}_X^*)$, which in turn is given by functions $f_i \in \mathcal{K}_X^*(U_i)$ for an open covering $X = \bigcup U_i$, then we define $\mathcal{O}(D) \in \operatorname{Pic}(X)$ as the line bundle with transition function $\psi_{ij} := f_i \cdot f_j^{-1} \in H^0(U_i \cap U_j, \mathcal{O}_X^*)$. By the very definition $\{(U_i, \psi_{ij})\}$ satisfies the cocycle condition, i.e. $\{(U_i, \psi_{ij})\} \in \check{H}^1(\{U_i\}, \mathcal{O}_X^*)$ and thus we obtain an element in $H^1(X, \mathcal{O}_X^*) = \operatorname{Pic}(X)$.

If $D, D' \in H^0(X, \mathcal{K}_X^*/\mathcal{O}_X^*)$ are given by $\{f_i\}$ and $\{f_i'\}$, respectively, then $D + D'$ corresponds to $\{f_i \cdot f_i'\}$. (Note that we may always pass to a refinement of a given open covering, so that we can assume that $\{f_i\}$ and $\{f'_j\}$ are defined with respect to the same open covering.)

By definition $\mathcal{O}(D + D')$ is described by $\{(f_i \cdot f'_i) \cdot (f_j \cdot f'_j)^{-1}\} = \{\psi_{ij} \cdot \psi'_{ij}\}$. Hence, $\mathcal{O}(D + D') = \mathcal{O}(D) \otimes \mathcal{O}(D')$. Clearly, $\mathcal{O}(0) \cong \mathcal{O}$ and thus $\mathcal{O}(-D) \cong \mathcal{O}(D)^*$. $\qquad\square$

Remark 2.3.11 Suppose $Y \subset X$ is a smooth hypersurface of a compact complex manifold X of dimension n. In particular, Y is a real codimension two submanifold of a compact manifold and thus defines an element $[Y] \in H^2(X, \mathbb{R})$, its *fundamental class*. Indeed, by Poincaré duality (cf. Proposition A.0.6) the linear map $\alpha \mapsto \int_Y \alpha$ on $H^{2n-2}(X, \mathbb{R})$ defines an element in $H^2(X, \mathbb{R})$.

On the other hand, we have introduced the first Chern class $c_1(\mathcal{O}(Y)) \in H^2(X, \mathbb{R})$ of the line bundle $\mathcal{O}(Y)$ (see Definition 2.2.13). *A priori*, it is not clear that $c_1(\mathcal{O}(Y)) = [Y]$ and we have to refer to Section 4.4 for a proof of this.

The homomorphism $\mathrm{Div}(X) \to \mathrm{Pic}(X)$ is compatible with pull-back. However, the pull-back of a divisor D under a morphism $f : X \to Y$ is not always well-defined, one has to assume that the image of f is not contained in the support of D. Thus, one usually considers only *dominant* morphisms, i.e. morphisms with dense image. There are two ways to define the pull-back of a divisor:

i) Let $f : X \to Y$ be a holomorphic map and let $Z \subset Y$ be an irreducible hypersurface such that no component of $f(X)$ is contained in Z. Then the preimage $f^{-1}(Z)$ of Z is again a hypersurface, although in general not irreducible. Indeed, if Z is locally the zero set of a holomorphic function g then $f^{-1}(Z)$ is the zero set of $g \circ f$.

Let $Y_i \subset X$, $i = 1, \ldots, k$; be the irreducible components of $f^{-1}(Z)$. The pull-back of Z considered as a divisor is the linear combination $f^*Z = \sum n_i[Y_i]$ where the coefficients are determined as follows. Consider a smooth point $y_i \in Y_i$ and its image $z = f(y_i)$. Then locally near z the hypersurface Z is defined by a holomorphic function g, the pull-back of which can be decomposed into its irreducible factors $g \circ f = \prod g_j^{n_j}$, where g_i is a local equation of Y_i near y_i. This determines the coefficient n_i in $f^*Z = \sum n_i[Y_i]$. (Note that one has to pay attention to the fact that an irreducible hypersurface may very well be reducible at singular points.)

More generally, one defines the pull-back of a divisor $D = \sum a_i[Z_i] \in \mathrm{Div}(Y)$ such that no component of $f(X)$ is contained in any of the hypersurfaces Z_i as $f^*D = \sum_i a_i f^*(Z_i)$.

ii) We can also use Proposition 2.3.9 to give an alternative definition of the pull-back of a divisor under a dominant holomorphic map. If $f : X \to Y$ is a dominant holomorphic map and the divisor $D \in \mathrm{Div}(Y)$ corresponds to the Cartier divisor $\{(U_i, f_i \in \mathcal{K}^*_X(U_i))\}$ then f^*D is the divisor given by $\{(f^{-1}(U_i), f_i \circ f \in \mathcal{K}^*_X(f^{-1}(U_i)))\}$.

The two definitions are indeed equivalent, as follows easily from an inspection of the proof of Proposition 2.3.9.

The pull-back as defined above yields a group homomorphism between the divisor groups of Y and X. More precisely, one has

Proposition 2.3.12 *Let* $f : X \to Y$ *be a holomorphic map of connected complex manifolds and suppose that* f *is dominant, i.e.* $f(X)$ *is dense in* Y. *Then the pull-back defines a group homomorphism*

$$f^* : \mathrm{Div}(Y) \longrightarrow \mathrm{Div}(X).$$

□

This description ii) of the pull-back shows that it is compatible with the group homomorphism $\mathrm{Div}(X) \to \mathrm{Pic}(X)$ and the pull-back of line bundles (cf. Corollary 2.2.11).

Corollary 2.3.13 *Let* $f : X \to Y$ *be a holomorphic map. If* $D \in \mathrm{Div}(Y)$ *is a divisor such that* f^*D *is defined then* $\mathcal{O}(f^*D) \cong f^*\mathcal{O}(D)$. □

Thus, if X is connected and f is dominant one obtains a commutative diagram of abelian groups

$$
\begin{array}{ccc}
\mathrm{Div}(Y) & \longrightarrow & \mathrm{Pic}(Y) \\
{\scriptstyle f^*}\big\downarrow & & \big\downarrow{\scriptstyle f^*} \\
\mathrm{Div}(X) & \longrightarrow & \mathrm{Pic}(X)
\end{array}
$$

Lemma 2.3.14 *A divisor* $D \in \mathrm{Div}(X)$ *is principal if and only if* $\mathcal{O}(D) \cong \mathcal{O}$.

Proof. If D is the principal divisor (f), then we may take as the corresponding section of $\mathcal{K}_X^*/\mathcal{O}_X^*$ the image of $f \in K(X)$ under the natural map $K(X)^* = H^0(X, \mathcal{K}_X^*) \to H^0(X, \mathcal{K}_X^*/\mathcal{O}_X^*)$. But then the associated cocycle $\{\psi_{ij}\}$ is the trivial one, i.e. $\psi_{ij} \equiv 1$. This shows that $\mathcal{O}(D) \cong \mathcal{O}$.

Conversely, let $\mathcal{O}(D)$ be trivial. Then, we may find an open covering $X = \bigcup U_i$ such that D corresponds to $\{f_i \in \mathcal{K}_X^*(U_i)\}$ and such that there exist holomorphic units $g_i \in \mathcal{O}_X^*(U_i)$ with $f_i \cdot f_j^{-1} = \psi_{ij} = g_i \cdot g_j^{-1}$. Hence, $f_i \cdot g_i^{-1} = f_j \cdot g_j^{-1}$ on $U_i \cap U_j$, i.e. there exists a global meromorphic function $f \in K(X)$ with $f|_{U_i} = f_i \cdot g_i^{-1}$ for all i. Since g_i are nowhere vanishing, one finds $(f) = D$. □

Remark 2.3.15 In fact, the above discussion is an application of the general machinery of long exact cohomology sequences to the case of the short exact sequence

$$0 \longrightarrow \mathcal{O}_X^* \longrightarrow \mathcal{K}_X^* \longrightarrow \mathcal{K}_X^*/\mathcal{O}_X^* \longrightarrow 0.$$

In particular, the map $\text{Div}(X) \to \text{Pic}(X)$ is nothing but the boundary map $H^0(X, \mathcal{K}_X^*/\mathcal{O}_X^*) \to H^1(X, \mathcal{O}_X^*)$, the kernel of which coincides with the image of $K(X)^* = H^0(X, \mathcal{K}_X^*) \to H^0(X, \mathcal{K}_X^*/\mathcal{O}_X^*)$. The latter is by definition the set of principal divisors.

Definition 2.3.16 Two divisors $D, D' \in \text{Div}(X)$ are called *linearly equivalent*, $D \sim D'$, if $D - D'$ is a principal divisor.

Thus, the lemma shows that the group homomorphism $\text{Div}(X) \to \text{Pic}(X)$ factorizes over an injection

$$\text{Div}(Y)/_\sim \hookrightarrow \text{Pic}(X)$$

In general, the inclusion is strict. But we will see that as soon as a line bundle admits a global section, it is contained in the image. In order to see this we shall construct a canonical map

$$H^0(X, L) \setminus \{0\} \longrightarrow \text{Div}(X), \quad s \longmapsto Z(s).$$

(Here, we assume that X is connected. Otherwise we have to take out all sections $s \in H^0(X, L)$ vanishing on some connected component of X.)

The map is constructed as follows: Fix trivializations $\psi_i : L|_{U_i} \cong \mathcal{O}_{U_i}$. Thus, $L \in \text{Pic}(X) = H^1(X, \mathcal{O}_X^*)$ is given by the cocycle $\{(U_i, \psi_{ij} := \psi_i \circ \psi_j^{-1})\}$. If $s \in H^0(X, L) \setminus \{0\}$ we set $f_i := \psi_i(s|_{U_i}) \in \mathcal{O}(U_i)$. Then, by the identity theorem, $f_i \in \mathcal{K}_X^*(U_i)$. Moreover, $f_i \cdot f_j^{-1} = \psi_i(s|_{U_i \cap U_j}) \cdot (\psi_j(s|_{U_i \cap U_j}))^{-1} = \psi_{ij} \in H^0(U_i \cap U_j, \mathcal{O}_X^*)$. This way we obtain a (Cartier) divisor $f := \{(U_i, f_i)\} \in H^0(X, \mathcal{K}_X^*/\mathcal{O}_X^*)$. The corresponding element in $\text{Div}(X)$ is denoted $Z(s)$.

It is not hard to check that for sections $0 \neq s_1 \in H^0(X, L_1)$ and $0 \neq s_2 \in H^0(X, L_2)$ the divisor associated to $s_1 \otimes s_2 \in H^0(X, L_1 \otimes L_2)$ satisfies $Z(s_1 \otimes s_2) = Z(s_1) + Z(s_2)$.

Remarks 2.3.17 i) There is a slightly different description of $Z(s)$ for $0 \neq s \in H^0(X, L) \setminus \{0\}$. In fact, $Z(s) = \sum \text{ord}_Y(s)[Y]$, where $\text{ord}_Y(s)$ is defined as $\text{ord}_Y(\psi_i(s))$. The latter does not depend on the trivialization. In particular, this description shows that the divisor $Z(s)$ associated to a section $0 \neq s \in H^0(X, L)$ is effective. Also note that for $\lambda \in \mathbb{C}^*$ one has $Z(s) = Z(\lambda s)$. In fact, this holds true for any global holomorphic function λ without zeros.

ii) The reader should be aware of the fact that we use $Z(s)$ or $Z(f)$ in two different contexts. In the context of the present discussion $Z(s)$ is an element of $\text{Div}(X)$ and, therefore, it encodes more than just the vanishing locus of s, all the multiplicities are taken into account. Before, e.g. page 18, we denoted by $Z(f)$ just the zero set of a holomorphic function, i.e. in this case $Z(f)$ and $Z(f^2)$ would be the same thing. I hope that this will not lead to any confusion.

The next result explains the relation between $\mathcal{O}(Z(s))$ and L for a section $0 \neq s \in H^0(X, L)$. For simplicity we continue to assume that X is connected.

Proposition 2.3.18 i) *Let* $0 \neq s \in H^0(X, L)$. *Then the line bundle* $\mathcal{O}(Z(s))$ *is isomorphic to* L.

ii) *For any effective divisor* $D \in \mathrm{Div}(X)$ *there exists a section* $0 \neq s \in H^0(X, \mathcal{O}(D))$ *with* $Z(s) = D$.

Proof. Let $L \in \mathrm{Pic}(X)$ on the open subsets U_i of an open covering $X = \bigcup U_i$ be trivialized by $\psi_i : L|_{U_i} \cong \mathcal{O}_{U_i}$. The divisor $Z(s)$ associated to $0 \neq s \in H^0(X, L)$ is given by $f := \{f_i := \psi_i(s|_{U_i})\} \in H^0(X, \mathcal{K}_X^*/\mathcal{O}_X^*)$. By definition, the line bundle associated to $Z(s)$ is the line bundle that corresponds to the cocycle $\{(U_i, f_i \cdot f_j^{-1})\}$. The first assertion follows from this, as $f_i \cdot f_j^{-1} = \psi_i(s|_{U_i \cap U_j}) \cdot (\psi_j(s|_{U_i \cap U_j}))^{-1} = \psi_i \circ \psi_j^{-1}$.

Let $D \in \mathrm{Pic}(X)$ be given by $\{(U_i, f_i \in \mathcal{K}_X^*(U_i))\}$. Since D is effective, the meromorphic functions $f_i \in \mathcal{K}_X^*(U_i)$ are in fact holomorphic, i.e. $f_i \in \mathcal{O}(U_i)$. On the other hand, the line bundle $\mathcal{O}(D)$ is associated to the cocycle $\{(U_i, \psi_{ij} = f_i \cdot f_j^{-1})\} \in H^1(X, \mathcal{O}_X^*)$. The holomorphic functions $f_i \in \mathcal{O}(U_i)$ define a global section $s \in H^0(X, \mathcal{O}(D))$, as $\psi_{ij} \cdot f_j = f_i$. Moreover, $Z(s)|_{U_i} = Z(s|_{U_i}) = Z(f_i) = D \cap U_i$. Hence, $Z(s) = D$.

Note that the section s is not unique or canonical, e.g. we could change the f_i by some $\lambda \in \mathbb{C}^*$.   □

Corollary 2.3.19 *Non-trivial sections* $s_1 \in H^0(X, L_1)$ *and* $s_2 \in H^0(X, L_2)$ *define linearly equivalent divisors* $Z(s_1) \sim Z(s_2)$ *if and only if* $L_1 \cong L_2$.

Proof. This is a consequence of $\mathcal{O}(Z(s_i)) \cong L_i$ and Lemma 2.3.14.   □

Corollary 2.3.20 *The image of the natural map* $\mathrm{Div}(X) \to \mathrm{Pic}(X)$ *is generated by those line bundles* $L \in \mathrm{Pic}(X)$ *with* $H^0(X, L) \neq 0$.

Proof. Proposition 2.3.18 shows that any $L \in \mathrm{Pic}(X)$ with $H^0(X, L) \neq 0$ is contained in the image.

Conversely, any divisor $D = \sum a_i[Y_i] \in \mathrm{Div}(X)$ can be written as $D = \sum a_i^+[Y_i] - \sum a_j^-[Y_j]$ with $a_k^\pm \geq 0$. Hence, $\mathcal{O}(D) \cong \mathcal{O}(\sum a_i^+[Y_i]) \otimes \mathcal{O}(\sum a_j^-[Y_j])^*$ and the line bundles $\mathcal{O}(\sum a_i^+[Y_i])$ and $\mathcal{O}(\sum a_j^-[Y_j])$ are both associated to effective divisors and, therefore, admit non-trivial global sections.   □

Remark 2.3.21 Later we will see that for projective manifolds, i.e. complex manifolds that can be realized as closed submanifolds of $\mathbb{P}^n$ (Definition 2.1.17), the map $\mathrm{Div}(X) \to \mathrm{Pic}(X)$ is surjective (see Section 5.3). But note that even for very easy manifolds, e.g. a generic complex torus of dimension two, this is no longer the case.

Let $Y \subset X$ be an irreducible hypersurface. Then, the (non-unique) section $0 \neq s \in H^0(X, \mathcal{O}(Y))$ with $Z(s) = Y$ gives rise to a sheaf (not vector bundle) homomorphism $\mathcal{O}_X \to \mathcal{O}(D)$ and dually to $\mathcal{O}(-D) \to \mathcal{O}_X$.

Lemma 2.3.22 *The induced map $\mathcal{O}(-D) \to \mathcal{O}_X$ is injective and the image is the ideal sheaf $\mathcal{I}$ of $Y \subset X$ of holomorphic functions vanishing on Y.*

Proof. This is a local statement. Thus, we may assume that $\mathcal{O}(-Y)$ is trivial and that the map $\mathcal{O}(-Y) \to \mathcal{O}_X$ is given by multiplication with the equation defining Y. Clearly, this map is injective.

If $x \in X$ is either a smooth point of Y or not contained in Y at all, then the image is, locally near x, the ideal generated by the defining equation of Y, i.e. the ideal sheaf of Y (use Corollary 1.1.19). Since the ideal sheaf $\mathcal{I}$ of Y and the subsheaf $\mathcal{O}(-Y) \subset \mathcal{O}_X$ are both invertible and coincide on the complement of the codimension at least two subset $Y_{\mathrm{sing}} \subset X$, they are equal. (Use Exercise 2.2.7.) $\square$

More generally, for any effective divisor $D = \sum a_i[Y_i]$ we obtain a short exact sequence

$$0 \longrightarrow \mathcal{O}(-D) \longrightarrow \mathcal{O}_X \longrightarrow \mathcal{O}_D \longrightarrow 0,$$

where $\mathcal{O}_D$ is the quotient of $\mathcal{O}_X$ by all holomorphic functions vanishing of order at least a_i along Y_i. In the language of complex spaces, $\mathcal{O}_D$ is the structure sheaf of the (possibly non-reduced) subspace associated to D.

Definition 2.3.23 Let L be a holomorphic line bundle on X. A *meromorphic section* of L is a collection of meromorphic functions $f_i \in \mathcal{K}_X(U_i)$ on open subsets U_i of an open covering $X = \bigcup U_i$ such that $f_i = \psi_{ij} \cdot f_j$, where ψ_{ij} is the cocycle defined by trivializations $\psi_i : L|_{U_i} \cong \mathcal{O}_{U_i}$.

Remarks 2.3.24 i) Any holomorphic section $s \in H^0(X, L)$ is a meromorphic section, but usually, e.g. if X is projective, L admits many more meromorphic sections than holomorphic ones. In fact, later (Section 5.3) we shall see that on a projective manifold any line bundle has a non-trivial meromorphic section, but for a general complex manifold this is not true.

ii) It is easy to see that there exists a canonical divisor (s) associated to any meromorphic section s of a line bundle L. As before, $\mathcal{O}((s)) \cong L$.

iii) If $0 \neq s \in H^0(X, L)$ and $s' \in H^0(X, L')$, then there exists a meromorphic section s'/s of $L^* \otimes L'$, which locally is given by $\psi_i'(s'|_{U_i})/\psi_i(s|_{U_i})$. As a special case of this construction, we can associate meromorphic functions $s'/s \in K(X)$ to sections $0 \neq s, s' \in H^0(X, L)$ (see the discussion on page 74).

iv) It is not difficult to show that the image of the natural map $\mathrm{Div}(X) \to \mathrm{Pic}(X)$ is nothing but the set of all line bundles that admit at least one non-trivial meromorphic section. See Exercise 2.3.4.

v) The interplay between (holomorphic or meromorphic) sections of line bundles and the function field $K(X)$ is illustrated by Proposition 2.2.28.

A holomorphic function $s \in H^0(X, \mathcal{O}_X)$ on a complex manifold X defines a holomorphic map $X \to \mathbb{C}$. More generally, a collection of holomorphic functions $s_1, \ldots, s_N$ defines a holomorphic map $X \to \mathbb{C}^N$. However, often, e.g.

when X is compact, there exist only very few (or none at all) non-constant holomorphic functions. Replacing holomorphic functions by holomorphic sections $s \in H^0(X, L)$ of a holomorphic line bundle L gives us more flexibility, as one frequently, e.g. for projective manifolds, disposes of holomorphic line bundles with many sections. The prize one has to pay is that sections $s_0, \ldots, s_N \in H^0(X, L)$ of a holomorphic line bundle L only define a holomorphic map to some projective space $\mathbb{P}^N$ and, moreover, that this map might not be everywhere well-defined on X.

Definition 2.3.25 Let L be a holomorphic line bundle on a complex manifold X. A point $x \in X$ is a *base point* of L if $s(x) = 0$ for all $s \in H^0(X, L)$. The *base locus* $\mathrm{Bs}(L)$ is the set of all base points of L.

Clearly, if $s_0, \ldots, s_N \in H^0(X, L)$ is a basis of global sections then $\mathrm{Bs}(L) = Z(s_0) \cap \ldots \cap Z(s_N)$ is an analytic subvariety. Note that in general $H^0(X, L)$ could be infinite-dimensional. Later we will see that for compact X these spaces are actually finite-dimensional (cf. Theorem 4.1.13). Thus, the fact that $H^0(X, \mathcal{O}_X) = \mathbb{C}$ when X is compact is generalized to $\dim H^0(X, L) < \infty$.

Proposition 2.3.26 *Let L be a holomorphic line bundle on a complex manifold X and suppose that $s_0, \ldots, s_N \in H^0(X, L)$ is a basis. Then*

$$\varphi_L : X \setminus \mathrm{Bs}(L) \longrightarrow \mathbb{P}^N \ , \quad x \longmapsto (s_0(x) : \ldots : s_N(x))$$

defines a holomorphic map such that $\varphi_L^* \mathcal{O}_{\mathbb{P}^N}(1) \cong L|_{X \setminus \mathrm{Bs}(L)}$.

Proof. Since $\mathrm{Bs}(L) = Z(s_0) \cap \ldots \cap Z(s_N) \subset X$ is a closed subset, the subset $X \setminus \mathrm{Bs}(L)$ is an open submanifold of X. Thus, the notion of a holomorphic map makes sense.

Let us now explain the definition of φ_L. If $x \in X \setminus \mathrm{Bs}(L)$ and $\psi : L|_U \cong \mathcal{O}_U$ is a trivialization of L over an open neighbourhood U of $x \in X$, then $\varphi_L(x) = (s_0(x) : \ldots : s_N(x))$ denotes the point $(\psi(s_0(x)) : \ldots : \psi(s_N(x))) \in \mathbb{P}^N$. The definition is independent of the trivialization ψ, for any other trivialization is of the form $\lambda \cdot \psi$ for some $\lambda \in \mathcal{O}^*(U)$ and $(\lambda(x) \cdot \psi(s_0(x)) : \ldots : \lambda(x) \cdot \psi(s_N(x))) = (\psi(s_0(x)) : \ldots : \psi(s_N(x)))$. The local description of φ_L shows that it is well-defined and holomorphic on $X \setminus \mathrm{Bs}(L)$.

Now consider the section z_0 of $\mathcal{O}(1)$ on $\mathbb{P}^N$ and let D_0 be the associated divisor, i.e. the hyperplane $\{(0 : z_1 : \ldots : z_N)\} \subset \mathbb{P}^n$. Then $f^* D_0$ is the divisor $Z(s_0)$. Thus, by Corollary 2.3.13 and Proposition 2.3.18 one has $L \cong \mathcal{O}(Z(s_0)) \cong \mathcal{O}(f^* D_0) \cong f^* \mathcal{O}(D_0) \cong f^* \mathcal{O}(1)$ on the open subset $X \setminus \mathrm{Bs}(L)$. $\square$

Remarks 2.3.27 i) The map φ_L does depend on the choice of the base, but for two different choices the induced maps differ by a linear transformation of $\mathbb{P}^N$. If we do not want to choose any base at all then we still can define a holomorphic map

$$\varphi_L : X \setminus \mathrm{Bs}(L) \longrightarrow \mathbb{P}(H^0(X, L)^*)$$

by associating to $x \in X \setminus \mathrm{Bs}(L)$ the linear map $H^0(X, L) \to \mathbb{C}, s \mapsto s(x) \in$

$L(x) \cong \mathbb{C}$. As this linear map depends on the isomorphism $L(x) \cong \mathbb{C}$, we only obtain this way a point in $\mathbb{P}(H^0(X, L)^*)$. Clearly, the two description of φ_L are compatible, i.e. after fixing a basis of $H^0(X, L)$ the two morphisms coincides under the induced isomorphism $\mathbb{P}(H^0(X, L)^*) \cong \mathbb{P}^N$.

ii) It is easy to generalize the above construction as follows. Instead of taking a basis $s_0, \ldots, s_N$ one could just take any collection of sections, neither necessarily linearly independent nor generating $H^0(X, L)$. This yields a holomorphic map on the complement of the base locus $\mathrm{Bs}(L, s_0, \ldots, s_N) := Z(s_0) \cap \ldots \cap Z(s_N)$ to $\mathbb{P}((\bigoplus \mathbb{C} s_i)^*)$. The map φ_L as in the proposition is said to be associated to the *complete linear system* $H^0(X, L)$, whereas a subspace of $H^0(X, L)$ is simply called a *linear system* of L. Sometimes, one denotes the complete linear system $\mathbb{P}(H^0(X, L))$ also by $|L|$. One says that L is *globally generated* by the sections $s_0, \ldots, s_N$ if $\mathrm{Bs}(L, s_0, \ldots, s_N) = \emptyset$.

Definition 2.3.28 A line bundle L on a complex manifold is called *ample* if for some $k > 0$ and some linear system in $H^0(X, L^k)$ the associated map φ is an embedding.

By definition, a compact complex manifold is projective if and only if it admits an ample line bundle.

As examples of the above construction we shall discuss the Veronese map and the Segre map.

Veronese map. Let $X = \mathbb{P}^N$ and let L be the line bundle $\mathcal{O}(d)$. We have seen (Exercise 2.2.8) that homogeneous polynomials of degree d can be viewed as sections of L. Indeed, in Proposition 2.4.1 we will show that all sections of $\mathcal{O}(d)$ are of this type. Let $N := \dim H^0(\mathbb{P}^n, \mathcal{O}(d)) - 1$ and choose the monomials of degree d as a basis of $H^0(\mathbb{P}^n, \mathcal{O}(d))$. The induced map $\varphi_{\mathcal{O}(d)}$ is the *Veronese embedding*.

More explicitly, $\varphi_{\mathcal{O}(d)}$ is given by

$$\mathbb{P}^n \longrightarrow \mathbb{P}^N, \quad (z_0 : \ldots : z_n) \longmapsto (z_0^{i_0} \cdot \ldots \cdot z_n^{i_n})_{(i_0, \ldots, i_n)},$$

where $(i_0, \ldots, i_n)$ runs through all multi-indices with $\sum_{j=0}^n i_j = d$. Obviously, $\mathrm{Bs}(\mathcal{O}(d)) = \emptyset$ and, therefore, φ_L is everywhere defined. The reader may verify that the Veronese map is indeed an embedding (cf. Exercise 2.3.5).

Let us consider the case $X = \mathbb{P}^1$ and $d = 2$. Then the Veronese map defines an isomorphism of $\mathbb{P}^1$ with the hypersurface $Z(x_0 x_2 - x_1^2)$ of $\mathbb{P}^2$ via $(z_0 : z_1) \mapsto (z_0^2 : z_0 z_1 : z_1^2)$.

Segre map. Here we let $X = \mathbb{P}^{n_1} \times \mathbb{P}^{n_2}$. The two projections are denoted by $p_1 : X \to \mathbb{P}^{n_1}$ and $p_2 : X \to \mathbb{P}^{n_2}$. Let $L \in \mathrm{Pic}(X)$ be the line bundle $p_1^*(\mathcal{O}(1)) \otimes p_2^*(\mathcal{O}(1))$. Pulling-back (Remark 2.2.22) and taking tensor product (see page 73) defines a bilinear map

$$H^0(\mathbb{P}^{n_1}, \mathcal{O}(1)) \otimes H^0(\mathbb{P}^{n_2}, \mathcal{O}(1))$$

$$\downarrow$$

$$H^0(X, p_1^*(\mathcal{O}(1))) \otimes H^0(X, p_2^*(\mathcal{O}(1))) \longrightarrow H^0(X, L).$$

If $z_0, \ldots, z_{n_1}$ and $z_0', \ldots, z_{n_2}'$ denote the linear coordinates on $\mathbb{P}^{n_1}$ respectively $\mathbb{P}^{n_2}$, then we obtain this way sections $z_i z_j'$ of L. One may verify that these sections are linearly independent (cf. Exercise 2.3.6) and that they indeed form a basis (cf. Proposition 2.4.1). The *Segre map* is the induced map

$$\mathbb{P}^{n_1} \times \mathbb{P}^{n_2} \longrightarrow \mathbb{P}^{(n_1+1)(n_2+1)-1}$$

$$((z_0 : \ldots : z_{n_1}), (z_0' : \ldots : z_{n_2}')) \longmapsto (z_0 z_0' : z_0 z_1' : \ldots : z_{n_1} z_{n_2}')$$

In the case $n_1 = n_2 = 1$ this yields the map

$$\mathbb{P}^1 \times \mathbb{P}^1 \longrightarrow \mathbb{P}^3, \quad ((z_0 : z_1), (z_0' : z_1')) \longmapsto (z_0 z_0' : z_0 z_1' : z_1 z_0' : z_1 z_1'),$$

which in fact defines an isomorphism of $\mathbb{P}^1 \times \mathbb{P}^1$ with the quadric $Z(x_0 x_3 - x_1 x_2)$.

In both examples the use of the z_i's might be confusing. They denote sections of $\mathcal{O}(1)$, i.e. linear coordinates on $\mathbb{P}^n$, on the one hand, but they are also used to write a point in $\mathbb{P}^n$ as $(z_0 : \ldots : z_n)$ on the other. This is completely analogous to the use of the standard coordinates, say x_i of $\mathbb{R}^n$, in linear algebra.

To conclude this section we discuss the case of **line bundles and divisors on curves**. For the rest of this section we assume that X is a compact connected curve. Thus, any point $x \in X$ is an irreducible hypersurface and thus defines a divisor $[x] \in \mathrm{Div}(X)$. In fact, any divisor $D \in \mathrm{Div}(X)$ is of the form $D = \sum n_i [x_i]$ with $x_i \in X$.

To simplify we write $D = \sum n_i x_i$ and $\mathcal{O}(x)$ for the line bundle associated to the divisor $[x]$.

Definition 2.3.29 The *degree* of a divisor $D = \sum n_i [x_i]$ on a curve X is the integer

$$\deg(D) = \sum_i n_i.$$

Clearly, the degree defines a surjective group homomorphism

$$\deg : \mathrm{Div}(X) \longrightarrow \mathbb{Z}.$$

Proposition 2.3.30 *If D is a principal divisor on a compact curve then* $\deg(D) = 0$.

In order to prove the proposition we will need a lemma. Before stating it let us recall that any meromorphic function $f \in K(X)$ defines a holomorphic map $f : X \setminus P(f) \to \mathbb{C}$. By abuse of notation we denote by $P(f)$ not only the pole divisor, but also the support of it. Thus, $X \setminus P(f)$ is just the open subset where f is holomorphic.

Lemma 2.3.31 *Let $f \in K(X)$ be a meromorphic function on a curve X. Then the induced map $f : X \setminus P(f) \to \mathbb{C}$ extends naturally to a holomorphic map $X \to \mathbb{P}^1$.*

Proof. Let $X = \bigcup U_i$ be an open covering such that $f|_{U_i}$ is given as g_i/h_i with $g_i, h_i \in \mathcal{O}(U_i)$. We may assume that g_i and h_i have no common zero. Indeed, if $g, h : U \to \mathbb{C}$ are two holomorphic functions on an open subset $U \subset \mathbb{C}$ with $g(z_0) = h(z_0) = 0$ for some $z_0 \in U$, then $g/h = (g \cdot (z - z_0)^{-1})/(h \cdot (z - z_0)^{-1})$ and $g \cdot (z - z_0)^{-1}$ and $h \cdot (z - z_0)^{-1}$ are still holomorphic.

Thus, $U_i \to \mathbb{P}^1$, $x \mapsto (g_i(x) : h_i(x))$ is well-defined. Moreover, on $U_i \setminus P(f)$ it coincides with $x \mapsto (f(x) : 1)$. Either by a direct argument or by invoking the identity principle we see that the different maps $U_i \to \mathbb{P}^1$ glue to a holomorphic map $X \to \mathbb{P}^1$ which extends $f : X \setminus P(f) \to \mathbb{C}$, where the embedding $\mathbb{C} \subset \mathbb{P}^1$ is given by $z \mapsto (z : 1)$. $\square$

Proof of the Proposition. By definition, a principal divisor D is of the form $D = (f) = Z(f) - P(f)$ for some meromorphic function $f \in K(X)$. By the lemma this yields a holomorphic map $f : X \to \mathbb{P}^1$. By definition $Z(f) = f^*0$ and $P(f) = f^*\infty$.

Over the regular values, which form a connected set, the map f is a topological cover. Thus, any fibre contains the same number d of points.

In a neighbourhood of a critical $x \in \mathbb{P}^1$ the map is of the form $\bigsqcup z_i^{d_i}$: $\bigsqcup B_1(0) \to B_1(0)$ with $\sum d_i = d$. This allows to compare the degree of the pull-back divisors of different (critical) points. In particular, $\deg(Z(f)) = d = \deg(P(f))$. Hence, $\deg(D) = \deg(Z(f)) - \deg(P(f)) = 0$. $\square$

Corollary 2.3.32 *If D_1 and D_2 are two linearly equivalent divisors then* $\deg(D_1) = \deg(D_2)$, *i.e. the degree factorizes over the image of* $\mathrm{Div}(X) \to \mathrm{Pic}(X)$. $\square$

Once we know that $\mathrm{Div}(X) \to \mathrm{Pic}(X)$ is surjective (see i) of Examples 5.3.6 and Corollary 5.3.7), the corollary shows that there is a commutative diagram

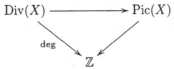

For $\mathcal{O}(1) \in \mathrm{Pic}(\mathbb{P}^1)$, which is in the image of $\mathrm{Div}(\mathbb{P}^1) \to \mathrm{Pic}(\mathbb{P}^1)$, one has $\deg \mathcal{O}(1) = 1$.

Definition 2.3.33 The *Abel–Jacobi map* of a compact connected curve X is the map
$$X \longrightarrow \mathrm{Pic}(X), \quad x \longmapsto \mathcal{O}(x - x_0),$$
which depends on a chosen point $x_0 \in X$.

Proposition 2.3.34 *For a compact connected curve X the following conditions are equivalent:*
   i) *The Abel–Jacobi map is not injective.*
   ii) *The curve X is isomorphic to $\mathbb{P}^1$.*
   iii) *There exist points $x_1 \neq x_2 \in X$ such that $\mathcal{O}(x_1) \cong \mathcal{O}(x_2)$.*

Proof. If $\mathcal{O}(x_1) \cong \mathcal{O}(x_2)$, then the line bundle $L := \mathcal{O}(x_i)$ admits two holomorphic sections $s_1, s_2 \in H^0(X, L)$ with $Z(s_i) = x_i$. Thus, the induced holomorphic map $\varphi : X \to \mathbb{P}^1$, $z \mapsto (s_1(x) : s_2(x))$ is everywhere defined. Furthermore, φ is injective in a neighbourhood of x_i and therefore injective everywhere. This shows that φ is an isomorphism. Hence, we have proven that iii) implies ii).

The converse follows from the observation that any point in $\mathbb{P}^1$ is the zero locus of a linear homogeneous polynomial. Since linear homogeneous polynomials are all regarded as global sections of $\mathcal{O}(1)$, this shows that for two arbitrary points $x_1, x_2 \in \mathbb{P}^1$ one always has $\mathcal{O}(x_1) \cong \mathcal{O}(x_2)$.

The equivalence of i) and iii) is obvious.   □

Exercises

2.3.1 Show that the natural map $\mathrm{Div}(X) \to \mathrm{Pic}(X)$ is not injective if and only if $a(X) > 0$.

2.3.2 Let $Y \subset X$ be a smooth hypersurface defined by a section $s \in H^0(X, L)$ for some holomorphic line bundle $L \in \mathrm{Pic}(X)$. Show that the normal bundle $\mathcal{N}_{Y/X}$ is isomorphic to $L|_Y$. (You may use Exercise 2.2.12 or the cocycle description of the normal bundle. A complete proof of this assertion will be given in the next section, but give it a try yourself before looking at the proof there.)

2.3.3 Determine the normal bundle of a complete intersection $X \subset \mathbb{P}^N$ defined by irreducible homogeneous polynomials $f_1, \ldots, f_k$ of degree $d_1, \ldots, d_k$. (Together with Proposition 2.4.3 of the next section this yields a description of the canonical bundles of any complete intersection in $\mathbb{P}^n$.)

2.3.4 Show that the image of $\mathrm{Div}(X) \to \mathrm{Pic}(X)$ consists of those line bundles admitting non-trivial meromorphic sections

2.3.5   Prove that the Veronese map $\mathbb{P}^n \to \mathbb{P}^N$ is an embedding.

2.3.6   Let L_1 and L_2 be holomorphic line bundles on complex manifolds X_1 and X_2, respectively. To two sections $s_i \in H^0(X_i, L_i)$, $i = 1, 2$, one associates a section $s_1 \cdot s_2 \in H^0(X_1 \times X_2, p_1^*(L_1) \otimes p_2^*(L_2))$. Show that if $s_1^1, \ldots, s_1^k$ and $s_2^1, \ldots, s_2^\ell$ are linearly independent sections of L_1 and L_2, respectively, then $s_1^i s_2^j$ form linearly independent sections of $p_1^*(L_1) \otimes p_2^*(L_2)$. Here, p_1 and p_2 are the two projections.

2.3.7   Let $x \in \mathbb{P}^n$ and consider the linear system

$$\{s \in H^0(\mathbb{P}^n, \mathcal{O}(1)) \mid s(x) = 0\}.$$

Show that it defines a holomorphic map $\varphi : \mathbb{P}^n \setminus \{x\} \to \mathbb{P}^{n-1}$. Describe this map geometrically.

2.3.8   (Bézout's theorem) Let $C \subset \mathbb{P}^2$ be a smooth curve defined by a homogeneous polynomial f of degree d. Show that the line bundle $\mathcal{O}(1)$ restricted to C is of degree d. Let $D \subset \mathbb{P}^2$ be a second smooth curve different from C defined by a homogeneous polynomial g of degree e. Show that

$$d \cdot e = \sum_{p \in C \cap D} \dim \mathcal{O}_{\mathbb{P}^2, p}/(f, g).$$

2.3.9   Show that the image of $\varphi_{\mathcal{O}(3)} : \mathbb{P}^1 \to \mathbb{P}^3$ is not a complete intersection.

2.3.10   Let $C = \varphi_{\mathcal{O}(2)}(\mathbb{P}^1) \subset \mathbb{P}^2$ and consider the restriction of the linear system in Exercise 2.3.7 to C. Study the induced map $C \setminus \{x\} \to \mathbb{P}^1$. (There are two cases to be considered: $x \in C$ and $x \notin C$.)

2.3.11   Show that on a compact curve $X \not\cong \mathbb{P}^1$ there always exist divisors D with $\deg(D) = 0$, but which are not principal. (Thus, the converse of Proposition 2.3.30 is false in general.)

Comments: - The relation between line bundles and hypersurfaces as presented is well understood. It should be clear, and we leave this to the reader, that complete intersections can be treated quite analogously. This yields a relation between complete intersections of codimension d and holomorphic vector bundles of the form $L_1 \oplus \ldots \oplus L_d$ with $L_i \in \mathrm{Pic}(X)$. The situation is more complicated for arbitrary submanifolds. One could hope that those can be studied in terms of non-split vector bundles, but, unfortunately, this is not always possible.

- Some parts of the above discussion would be more natural if formulated in the language of complex spaces, e.g. we could speak of non-reduced analytic hypersurfaces associated to a divisor $a[Y]$ with $a > 1$.

- For more elementary and sometimes not so elementary examples illustrating Veronese and Segre maps, we recommend [65].

- Of course, the discussion of curves is very sketchy. We will come back to certain aspects, but for a more complete discussion we have to refer to one of the many available textbooks on the subject.

2.4 The Projective Space

The aim of this section is to present certain explicit calculations for the projective space $\mathbb{P}^n$. We shall describe sections of the holomorphic line bundles $\mathcal{O}(k)$ on $\mathbb{P}^n$ and show that the canonical bundle $K_{\mathbb{P}^n}$ of the projective space $\mathbb{P}^n$ is isomorphic to the line bundle $\mathcal{O}(-n-1)$. It will turn out that everything can be expressed in polynomials, which turns complex geometry for $\mathbb{P}^n$ into complex algebraic geometry.

In order to determine all holomorphic sections of $\mathcal{O}(k)$, let us first recall that $\mathcal{O}(-1)$ is the holomorphic line bundle which is naturally embedded into the trivial vector bundle $\mathcal{O}^{\oplus n+1}$ such that the fibre of $\mathcal{O}(-1)$ over $\ell \in \mathbb{P}^n$ is $\ell \subset \mathbb{C}^{n+1}$ (cf. Proposition 2.2.6).

We will also use the following observation. If E is a holomorphic subbundle of F, then the inclusion $E \subset F$ naturally defines an inclusion of all tensor products $E^{\otimes k} \subset F^{\otimes k}$. In particular, for $k > 0$ the line bundle $\mathcal{O}(-k)$ is a subbundle of $\mathbb{P}^n \times (\mathbb{C}^{n+1})^{\otimes k}$.

Any homogeneous polynomial $s \in \mathbb{C}[z_0, \ldots, z_n]_k$ of degree k defines a linear map $(\mathbb{C}^{n+1})^{\otimes k} \to \mathbb{C}$. This gives rise to a holomorphic map $\mathbb{P}^n \times (\mathbb{C}^{n+1})^{\otimes k} \to \mathbb{C}$ which is linear on any fibre of the projection to $\mathbb{P}^n$. The restriction to $\mathcal{O}(-k)$ thus provides a holomorphic section of $\mathcal{O}(k)$. This way we associate to any homogeneous polynomial s of degree k a global holomorphic section of $\mathcal{O}(k)$, which will also be called s (cf. Exercise 2.2.8).

Proposition 2.4.1 *For $k \geq 0$ the space $H^0(\mathbb{P}^n, \mathcal{O}(k))$ is canonically isomorphic to the space $\mathbb{C}[z_0, \ldots, z_n]_k$ of all homogeneous polynomials of degree k.*

Proof. Note that the section associated to a non-trivial polynomial $0 \neq s \in \mathbb{C}[z_0, \ldots, z_n]_k$ is not trivial. Indeed, the composition $\mathcal{O}(-1) \subset \mathbb{P}^n \times \mathbb{C}^{n+1} \to \mathbb{C}^{n+1}$ is surjective and, therefore, any polynomial s that induces the trivial section of $\mathcal{O}(k)$ defines a trivial map $\mathbb{C}^{n+1} \to \mathbb{C}$, i.e. $s = 0$. Since the map $\mathbb{C}[z_0, \ldots, z_n]_k \to H^0(\mathbb{P}^n, \mathcal{O}(k))$ is clearly linear, it suffices to show surjectivity.

Let $0 \neq t \in H^0(\mathbb{P}^n, \mathcal{O}(k))$. Choose a non-trivial $0 \neq s_0 \in H^0(\mathbb{P}^n, \mathcal{O}(k))$ induced by a homogeneous polynomial of degree k and consider the meromorphic function $F := t/s_0 \in K(\mathbb{P}^n)$ (cf. Remark 2.3.24). Composing with the projection $\mathbb{C}^{n+1} \setminus \{0\} \to \mathbb{P}^n$ yields a meromorphic map $\tilde{F} \in K(\mathbb{C}^{n+1} \setminus \{0\})$. Moreover, $G := s_0 \cdot \tilde{F}$ is a holomorphic function $\mathbb{C}^{n+1} \setminus \{0\} \to \mathbb{C}$, which, due to Hartogs' theorem 1.1.4, can be extended to a holomorphic function $G : \mathbb{C}^{n+1} \to \mathbb{C}$. By definition of $\tilde{F}$ and G, this function is homogeneous of degree k, i.e. $G(\lambda \cdot z) = \lambda^k \cdot G(z)$ for $\lambda \in \mathbb{C}$ and $z \in \mathbb{C}^{n+1}$.

Using the power series expansion of G in $z = 0$ this proves that G is a homogeneous polynomial s of degree k. Clearly, the section of $\mathcal{O}(k)$ induced by G equals $t \in H^0(\mathbb{P}^n, \mathcal{O}(k))$. $\qquad\square$

It should be obvious from the proof that the isomorphism described above is compatible with the multiplicative structure, i.e. we obtain a ring isomorphism

$$R(\mathbb{P}^n, \mathcal{O}(1)) = \bigoplus_{k \geq 0} H^0(\mathbb{P}^n, \mathcal{O}(k)) \cong \mathbb{C}[z_0, \ldots, z_n].$$

Applying Exercise 2.2.5 the proposition also yields

Corollary 2.4.2 *For $k < 0$ the line bundle $\mathcal{O}(k)$ admits no global holomorphic sections, i.e. $H^0(\mathbb{P}^n, \mathcal{O}(k)) = 0$.* $\square$

The higher cohomology groups of $\mathcal{O}(k)$ will be computed in Example 5.2.5 by applying the Kodaira vanishing theorem.

Let us next construct an isomorphism $K_{\mathbb{P}^n} \cong \mathcal{O}(-n-1)$. This can be done by a direct cocycle calculation or by using the Euler sequence. We shall discuss both. Using the adjunction formula, the isomorphism $K_{\mathbb{P}^n} \cong \mathcal{O}(-n-1)$ will enable us to compute the Kodaira dimension of hypersurfaces and complete intersections.

First, let us recall the standard open covering of $\mathbb{P}^n$ (see page 56). Let U_i be the open set

$$U_i := \{(z_0 : \ldots : z_n) \mid z_i \neq 0\} \subset \mathbb{P}^n.$$

Then $\mathbb{P}^n = \bigcup_{i=0}^n U_i$ and the maps

$$\varphi_i : U_i \longrightarrow \mathbb{C}^n, \quad (z_0 : \ldots : z_n) \longmapsto (\tfrac{z_0}{z_i}, \ldots, \tfrac{\widehat{z_i}}{z_i}, \ldots, \tfrac{z_n}{z_i})$$

are biholomorphic. For $i > j$ the transition map

$$\varphi_{ij} = \varphi_i \circ \varphi_j^{-1} : \{(w_1, \ldots, w_n) \mid w_i \neq 0\} \longmapsto \{(w_1, \ldots, w_n) \mid w_j \neq 0\}$$

is the composition of

$$\tilde{\varphi}_{ij}(w_1, \ldots, w_n) = (w_i^{-1}w_1, \ldots, w_i^{-1}w_{i-1}, w_i^{-1}, w_i^{-1}w_{i+1}, \ldots, w_i^{-1}w_n)$$

and the permutation $(j+1, i)$, the signature of which is $(-1)^{i-j-1}$.

Proposition 2.4.3 *The canonical bundle $K_{\mathbb{P}^n}$ is isomorphic to $\mathcal{O}(-n-1)$.*

Proof. By definition $K_{\mathbb{P}^n} \cong \det(T_{\mathbb{P}^n}^*)$, where the holomorphic tangent bundle $T_{\mathbb{P}^n}$ is uniquely determined by its cocycle $\{J(\varphi_{ij}) \circ \varphi_j : U_i \cap U_j \to \mathrm{Gl}(n, \mathbb{C})\}$. Thus, it suffices to show that the line bundle associated to the cocycle $\{\det(J(\varphi_{ij})) \circ \varphi_j\}$ is isomorphic to $\mathcal{O}(n+1)$. Using the above remark, we have

$$\det(J(\varphi_{ij})) = (-1)^{i-j-1} \det \left(\frac{\partial \tilde{\varphi}_{ij}^k}{\partial w_\ell} \right)_{k, \ell},$$

where $\tilde{\varphi}_{ij} = (\tilde{\varphi}_{ij}^1, \ldots, \tilde{\varphi}_{ij}^n)$.

The matrix $\left(\dfrac{\partial \tilde{\varphi}_{ij}^k}{\partial w_\ell}\right)_{k,\ell}$ is easily identified as

$$\begin{pmatrix} w_i^{-1} & 0 & \cdot\cdot & 0 & -\frac{w_1}{w_i^2} & 0 & \cdot & \cdot & 0 \\ 0 & w_i^{-1} & \cdot\cdot & 0 & -\frac{w_2}{w_i^2} & 0 & \cdot & \cdot & 0 \\ \cdot & & & & & & & & \\ \cdot & & & & & & & & \\ 0 & & \cdot\cdot & w_i^{-1} & -\frac{w_{i-1}}{w_i^2} & 0 & \cdot & \cdot & 0 \\ & & & & -\frac{1}{w_i^{-2}} & 0 & \cdot & \cdot & 0 \\ 0 & & \cdot\cdot & 0 & -\frac{w_{i+1}}{w_i^2} & w_i^{-1} & \cdot & \cdot & 0 \\ & & & & & & & & \\ 0 & & \cdot\cdot & 0 & -\frac{w_n}{w_i^2} & 0 & \cdot & \cdot & w_i^{-1} \end{pmatrix}$$

Hence, $\det\left(\dfrac{\partial \tilde{\varphi}_{ij}^k}{\partial w_\ell}\right) = -w_i^{-(n+1)}$ and, therefore,

$$\det(J(\varphi_{ij})) \circ \varphi_j = (-1)^{i-j} w_i^{-(n+1)} \circ \varphi_j = (-1)^{i-j} (z_i/z_j)^{-(n+1)}.$$

On the other hand, $\mathcal{O}(-1)$ is the line bundle $\{(\ell, z) \mid z \in \ell\} \subset \mathbb{P}^n \times \mathbb{C}^{n+1}$, to which is associated the cocycle $\{z_i/z_j\}$ (cf. Proposition 2.2.6). Thus, $\mathcal{O}(n+1)$ corresponds to the cocycle $\{(z_j/z_i)^{n+1}\}$. The two cocyles $\{(z_j/z_i)^{n+1}\}$ and $\{(-1)^{i-j}(z_i/z_j)^{-(n+1)}\}$ define isomorphic line bundles, for they only differ by the boundary of the cocycle $\{\eta_i \equiv (-1)^i : U_i \to \mathbb{C}^* \subset \mathcal{O}^*\}$.   □

Recall that any short exact sequence $0 \to E \to F \to G \to 0$ of holomorphic vector bundles (or locally free sheaves), induces a canonical isomorphism $\det(F) \cong \det(E) \otimes \det(G)$ (cf. viii) of Example 2.2.4). Thus, the above proposition is also a consequence of the following one.

Proposition 2.4.4 (Euler sequence) *On $\mathbb{P}^n$ there exists a natural short exact sequence of holomorphic vector bundles*

$$0 \longrightarrow \mathcal{O} \longrightarrow \bigoplus_{j=0}^{n} \mathcal{O}(1) \longrightarrow T_{\mathbb{P}^n} \longrightarrow 0.$$

Proof. The inclusion $\mathcal{O} \subset \bigoplus_{j=0}^{n} \mathcal{O}(1)$ is obtained from the natural inclusion $\mathcal{O}(-1) \subset \bigoplus_{j=0}^{n} \mathcal{O}$ (cf. Proposition 2.2.6) by twisting with $\mathcal{O}(1)$ (cf. Exercise 2.2.4). It suffices to show that the kernel of the dual map $\bigoplus_{j=0}^{n} \mathcal{O}(-1) \to \mathcal{O}$ is canonically isomorphic to $\Omega_{\mathbb{P}^n}$. (Here we use that a sequence of holomorphic vector bundles $0 \to E \to F \to G \to 0$ is exact if and only if $0 \to G^* \to F^* \to E^* \to 0$ is exact.)

The proof of the identification of $\Omega_{\mathbb{P}^n}$ with the kernel is a little tedious. Using the standard trivialization of $\bigoplus_{j=0}^{n} \mathcal{O}(-1)|_{U_i}$ we will identify $\Omega_{\mathbb{P}^n}|_{U_i}$ in a natural way with the kernel of the surjection $\bigoplus_{j=0}^{n} \mathcal{O}(-1) \to \mathcal{O}$ over U_i. The proof concludes by showing that these inclusions glue.

First note that for any local holomorphic function $f : U \to \mathbb{C}$, where $U \subset X$ is an open subset of an arbitrary manifold X the differential $df = \partial f$ is a section of $\Omega_X|_U$ (cf. Exercise 2.2.10 or Section 2.6). Moreover, if $\varphi : U_i \cong \varphi(U_i) \subset \mathbb{C}^n$ is a local coordinate system, then $\Omega_X|_U = \bigoplus_{k=1}^{n} (d\varphi^k) \cdot \mathcal{O}_U$. In our situation $\Omega_{\mathbb{P}^n}|_{U_i} = \bigoplus_{i \neq k=0}^{n} d\left(\frac{z_k}{z_i}\right) \cdot \mathcal{O}_{U_i}$.

Since $\mathcal{O}(-1)|_{U_i}$ is canonically trivialized by $(\ell, z) \mapsto z_i$, one has a natural diagonal trivialization $\psi_i(\ell, z^0, \ldots, z^n) = (\ell, z_i^0, \ldots, z_i^n)$ of $\bigoplus_{k=0}^{n} \mathcal{O}(-1)|_{U_i}$. In this trivialization the surjection $\varphi : \bigoplus \mathcal{O}(-1) \to \mathcal{O}$ is the map

$$\varphi_i(\ell, w_0, \ldots, w_n) = \sum_{k=0}^{n} w_k \frac{z_k}{z_i}.$$

Hence, the map

$$\phi_i : \Omega_{\mathbb{P}^n}|_{U_i} = \bigoplus_{k=0}^{n} d\left(\frac{z_k}{z_i}\right) \cdot \mathcal{O}_{U_i} \longrightarrow \bigoplus_{k=0}^{n} \mathcal{O}_{U_i}, \quad d\left(\frac{z_k}{z_i}\right) \longmapsto e_k - \frac{z_k}{z_i} e_i,$$

where e_k is the k-th unit vector in $\bigoplus_{k=0}^{n} \mathcal{O}$, identifies $\Omega_{\mathbb{P}^n}|_{U_i}$ with the kernel of φ_i.

Let us now check that $\phi_i(d\left(\frac{z_k}{z_i}\right)) = (\psi_i \psi_j^{-1})(\phi_j(d\left(\frac{z_k}{z_i}\right)))$. Here we use

$$d\left(\frac{z_k}{z_i}\right) = d\left(\frac{z_k}{z_j}\right) \frac{z_j}{z_i} - \frac{z_k}{z_i} \frac{z_j}{z_i} d\left(\frac{z_i}{z_j}\right).$$

Thus,

$$\phi_j\left(d\left(\frac{z_k}{z_i}\right)\right) = \frac{z_j}{z_i}\left(e_k - \frac{z_k}{z_j} e_j\right) - \frac{z_k}{z_i} \frac{z_j}{z_i}\left(e_i - \frac{z_i}{z_j} e_j\right) = \frac{z_j}{z_i}\left(e_k - \frac{z_k}{z_i} e_i\right).$$

Since $\psi_i \psi_j^{-1}$ is given by multiplication with $\frac{z_i}{z_j}$, this concludes the proof. $\square$

Remark 2.4.5 The dual of the Euler sequence twisted by $\mathcal{O}(1)$ takes the form

$$0 \longrightarrow \Omega_{\mathbb{P}^n}(1) \longrightarrow \bigoplus_{j=0}^{n} \mathcal{O} \longrightarrow \mathcal{O}(1) \longrightarrow 0. \qquad (2.1)$$

Let us think of $\mathbb{P}^n$ more invariantly as $\mathbb{P}(V)$ with V a complex vector space of dimension $n + 1$. In particular, V^* is naturally identified with the space of homogeneous linear forms on V and, therefore, $V^* = H^0(\mathbb{P}(V), \mathcal{O}(1))$. The exact sequence (2.1) can in this context be written as

$$0 \longrightarrow \Omega_{\mathbb{P}(V)}(1) \longrightarrow V^* \otimes \mathcal{O} \longrightarrow \mathcal{O}(1) \longrightarrow 0,$$

where the surjection $V^* \otimes \mathcal{O} \to \mathcal{O}(1)$ is the evaluation map.

The advantage of writing the Euler sequence coordinate free is that it lends itself to important generalizations in the relative context, but we only mention this in passing:

If E is a holomorphic vector bundles over X, any fibre $\pi^{-1}(x)$ of the projection $\pi : \mathbb{P}(E) \to X$ is a complex projective space $\mathbb{P}(E(x))$ which comes with the Euler sequence

$$0 \longrightarrow \Omega_{\mathbb{P}(E(x))}(1) \longrightarrow E(x)^* \otimes \mathcal{O} \longrightarrow \mathcal{O}(1) \longrightarrow 0.$$

It turns out that these sequences can be glued to the *relative Euler sequence*

$$0 \longrightarrow \Omega_\pi \otimes \mathcal{O}_\pi(1) \longrightarrow \pi^* E^* \longrightarrow \mathcal{O}_\pi(1) \longrightarrow 0,$$

where $\mathcal{O}_\pi(1)$ is the (dual of the) relative tautological bundle (see Exercise 2.4.9) and $\mathcal{T}_\pi = \Omega_\pi^*$ is the relative tangent bundle defined as the kernel of $\mathcal{T}_{\mathbb{P}(E)} \to \pi^* \mathcal{T}_X$.

Corollary 2.4.6 *One has* $\mathrm{kod}(\mathbb{P}^n) = -\infty$.

Proof. Since the line bundles $\mathcal{O}(d)$ have global sections for $d \geq 0$, none of the powers $K_{\mathbb{P}^n}^{\otimes m} = \mathcal{O}(-m(n+1))$ for $m > 0$ admits non-trivial global sections (Corollary 2.4.2). Thus, the canonical ring $R(\mathbb{P}^n)$ is isomorphic to $\mathbb{C}$. $\qquad \square$

This yields in particular an example of a complex compact manifold with $\mathrm{kod}(X) < a(X)$. Other examples are provided by projective complex tori.

In order to compute the canonical bundle of hypersurfaces in $\mathbb{P}^n$, or more generally of complete intersections, we will need one more result. The following works for any complex manifold Y and not just the projective space.

Proposition 2.4.7 *Let X be a smooth hypersurface of a complex manifold Y defined by a section $s \in H^0(Y, L)$ of some holomorphic line bundle L on Y. Then $\mathcal{N}_{X/Y} \cong L|_X$ and thus $K_X \cong (K_Y \otimes L)|_X$.*

Proof. First observe the following fact: Let $\varphi = (\varphi^1, \ldots, \varphi^n)$ be a holomorphic map $U \to \mathbb{C}^n$ of some connected open neighbourhood $U \subset \mathbb{C}^n$ of the origin, such that $\varphi^n(z_1, \ldots, z_{n-1}, 0) = 0$ for all $(z_1, \ldots, z_{n-1}, 0) \in U$. Then $\varphi^n(z) = z_n \cdot h(z_1, \ldots, z_n)$, where $h(z_1, \ldots, z_n)$ is a power series in $z_1, \ldots, z_n$. Hence,

$$\frac{\partial \varphi^n}{\partial z_k}(z_1, \ldots, z_{n-1}, 0) = \begin{cases} 0 & \text{for } k = 1, \ldots, n-1 \\ h(z_1, \ldots z_{n-1}, 0) & \text{for } k = n \end{cases}$$

Let us apply this to our situation. We may fix local coordinates $\varphi_i : U_i \cong \varphi_i(U_i) \subset \mathbb{C}^n$, such that $\varphi_i(U_i \cap X) = \{(z_1, \ldots, z_n) \in \varphi_i(U_i) \mid z_n = 0\}$. Thus, the transition maps $\varphi_{ij} : \varphi_j(U_i \cap U_j) \cong \varphi_i(U_i \cap U_j)$ have the above property.

This yields

$$J(\varphi_{ij})|_X = \begin{pmatrix} J(\varphi_{ij}|_X) & * \\ 0 & \frac{\partial \varphi_{ij}^n}{\partial z_n}|_X \end{pmatrix}$$

By Proposition 2.2.17, the normal bundle $\mathcal{N}_{X/Y}$ corresponds to the cocycle $\{\frac{\partial \varphi_{ij}^n}{\partial z_n} \circ \varphi_j|_X\}$.

On the other hand, due to Proposition 2.3.18, one has $L \cong \mathcal{O}(X)$ and, therefore, L corresponds to the cocycle $\{\frac{s_i}{s_j}\}$, where s_i and s_j are local equations of X on U_i respectively U_j. Thus, $s_i = \varphi_i^n$. Hence, for $x \in X$ with $\varphi_j(x) = (z_1, \ldots z_{n-1}, 0)$ one has

$$\frac{s_i}{s_j}(x) = \frac{\varphi_i^n}{\varphi_j^n}(x) = \frac{(\varphi_{ij} \circ \varphi_j)^n}{\varphi_j^n}(x)$$

$$= \left(\frac{\varphi_{ij}^n}{z_n} \circ \varphi_j\right)(x) = h(z_1, \ldots, z_{n-1}, 0)$$

$$= \frac{\partial \varphi_{ij}^n}{\partial z_n}(z_1, \ldots, z_{n-1}, 0) = \left(\frac{\partial \varphi_{ij}^n}{\partial z_n} \circ \varphi_j\right)(x).$$

Therefore, both line bundles $\mathcal{O}(X)|_X$ and $\mathcal{N}_{X/Y}$ are given by the same cocycle. The last assertion follows from the adjunction formula 2.2.17. □

Remark 2.4.8 Note that the isomorphism $\mathcal{N}_{Y/X} \cong \mathcal{O}(X)|_X$ can locally be described as the map induced by $T_Y \twoheadrightarrow \mathcal{O}(X)|_X$, $\frac{\partial}{\partial z} \mapsto \frac{\partial s}{\partial z_i}|_X$.

Corollary 2.4.9 If $X \subset \mathbb{P}^n$ is a smooth hypersurface of degree d, i.e. defined by a section $s \in H^0(\mathbb{P}^n, \mathcal{O}(d))$, then $K_X \cong \mathcal{O}(d - n - 1)|_X$.   □

Exercises

2.4.1   Show that the canonical bundle K_X of a complete intersection $X = Z(f_1) \cap \ldots \cap Z(f_k) \subset \mathbb{P}^n$ is isomorphic to $\mathcal{O}(\sum \deg(f_i) - n - 1))|_X$. What can you deduce from this for the Kodaira dimension of X?

2.4.2   Are there holomorphic vector fields on $\mathbb{P}^n$, i.e. global sections of $T_{\mathbb{P}^n}$, which vanish only in a finite number of points? If yes, in how many?

2.4.3   Compute the Kodaira dimension of the following smooth hypersurfaces:
i) $Z\left(\sum_{i=0}^2 x_i^2\right) \subset \mathbb{P}^2$, ii) $Z\left(\sum_{i=0}^2 x_i^3\right) \subset \mathbb{P}^2$, iii) $Z\left(\sum_{i=0}^3 x_i^3\right) \subset \mathbb{P}^3$, and
iv) $Z\left(\sum_{i=0}^4 x_i^5\right) \subset \mathbb{P}^4$.

2.4.4   Show that $H^0(\mathbb{P}^n, \Omega_{\mathbb{P}^n}^q) = 0$ for $q > 0$.

2.4.5   The surface $\Sigma_n = \mathbb{P}(\mathcal{O}_{\mathbb{P}^1} \oplus \mathcal{O}_{\mathbb{P}^1}(n))$ is called the n-th *Hirzebruch surface*. Show that Σ_n is isomorphic to the hypersurface $Z(x_0^n y_1 - x_1^n y_2) \subset \mathbb{P}^1 \times \mathbb{P}^2$, where $(x_0 : x_1)$ and $(y_0 : y_1 : y_2)$ are the homogeneous coordinates of $\mathbb{P}^1$ respectively $\mathbb{P}^2$.

2.4.6   Describe the tangent, cotangent, and canonical bundle of $\mathbb{P}^n \times \mathbb{P}^m$.

2.4.7   Determine all complete intersections $X = H_1 \cap \ldots \cap H_{n+m-2} \subset \mathbb{P}^n \times \mathbb{P}^m$ with $H_i \in |p_1^* \mathcal{O}(d_i) \otimes p_2^* \mathcal{O}(e_i)|$, $d_i, e_i > 0$ with trivial canonical bundle.

2.4.8   Let $Y \subset X$ be a smooth hypersurface in a complex manifold X of dimension n and let α be a meromorphic section of K_X with at most simple poles along Y. Locally one can write $\alpha = h \cdot \frac{dz_1}{z_1} \wedge dz_2 \wedge \ldots \wedge dz_n$ with z_1 defining Y. One sets $\mathrm{Res}_Y(\alpha) = (h \cdot dz_2 \wedge \ldots \wedge dz_n)|_Y$.
a) Show that $\mathrm{Res}_Y(\alpha)$ is well defined and that it yields an element in $H^0(Y, K_Y)$.
b) Consider α as an element in $H^0(X, K_X \otimes \mathcal{O}(Y))$ and compare the definition of the residue with the adjunction formula: $K_Y \cong (K_X \otimes \mathcal{O}(Y))|_Y$.
c) Consider a smooth hypersurface $Y \subset \mathbb{P}^n$ defined by a homogeneous polynomial $f \in H^0(\mathbb{P}^n, \mathcal{O}(n+1))$. Prove that $\alpha := \sum (-1)^i z_i f^{-1} dz_0 \wedge \ldots \wedge \widehat{dz_i} \wedge \ldots \wedge dz_n$ is a meromorphic section of $K_{\mathbb{P}^n}$ with simple poles along Y. Furthermore, show that $\mathrm{Res}_Y(\alpha) \in H^0(Y, K_Y)$ defines a holomorpic volume form on Y, i.e. a trivializing section of K_Y.

2.4.9   Let E be a holomorphic vector bundle on a complex manifold X. Construct the relative tautological holomorphic line bundle $\mathcal{O}_\pi(-1)$ on $\mathbb{P}(E)$. Here, $\pi : \mathbb{P}(E) \to X$ denotes the projection and the fibre $\mathcal{O}_\pi(-1)(\ell)$ for a line $\ell \subset E(x)$ representing a point $\ell \in \pi^{-1}(x)$ should be, by construction, identified with ℓ.

2.4.10   This exercise generalizes the Euler sequence to Grassmannians. Show that on $\mathrm{Gr}_k(V)$ there exists a natural short exact sequence (the tautological sequence) of holomorphic vector bundles

$$0 \longrightarrow S \longrightarrow \mathcal{O} \otimes V \longrightarrow Q \longrightarrow 0$$

such that over a point $w \in \mathrm{Gr}_k(V)$ corresponding to $W \subset V$ the inclusion $S(w) \subset (\mathcal{O} \otimes V)(w) = V$ is just $W \subset V$. Moreover, prove that $\mathrm{Hom}(S, Q)$ is isomorphic to the holomorphic tangent bundle of $\mathrm{Gr}_k(V)$. Observe that this generalize what has been said about the Euler sequence on $\mathbb{P}^n$ which corresponds to the case $k = 1$

2.4.11   Let X be a complex manifold and let $\mathcal{O}_X \otimes V \twoheadrightarrow E$ be a surjection of vector bundles where V is a vector space and E is a vector bundle of rank k. Show that there exists a natural morphism $\varphi : X \to \mathrm{Gr}_k(V)$ with $\varphi^*(\mathcal{O} \otimes V \to Q) = \mathcal{O}_X \otimes V \to E$. Proposition 2.3.26 is the special case $k = 1$. Compare the Euler sequence with the tautological sequence introduced in the previous exercise via the Plücker embedding.

2.5 Blow-ups

So far, we have seen quite a few examples of compact complex manifolds, but no classification scheme has been proposed. In fact, classifying all compact complex manifolds is almost impossible, except maybe for the case of curves, as it is very easy to produce many new complex manifolds from any given one. One way to do so is by blowing up a given manifold X along a submanifold $Y \subset X$, which might very well be just a point. The particular feature of this construction is that it leaves the complement of Y unchanged.

Thus, classifying compact complex manifolds usually just means classifying those that are minimal, i.e. those which are not obtained as a blow-up of another manifold. (That a suitable minimal model of a given manifold can be defined is not clear at all and, for the time being, a 'clean' theory exists only up to dimension two.) These minimal ones can then be studied by certain numerical invariants, like the Kodaira dimension. In fact, the extent to which a classification can be achieved depends very much on the Kodaira dimension. A more or less complete list is only hoped for for compact manifolds of Kodaira dimension ≤ 0.

Besides classification theory, blow-ups are a useful tool in many other situations. We shall see how they can be used to resolve linear systems that are not base point free and in Section 5.3 they will be essential in proving the Kodaira embedding theorem.

Let X be a complex manifold and let $Y \subset X$ be a closed submanifold. We shall construct the blow-up of X along Y which is a complex manifold $\hat{X} = \mathrm{Bl}_Y(X)$ together with a proper holomorphic map $\sigma : \hat{X} \to X$. This map has two characteristic properties which will be immediate consequences of the construction, see Proposition 2.5.3.

We shall first discuss the construction in two special cases. The general situation will then be reduced to this.

Example 2.5.1 **Blow-up of a point.** Recall that the line bundle $\mathcal{O}(-1)$ on $\mathbb{P}^n$ is given as the incidence variety

$$
\begin{array}{ccccccc}
\mathbb{P}^n & \lhook\joinrel\longrightarrow & \mathcal{O}(-1) & \lhook\joinrel\longrightarrow & \mathbb{P}^n \times \mathbb{C}^{n+1} & \longrightarrow & \mathbb{P}^n \\
\downarrow & & {\scriptstyle \sigma}\downarrow & & & & \downarrow \\
\{0\} & \lhook\joinrel\longrightarrow & \mathbb{C}^{n+1} & =\!=\!=\!= & \mathbb{C}^{n+1} & &
\end{array}
$$

Thus, the fibre of the projection $\pi : \mathcal{O}(-1) \to \mathbb{P}^n$ over a line $\ell \in \mathbb{P}^n$ is just the line ℓ itself. Let us consider the other projection $\sigma : \mathcal{O}(-1) \to \mathbb{C}^{n+1}$. For $z \neq 0$ the pre-image $\sigma^{-1}(z)$ consists of the unique line ℓ_z passing through $z \in \mathbb{C}^{n+1}$. For $z = 0$ however the pre-image is $\sigma^{-1}(0) = \mathbb{P}^n$, as any line in $\mathbb{C}^{n+1}$ contains the origin $0 \in \mathbb{C}^{n+1}$. In fact, $\sigma^{-1}(0)$ is nothing but the zero section of $\mathcal{O}(-1) \to \mathbb{P}^n$.

The *blow-up* $\sigma : \mathrm{Bl}_0(\mathbb{C}^{n+1}) \to \mathbb{C}^{n+1}$ of $\mathbb{C}^{n+1}$ along the zero-dimensional submanifold $\{0\}$ is by definition the holomorphic line bundle $\mathcal{O}(-1)$ together with the natural projection $\sigma : \mathcal{O}(-1) \to \mathbb{C}^{n+1}$.

Clearly, the construction is independent of any coordinates, i.e. we could have worked with the abstract projective space $\mathbb{P}(V)$. But using the coordinates explicitly, the incidence variety $\mathcal{O}(-1) \subset \mathbb{P}^n \times \mathbb{C}^{n+1}$ can also be described as the subset of $\mathbb{P}^n \times \mathbb{C}^{n+1}$ satisfying the equations $z_i \cdot x_j = z_j \cdot x_i$, $i, j = 0, \ldots, n$, where $(z_0, \ldots, z_n)$ and $(x_0 : \ldots : x_n)$ are the coordinates of $\mathbb{C}^{n+1}$ and $\mathbb{P}^n$, respectively. Also note that the normal bundle of $0 \in \mathbb{C}^{n+1}$, which is just a vector space, is naturally isomorphic to $\mathbb{C}^{n+1}$.

Example 2.5.2 **Blow-up along a linear subspace.** Let $\mathbb{C}^m \subset \mathbb{C}^n$ be the linear subspace satisfying $z_{m+1} = \ldots = z_n = 0$ and denote by $(x_{m+1} : \ldots : x_n)$ the homogeneous coordinates of $\mathbb{P}^{n-m-1}$. We define

$$\mathrm{Bl}_{\mathbb{C}^m}(\mathbb{C}^n) := \{(x, z) \mid z_i \cdot x_j = z_j \cdot x_i, \ i, j = m+1, \ldots, n\} \subset \mathbb{P}^{n-m-1} \times \mathbb{C}^n.$$

In other words, $\mathrm{Bl}_{\mathbb{C}^m}(\mathbb{C}^n)$ is the incidence variety $\{(\ell, z) \mid z \in \langle \mathbb{C}^m, \ell\rangle\}$, where $\ell \in \mathbb{P}^{n-m-1}$ is a line in the complement $\mathbb{C}^{n-m}$ of $\mathbb{C}^m \subset \mathbb{C}^n$ and $\langle \mathbb{C}^m, \ell\rangle$ is the span of $\mathbb{C}^m$ and the line ℓ.

Using the projection $\pi : \mathrm{Bl}_{\mathbb{C}^m}(\mathbb{C}^n) \to \mathbb{P}^{n-m-1}$ one realizes $\mathrm{Bl}_{\mathbb{C}^m}(\mathbb{C}^n)$ as a $\mathbb{C}^{m+1}$-bundle over $\mathbb{P}^{n-m-1}$. The fibre over $\ell \in \mathbb{P}^{n-m-1}$ is just $\pi^{-1}(\ell) = \langle \mathbb{C}^m, \ell\rangle$. Thus, $\mathrm{Bl}_{\mathbb{C}^m}(\mathbb{C}^n)$ is a complex manifold.

Moreover, the projection $\sigma : \mathrm{Bl}_{\mathbb{C}^m}(\mathbb{C}^n) \to \mathbb{C}^n$ is an isomorphism over $\mathbb{C}^n \setminus \mathbb{C}^m$ and $\sigma^{-1}(\mathbb{C}^m)$ is canonically isomorphic to $\mathbb{P}(\mathcal{N}_{\mathbb{C}^m/\mathbb{C}^n})$, where the normal bundle $\mathcal{N}_{\mathbb{C}^m/\mathbb{C}^n}$ is canonically isomorphic to the trivial bundle $\mathbb{C}^m \times \mathbb{C}^{n-m}$ over $\mathbb{C}^m$.

Let us now construct the blow-up of an arbitrary complex manifold X of dimension n along an arbitrary submanifold $Y \subset X$ of dimension m. In order to do so, we choose an atlas $X = \bigcup U_i$, $\varphi_i : U_i \cong \varphi_i(U_i) \subset \mathbb{C}^n$ such that $\varphi_i(U_i \cap Y) = \varphi(U_i) \cap \mathbb{C}^m$.

Let $\sigma : \mathrm{Bl}_{\mathbb{C}^m}(\mathbb{C}^n) \to \mathbb{C}^n$ be the blow-up of $\mathbb{C}^n$ along $\mathbb{C}^m$ as constructed in the example above and denote by $\sigma_i : Z_i \to \varphi_i(U_i)$ its restriction to the open subset $\varphi_i(U_i) \subset \mathbb{C}^n$, i.e. $Z_i = \sigma^{-1}(\varphi_i(U_i))$ and $\sigma_i = \sigma|_{Z_i}$. We shall prove that all the blow-ups on the various charts $\varphi_i(U_i)$ naturally glue.

Consider arbitrary open subsets $U, V \subset \mathbb{C}^n$ and a biholomorphic map $\phi : U \cong V$ with the property that $\phi(U \cap \mathbb{C}^m) = V \cap \mathbb{C}^m$. Write $\phi = (\phi^1, \ldots, \phi^n)$. Then for $k > m$ one has $\phi^k = \sum_{j=m+1}^n z_j \phi_{k,j}$ (see the proof of Proposition 2.4.7). Let us define the biholomorphic map $\hat{\phi} : \sigma^{-1}(U) \cong \sigma^{-1}(V)$ as

$$\hat{\phi}(x, z) := \left((\phi_{k,j}(z))_{k,j=m+1,\ldots,n} \cdot x, \phi(z)\right).$$

It is straightforward to check that $\hat{\phi}(x, z)$ is indeed contained in the incidence variety. In order to obtain the global blow-up $\sigma : \mathrm{Bl}_{\mathbb{C}^m}(\mathbb{C}^n) \to X$ we have to ensure that these gluings are compatible. This is clear over $X \setminus Y$. Over Y the matrices we obtain for every $\phi_{ij}|_{\mathbb{C}^m}$ are by definition the cocycle of the normal bundle $\mathcal{N}_{Y/X}$. Thus, they do satisfy the cocycle condition. At the same time this proves that $\sigma^{-1}(Y) \cong \mathbb{P}(\mathcal{N}_{Y/X})$.

The discussion is summarized by the following

Proposition 2.5.3 *Let Y be a complex submanifold of X. Then there exists a complex manifold $\hat{X} = \mathrm{Bl}_Y(X)$, the blow-up of X along Y, together with a holomorphic map $\sigma : \hat{X} \to X$ such that $\sigma : \hat{X} \setminus \sigma^{-1}(Y) \cong X \setminus Y$ and $\sigma : \sigma^{-1}(Y) \to Y$ is isomorphic to $\mathbb{P}(\mathcal{N}_{Y/X}) \to Y$.* □

Definition 2.5.4 The hypersurface $\sigma^{-1}(Y) = \mathbb{P}(\mathcal{N}_{Y/X}) \subset \mathrm{Bl}_Y(X)$ is called the *exceptional divisor* of the blow-up $\sigma : \mathrm{Bl}_Y(X) \to X$.

Note that blowing-up along a smooth divisor $Y \subset X$ does not change X, i.e. in this case $\hat{X} = X$.

Let us now come back to the case that Y is just a point x in an n-dimensional manifold X. Then the blow-up $\sigma : \hat{X} := \mathrm{Bl}_x(X) \to X$ replaces $x \in X$ by $\mathbb{P}^{n-1}$. The exceptional divisor $\sigma^{-1}(x)$ is denoted E.

Proposition 2.5.5 *The canonical bundle $K_{\hat{X}}$ of the blow-up $\hat{X}$ is isomorphic to $\sigma^* K_X \otimes \mathcal{O}_{\hat{X}}((n-1)E)$.*

Proof. The morphism $\sigma : \hat{X} \to X$ induces a sheaf homomorphism $T_{\hat{X}} \to \sigma^* T_X$ (either use Exercise 2.2.10 or the discussion in Section 2.6). Taking the determinant and dualizing yields an injection $\sigma^* K_X \subset K_{\hat{X}}$ of sheaves (not of vector bundles!). In fact, over $\hat{X} \setminus E$ it is an isomorphism. It then suffices to show that the cokernel is $\mathcal{O}_{(n-1)E}$.

This is a local calculation and we can therefore assume that $X = \mathbb{C}^n$. Then we can conclude by the following cocycle calculation. Choose an open covering $\hat{X} = \bigcup V_i$, where $V_i := \{(x, z) \mid x_i \neq 0\}$, $i = 1, \ldots, n$, and coordinates

$$\varphi_i : V_i \cong \mathbb{C}^n \subset \mathbb{C}^{n+1}, \ (x, z) \mapsto \left(\frac{x_1}{x_i}, \ldots, \frac{x_i}{x_i}, \ldots, \frac{x_n}{x_i}, z_i \right).$$

Thus, $\mathbb{C}^n \subset \mathbb{C}^{n+1}$ is the affine subspace defined by $u_i = 1$. The transition functions are given by

$$\varphi_{ij}(u_1, \ldots, u_{n+1}) = (u_i^{-1} \cdot u_1, \ldots, u_i^{-1} \cdot u_n, u_i \cdot u_{n+1}).$$

Hence, $\det J(\varphi_{ij}) = \psi_{ij} \cdot u_i$, where $\{\psi_{ij}\}$ is the corresponding cocycle on $\mathbb{P}^{n-1}$, i.e. $\psi_{ij} = u_i^{-n}$ (see the proof of Proposition 2.4.3). Hence, $K_{\hat{X}}$ is given by the cocycle $u_i^{-(n-1)} = \left(\frac{x_i}{x_j} \right)^{-(n-1)}$. (In fact, as in the proof of Proposition 2.4.3 there is an extra boundary $(-1)^{i-j}$ which we have suppressed here.)

On the other hand, the divisor $E \subset \hat{X}$ is defined as the vanishing locus of all functions $z_1, \ldots, z_n$. Using $z_k \cdot x_\ell = z_\ell \cdot x_k$ this can be described on the open subset V_i by the single equation $z_i = 0$. Thus, the cocycle associated to E that describes the line bundle $\mathcal{O}(E)$ is $\{V_{ij}, u_i^{-1}\}$. (Recall that, if a divisor D on the open sets $D \cap V_i$ is given by f_i, then $\mathcal{O}(D)$ is the line bundle determined by the cocycle $\{V_i, \frac{f_j}{f_i}\}$. See Corollary 2.3.10.)   □

Corollary 2.5.6 *For the exceptional divisor* $E = \mathbb{P}^{n-1} \subset \hat{X} \to X$ *one has* $\mathcal{O}(E)|_E \cong \mathcal{O}(-1)$.

Proof. Indeed, Propositions 2.4.3 and 2.4.7 yield $\mathcal{O}(-n) \cong K_{\mathbb{P}^{n-1}} \cong (K_{\hat{X}} \otimes \mathcal{O}(E))|_E$. Hence, $\mathcal{O}(nE)|_E \cong \mathcal{O}(-n)$. Since $\mathrm{Pic}(\mathbb{P}^{n-1})$ is torsion free (cf. Exercise 3.2.11), this proves the assertion.

Without using this information on the Picard group, one has to argue as follows. The assertion is clearly local. So we may assume $X = \mathbb{C}^n$. The line bundle $\mathcal{O}(E)$ in this case can in fact be identified with the line bundle $\bigcup_{(\ell,z)\in\hat{X}} \ell$. Indeed, the latter admits the section $t(\ell, z) = ((\ell, z), z)$ which vanishes along E with multiplicity one. In other words, $\mathcal{O}(E)$ is isomorphic to the pull-back $\pi^*\mathcal{O}(-1)$ under the second projection $\pi : \hat{X} \to \mathbb{P}^{n-1}$. This yields the assertion $\mathcal{O}(E)|_E \cong \mathcal{O}(-1)$.   □

We will conclude by giving a more differential geometric description of the blow-up of a point. This is done by using a standard gluing operation known as the *connected sum* of two differentiable manifolds.

Let M and M' be differentiable manifolds both of dimension m. By $D \subset \mathbb{R}^m$ we shall denote the open ball of radius one. Choose open subsets $U \subset M$ and $U' \subset M'$ and diffeomorphisms $\eta : D \cong U$ and $\eta' : D \cong U'$. Furthermore we consider the diffeomorphim

$$\xi : D \setminus \left(\tfrac{1}{2}\bar{D}\right) \xrightarrow{\ \sim\ } D \setminus \left(\tfrac{1}{2}\bar{D}\right), \quad x \longmapsto \left(\tfrac{1}{2}\|x\|^{-2}\right) \cdot x$$

of the annulus $D \setminus (\tfrac{1}{2}D) = \{x \in \mathbb{R}^m \mid 1/2 < \|x\| < 1\}$.

Definition 2.5.7 The *connected sum*

$$M \# M' := \left(M \setminus \eta\left(\tfrac{1}{2}\bar{D}\right)\right) \bigcup\nolimits_\xi \left(M' \setminus \eta'\left(\tfrac{1}{2}\bar{D}\right)\right)$$

is obtained by gluing $M \setminus \eta(\tfrac{1}{2}\bar{D})$ and $M' \setminus \eta'(\tfrac{1}{2}\bar{D})$ over the open subsets $U \setminus \eta\left(\tfrac{1}{2}\bar{D}\right)$ and $U' \setminus \eta'\left(\tfrac{1}{2}\bar{D}\right)$ via the map ξ (or, more explicitly, via $\eta' \circ \xi \circ \eta^{-1}$).

It is not difficult to see that the diffeomorphism type of the connected sum does not depend neither on the chosen open subsets U and U' nor on η and η'.

The construction can easily be refined by taking possible orientations of M and M' into account. If M and M' are orientable and orientations have been chosen for both, then we require that η be orientation preserving and that η'

be orientation reversing. Here, use the standard orientation of $\mathbb{R}^m$. Since ξ is orientation reversing, we obtain a natural orientation on the connected sum $M \# M'$ which coincides with the given ones on the open subsets $M \setminus \eta(\frac{1}{2}\bar{D})$ and $M' \setminus \eta'(\frac{1}{2}\bar{D})$.

If M is oriented (the orientation is commonly suppressed in the notation), we shall denote by $\bar{M}$ the same differentiable manifold M, but with the reversed orientation. We will in particular be interested in $\bar{\mathbb{P}}^n$, the complex projective space with the orientation opposite to the one naturally induced by the complex structure. E.g. if $n = 2$ then $H^2(\bar{\mathbb{P}}^n, \mathbb{Z}) = \mathbb{Z}$ and its generator ℓ satisfies $\ell^2 = -1$.

Proposition 2.5.8 *Let $x \in X$ be a point in a complex manifold X. Then the blow-up $\mathrm{Bl}_x(X)$ is diffeomorphic as an oriented differentiable manifold to $X \# \bar{\mathbb{P}}^n$.*

Proof. The assertion is local, so we may assume that X is the unit disc $D = \{z \in \mathbb{C}^n \mid \|z\| < 1\}$ and $x = 0$. By definition $\hat{D} = \mathrm{Bl}_0(D) = \{(x, z) \in D \times \mathbb{P}^{n-1} \mid x_i \cdot z_j = x_j \cdot z_i\}$.

In order to specify the connected sum we choose $\eta = \mathrm{id}$ and $\eta' : D \to \bar{\mathbb{P}}^n$, $x \mapsto (1 : x)$, which is orientation reversing. Thus, $\mathbb{P}^n \setminus \eta'\left(\frac{1}{2}\bar{D}\right) = \{(x_0 : x) \mid |x_0| < 2 \cdot \|x\|\}$. In order to prove the assertion it suffices to find a orientation preserving diffeomorphism

$$ \mathbb{P}^n \setminus \eta'\left(\tfrac{1}{2}D\right) \longrightarrow \hat{D} \, . $$

which restricts to $\xi : \eta'(D) \setminus \eta'(\frac{1}{2}\bar{D}) \cong \{(x, z) \in \hat{D} \mid 1/2 < \|z\| < 1\} \cong D \setminus \left(\frac{1}{2}\bar{D}\right)$.

This map can explicitly be given as $(x_0 : x) \mapsto (\frac{x_0}{2} \cdot \|x\|^{-2} \cdot x, [x])$. The verification that it indeed defines a diffeomorphism is straightforward. $\square$

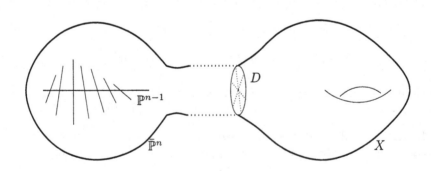

Remark 2.5.9 The connected sum with $\bar{\mathbb{P}}^2$ is often used in differential and symplectic topology of fourfolds. E.g. Taubes [104] has shown that any compact

differentiable manifold M of dimension four admits an anti-selfdual metric after taking the connected sum with $\bar{\mathbb{P}}^2$ sufficiently often. In the complex realm its says that after blowing-up a complex surfaces sufficiently often it can be endowed with an anti-selfdual metric.

Exercises

2.5.1 Let $\hat{X} \to X$ be the blow-up of a surface X in a point $x \in X$. Show that the pull-back of sections defines an isomorphism $H^0(X, K_X) \cong H^0(\hat{X}, K_{\hat{X}})$. More generally, one has $H^0(X, K_X^{\otimes m}) = H^0(\hat{X}, K_{\hat{X}}^{\otimes m})$.

2.5.2 Show that $\mathcal{O}(E)$ of the exceptional divisor $E = \mathbb{P}(\mathcal{N}_{Y/X})$ of a blow-up $\mathrm{Bl}_Y(X) \to X$ of a compact manifold X admits (up to scaling) only one section. (You might reduce to the case of the blow-up of a point.)

2.5.3 Consider the $\mathbb{Z}/2\mathbb{Z}$-action $z \mapsto -z$ on $\mathbb{C}^2$ which has one fixed point and whose quotient $\mathbb{C}^2/\pm$ is not smooth. Show that the action lifts to a $\mathbb{Z}/2\mathbb{Z}$-action on the blow-up $\mathrm{Bl}_0(\mathbb{C}^2)$ and prove that the quotient $\mathrm{Bl}_0(\mathbb{C}^2)/\pm$ is a manifold.

2.5.4 Let $C \subset \mathbb{C}^2$ be the reducible curve defined by $z_1 \cdot z_2 = 0$. Show that the closure of $C \setminus \{0\} \subset \mathbb{C} \setminus \{0\} = \mathrm{Bl}_0(\mathbb{C}^2) \setminus E$ in $\mathrm{Bl}_0(\mathbb{C}^2)$ is a smooth curve.

2.5.5 Let X be a K3 surface, i.e. X is a compact complex surface with $K_X \cong \mathcal{O}_X$ and $h^1(X, \mathcal{O}_X) = 0$. Show that X is not the blow-up of any other smooth surface.

Comments: - Blow-ups cannot only be useful in the study of complex manifolds, i.e. smooth geometric structures. In fact, Hironaka has shown that any singular complex variety can be made smooth by inductively blowing-up smooth subvarieties a finite number of times. This works in the analytic as well as in the algebraic setting. However, resolving singularities in characteristic p is still an open problem, but a weaker version, so called alterations, have been shown to exist by recent work of de Jong [30].

- As indicated in Exercise 2.5.4, blow-ups can be useful when one wants to desingularize curves embedded in surfaces (or more generally analytic subvarieties). We refer to [66] for the proof that this is always possible.

2.6 Differential Calculus on Complex Manifolds

In this section we will present a slightly different, although eventually equivalent approach to the notion of a complex manifold. We will see that a complex structure is determined by an almost complex structure on any tangent space, a purely linear algebra notion, satisfying a certain integrability condition. We thus combine the linear algebra of Section 1.1, the study of local differential forms in Section 1.3, and the notion of a complex manifold used so far. In passing, Dolbeault cohomology of complex manifolds and holomorphic vector bundles is introduced.

Definition 2.6.1 An *almost complex manifold* is a differentiable manifold X together with a vector bundle endomorphism

$$I : TX \longrightarrow TX, \text{ with } I^2 = -\text{id}.$$

Here, TX is the real tangent bundle of the underlying real manifold.

The endomorphism is also called the *almost complex structure* on the underlying differentiable manifold. If an almost complex structure exists, then the real dimension of X is even.

Proposition 2.6.2 *Any complex manifold X admits a natural almost complex structure.*

Proof. Cover X by holomorphic charts $\varphi : U_i \cong \varphi_i(U_i) \subset \mathbb{C}^n$ and use the almost complex structure defined in Section 1.3. By Proposition 1.3.2 it neither depends on the chart nor on the atlas in the equivalence class specified by the complex structure. Thus, we can define a natural almost complex structure on X depending just on its complex structure. $\square$

In order to distinguish explicitly between the (almost) complex manifold and the underlying real manifold one sometimes writes $X = (M, I)$, where M is just a real manifold and I is the additional structure.

Remark 2.6.3 Not every real manifold of even dimension admits an almost complex structure. The easiest example is provided by the four-dimensional sphere. See the comments at this end of this section.

Let X be an almost complex manifold. Then $T_{\mathbb{C}}X$ denotes the complexification of TX, i.e. $T_{\mathbb{C}}X = TX \otimes \mathbb{C}$. We emphasize that even for a complex manifold X the bundle $T_{\mathbb{C}}X$ is *a priori* just a complex vector bundle without a holomorphic structure.

Proposition 2.6.4 i) *Let X be an almost complex manifold. Then there exists a direct sum decomposition*

$$T_{\mathbb{C}}X = T^{1,0}X \oplus T^{0,1}X$$

of complex vector bundles on X, such that the $\mathbb{C}$-linear extension of I acts as multiplication by i on $T^{1,0}X$ respectively by $-i$ on $T^{0,1}X$.

ii) If X is a complex manifold, then $T^{1,0}X$ is naturally isomorphic (as a complex vector bundle) to the holomorphic tangent bundle $\mathcal{T}_X$.

Proof. i) One defines $T^{1,0}X$ and $T^{0,1}X$ as the kernel of $I - i \cdot \mathrm{id}$ respectively $I + i \cdot \mathrm{id}$. That these maps are vector bundle homomorphisms and that the direct sum decomposition holds, follows from the direct sum decomposition on all the fibres (cf. Lemma 1.2.5)

ii) As we have computed in Proposition 1.3.2, the Jacobian

$$df : T_x^{1,0}U \oplus T_x^{0,1}U \cong T_{f(x)}^{1,0}V \oplus T_{f(x)}^{0,1}V$$

of a biholomorphic map $f : U \cong V$, where $U, V \subset \mathbb{C}^n$ are open subsets, has the form $\begin{pmatrix} J(f) & 0 \\ 0 & \overline{J(f)} \end{pmatrix}$.

Let $X = \bigcup U_i$ be a covering by holomorphic charts $\varphi_i : U_i \cong \varphi(U_i) = V_i \subset \mathbb{C}^n$. Then $(\varphi_i^{-1})^*(T^{1,0}X|_{U_i}) \cong T^{1,0}V_i$ and the latter is canonically trivialized. With respect to these canonical trivializations the induced isomorphisms $T_{\varphi_j(x)}^{1,0}V_j \cong T_{\varphi_i(x)}^{1,0}V_i$ are given by $J(\varphi_i \circ \varphi_j^{-1}) \circ \varphi_j(x)$. Thus, both bundles $T^{1,0}X$ and $\mathcal{T}_X$ are associated to the same cocycle $\{J(\varphi_{ij}) \circ \varphi_j\}$. Hence, they are isomorphic. □

Definition 2.6.5 The bundles $T^{1,0}X$ and $T^{0,1}X$ are called the *holomorphic* respectively the *antiholomorphic tangent bundle* of the (almost) complex manifold X.

Remark 2.6.6 To any complex vector bundle $E \to X$ one can associate its complex conjugate $\bar{E} \to X$. As real vector spaces the fibres $E(x)$ and $\bar{E}(x)$ are naturally isomorphic, but multiplication with i differs by a sign. If E is described by the cocyle $\{\psi_{ij}\}$ then $\bar{E}$ is given by $\{\bar{\psi}_{ij}\}$. Note that the complex conjugate $\bar{E}$ of a holomorphic vector bundle E is *a priori* not holomorphic in general. Observe that $T^{0,1}X$ is canonically isomorphic to the complex conjugate $\overline{T^{1,0}X}$ of $T^{1,0}X$.

As in the local situation dealt with in Section 1.3, we are more interested in the dual bundles.

Definition 2.6.7 For an almost complex manifold X one defines the complex vector bundles

$$\bigwedge_{\mathbb{C}}^k X := \bigwedge^k (T_{\mathbb{C}}X)^* \quad \text{and} \quad \bigwedge^{p,q} X := \bigwedge^p (T^{1,0}X)^* \otimes_{\mathbb{C}} \bigwedge^q (T^{0,1}X)^*.$$

Their sheaves of sections are denoted by $\mathcal{A}_{X,\mathbb{C}}^k$ and $\mathcal{A}_X^{p,q}$, respectively. Elements in $\mathcal{A}^{p,q}(X)$, i.e. global sections of $\mathcal{A}_X^{p,q}$, are called forms of *type* (or *bidegree*) (p, q).

As in Section 1.3, we denote the projections $\mathcal{A}^*(X) \to \mathcal{A}^k(X)$ and $\mathcal{A}^* \to \mathcal{A}^{p,q}(X)$ by Π^k and $\Pi^{p,q}$, respectively.

Corollary 2.6.8 *There exists a natural direct sum decomposition*

$$\bigwedge_{\mathbb{C}}^k X = \bigoplus_{p+q=k} \bigwedge^{p,q} X \ \text{ and } \ \mathcal{A}_{X,\mathbb{C}}^k = \bigoplus_{p+q=k} \mathcal{A}_X^{p,q}.$$

Moreover, $\overline{\bigwedge^{p,q} X} = \bigwedge^{q,p} X$ and $\overline{\mathcal{A}_X^{p,q}} = \mathcal{A}_X^{q,p}$.

Proof. As Corollary 1.3.4, these assertions are immediate consequences of Proposition 1.2.8. □

Definition 2.6.9 Let X be an almost complex manifold. If $d : \mathcal{A}_{X,\mathbb{C}}^k \to \mathcal{A}_{X,\mathbb{C}}^{k+1}$ is the $\mathbb{C}$-linear extension of the exterior differential, then one defines

$$\partial := \Pi^{p+1,q} \circ d : \mathcal{A}_X^{p,q} \longrightarrow \mathcal{A}_X^{p+1,q}, \ \ \bar{\partial} := \Pi^{p,q+1} \circ d : \mathcal{A}_X^{p,q} \longrightarrow \mathcal{A}_X^{p,q+1}.$$

As in the proof of Lemma 1.3.6 the Leibniz rule for the exterior differential d implies the Leibniz rule for ∂ and $\bar{\partial}$, e.g. $\partial(\alpha \wedge \beta) = \partial(\alpha) \wedge \beta + (-1)^{p+q} \alpha \wedge \partial(\beta)$ for $\alpha \in \mathcal{A}^{p,q}(X)$.

Proposition 2.6.10 *Let $f : X \to Y$ be a holomorphic map between complex manifolds. Then the pull-back of differential forms respects the above decompositions, i.e. it induces natural $\mathbb{C}$-linear maps $f^* : \mathcal{A}^{p,q}(Y) \to \mathcal{A}^{p,q}(X)$. These maps are compatible with ∂ and $\bar{\partial}$.*

Proof. As for any differentiable map $f : X \to Y$ there exists the natural pull-back map $f^* : \mathcal{A}^k(Y) \to \mathcal{A}^k(X)$ which satisfies $f^* \circ d_Y = d_X \circ f^*$.

If f is holomorphic, then the pull-back f^* is compatible with the above decompositions (cf. Exercise 1.3.1). In particular, $f^*(\mathcal{A}^{p,q}(Y)) \subset \mathcal{A}^{p,q}(X)$ and $\Pi^{p+1,q} \circ f^* = f^* \circ \Pi^{p+1,q}$. Thus, for $\alpha \in \mathcal{A}^{p,q}(Y)$ one has

$$\partial_X(f^*\alpha) = \Pi^{p+1,q}(d_X(f^*(\alpha))) = \Pi^{p+1,q}(f^*(d_Y(\alpha)))$$
$$= f^*(\Pi^{p+1,q}(d_Y(\alpha))) = f^*(\partial_Y(\alpha)).$$

Analogously, one shows $\bar{\partial}_X \circ f^* = f^* \circ \bar{\partial}_Y$. □

By Proposition 2.6.4 we know that the complex vector bundles Ω_X^p and $\bigwedge^{p,0} X$ of a complex manifold X can be identified. In particular, any holomorphic section of Ω_X^p defines a section of $\bigwedge^{p,0} X$.

Proposition 2.6.11 *The space $H^0(X, \Omega_X^p)$ of holomorphic p-forms on a complex manifold X is the subspace $\{\alpha \in \mathcal{A}^{p,0}(X) \mid \bar{\partial}\alpha = 0\}$.*

Proof. This is a purely local statement. We use that locally the p-forms $dz_{i_1} \wedge \ldots \wedge dz_{i_p}$ provide a basis of the complex vector bundle $\bigwedge^{p,0} X$ as well as a holomorphic(!) basis of the holomorphic vector bundle Ω_X^p. This stems from the identification of $T^{1,0} = X$ with $\mathcal{T}_X$ given in the proof of Proposition 2.6.4.

A local section of the form $f dz_{i_1} \wedge \ldots \wedge dz_{i_p} \in \mathcal{A}^{p,0}(U)$ is a holomorphic section of Ω_U^p if and only if f is holomorphic, i.e. $\frac{\partial f}{\partial \bar{z}_i} = 0$ for $i = 1, \ldots, n$. But the latter is equivalent to

$$\bar{\partial}(f dz_{i_1} \wedge \ldots \wedge dz_{i_p}) = \sum \frac{\partial f}{\partial \bar{z}_i} d\bar{z}_i \wedge dz_{i_1} \wedge \ldots \wedge dz_{i_p} = 0.$$

$\square$

Corollary 2.6.12 *Let $f : X \to Y$ be a holomorphic map between complex manifolds. Then there exist the following natural maps:* i) *A sheaf homomorphism $\mathcal{T}_X \to f^* \mathcal{T}_Y$,* ii) *A sheaf homomorphism $f^* \Omega_Y \to \Omega_X$, and* iii) *A linear map $H^0(Y, \Omega_Y^p) \to H^0(X, \Omega_X^p)$.*

Proof. Assertion ii) implies the other two. Indeed, dualizing ii) yields i). Taking exterior powers of ii), passing to global sections, and composing with $H^0(Y, \Omega_Y^p) \to H^0(X, f^* \Omega_Y^p)$ yields iii).

In order to prove ii), we use the exact sequence $\Omega_Y \to \mathcal{A}_Y^{1,0} \xrightarrow{\bar{\partial}} \mathcal{A}_Y^{1,1}$. Since $f^* \circ \bar{\partial}_Y = \bar{\partial}_X \circ f^*$ the pull-back $f^{-1}(\mathcal{A}_Y^{1,0} \xrightarrow{\bar{\partial}} \mathcal{A}_Y^{1,1})$, maps to $\mathcal{A}_X^{1,0} \xrightarrow{\bar{\partial}} \mathcal{A}_X^{1,1}$. This yields a natural map $f^{-1} \Omega_Y \to \Omega_X$ and thus the desired map $f^* \Omega_Y \to \Omega_X$. $\square$

Definition 2.6.13 A holomorphic map $f : X \to Y$ is *smooth* at a point $x \in X$ if the induced map $\mathcal{T}_X(x) \to (f^* \mathcal{T}_Y)(x) = \mathcal{T}_Y(y)$ is surjective.

As an immediate consequence of Corollary 1.1.12 one finds:

Corollary 2.6.14 *Let $f : X \to Y$ be a holomorphic map and $y \in Y$. Assume that f is smooth in all points of the fibre $f^{-1}(y)$. Then the fibre $f^{-1}(y)$ is a smooth complex submanifold of X.* $\square$

We next wish to discuss what an almost complex structure needs in order to be induced by a complex one.

Proposition 2.6.15 *Let X be an almost complex manifold. Then the following two conditions are equivalent:*
i) $d\alpha = \partial(\alpha) + \bar{\partial}(\alpha)$ *for all* $\alpha \in \mathcal{A}^*(X)$.
ii) *On $\mathcal{A}^{1,0}(X)$ one has $\Pi^{0,2} \circ d = 0$.*
Both conditions hold true if X is a complex manifold.

Proof. The last assertion is easily proved by reducing to the local situation (cf. Section 1.3).

The implication i) $\Rightarrow$ ii) is trivial, since $d = \partial + \bar{\partial}$ clearly implies $\Pi^{0,2} \circ d = 0$ on $\mathcal{A}^{1,0}(X)$.

Conversely, $d = \partial + \bar{\partial}$ holds on $\mathcal{A}^{p,q}(X)$ if and only if $d\alpha \in \mathcal{A}^{p+1,q}(X) \oplus \mathcal{A}^{p,q+1}(X)$ for all $\alpha \in \mathcal{A}^{p,q}(X)$. Locally, $\alpha \in \mathcal{A}^{p,q}(X)$ can be written as a sum of terms of the form $f w_{i_1} \wedge \ldots \wedge w_{i_p} \wedge w'_{j_1} \wedge \ldots \wedge w'_{j_q}$ with $w_i \in \mathcal{A}^{1,0}(X)$ and $w'_j \in \mathcal{A}^{0,1}(X)$. Using Leibniz rule the exterior differential of such a form is computed in terms of df, dw_{i_k}, and dw'_{j_ℓ}. Clearly, $df \in \mathcal{A}^{1,0}(X) \oplus \mathcal{A}^{0,1}(X)$ and by assumption $dw_i \in \mathcal{A}^{2,0}(X) \oplus \mathcal{A}^{1,1}(X)$, and $dw'_j = \overline{dw'_j} \in \mathcal{A}^{1,1}(X) \oplus \mathcal{A}^{0,2}(X)$. For the latter we use that complex conjugating ii) yields $\Pi^{2,0} \circ d = 0$ on $\mathcal{A}^{0,1}$. Thus, $d\alpha \in \mathcal{A}^{p+1,q}(X) \oplus \mathcal{A}^{p,q+1}(X)$. $\qquad\square$

Definition 2.6.16 An almost complex structure I on X is called *integrable* if the condition i) or, equivalently, ii) in Proposition 2.6.15 is satisfied.

Here is another characterization of integrable almost complex structures.

Proposition 2.6.17 *An almost complex structure I is integrable if and only if the Lie bracket of vector fields preserves $T_X^{0,1}$, i.e. $[T_X^{0,1}, T_X^{0,1}] \subset T_X^{0,1}$.*

Proof. Let α be a $(1,0)$-form and let v, w be sections of $T^{0,1}$. Then, using the standard formula for the exterior differential (cf. Appendix A) and the fact that α vanishes on $T^{0,1}$, one finds

$$(d\alpha)(v,w) = v(\alpha(w)) - w(\alpha(v)) - \alpha([v,w]) = -\alpha([v,w]).$$

Thus, $d\alpha$ has no component of type $(0,2)$ for all α if and only if $[v,w]$ is of type $(0,1)$ for all v, w of type $(0,1)$. $\qquad\square$

Corollary 2.6.18 *If I is an integrable almost complex structure, then $\partial^2 = \bar{\partial}^2 = 0$ and $\partial\bar{\partial} = -\bar{\partial}\partial$. Conversely, if $\bar{\partial}^2 = 0$, then I is integrable.*

Proof. The first assertion follows directly from $d = \partial + \bar{\partial}$ (Proposition 2.6.15), $d^2 = 0$, and the bidegree decomposition.

Conversely, if $\bar{\partial}^2 = 0$ we show that $[T_X^{0,1}, T_X^{0,1}] \subset T_X^{0,1}$. For v, w local sections of $T_X^{0,1}$ we use again the formula $(d\alpha)(v,w) = v(\alpha(w)) - w(\alpha(v)) - \alpha([v,w])$, but this time for a $(0,1)$-form α. Hence, $(d\alpha)(v,w) = (\bar{\partial}\alpha)(v,w)$. If applied to $\alpha = \bar{\partial}f$ we obtain

$$
\begin{aligned}
0 = (\bar{\partial}^2 f)(v,w) &= v((\bar{\partial}f)(w)) - w((\bar{\partial}f)(v)) - (\bar{\partial}f)([v,w]) \\
&= v((df)(w)) - w((df)(v)) - (\bar{\partial}f)([v,w]), \text{ since } v,w \in T_X^{0,1} \\
&= (d^2 f)(v,w) + (df)([v,w]) - (\bar{\partial}f)([v,w]) \\
&= 0 + (\partial f)([v,w]), \quad \text{since } d = \partial + \bar{\partial} \text{ on } \mathcal{A}^0.
\end{aligned}
$$

Since at any given point $(1,0)$-forms of the type ∂f generate $\bigwedge^{1,0}$, this yields $[v, w] \in T_X^{0,1}$. $\square$

The importance of the concept of integrable almost complex structure stems from the following important result, the proof of which is highly non-trivial. For analytic manifolds an easy argument is given in [114].

Theorem 2.6.19 (Newlander–Nierenberg) *Any integrable almost complex structure is induced by a complex structure.* $\square$

Thus, complex manifolds and differentiable manifolds endowed with an integrable almost complex structure are describing the same geometrical object (see Exercise 2.6.1). Each definition has its advantages and it depends on the situation which one suits better.

Before stating a similar result for vector bundles, we introduce the Dolbeault cohomology of a complex manifold. This is the analogue of the cohomology group $H^q(X, \Omega_X^p)$, used to defines the Hodge numbers of a compact complex manifold, in the context of almost complex structures.

Definition 2.6.20 Let X be endowed with an integrable almost complex structure. Then the (p, q)-*Dolbeault cohomology* is the vector space

$$H^{p,q}(X) := H^q(\mathcal{A}^{p,\bullet}(X), \bar{\partial}) = \frac{\mathrm{Ker}\left(\bar{\partial} : \mathcal{A}^{p,q}(X) \to \mathcal{A}^{p,q+1}(X)\right)}{\mathrm{Im}\left(\bar{\partial} : \mathcal{A}^{p,q-1}(X) \to \mathcal{A}^{p,q}(X)\right)}$$

We have seen that $H^0(X, \Omega_X^p) = \ker(\bar{\partial} : \mathcal{A}^{p,0}(X) \to \mathcal{A}^{p,1}(X))$. Using the $\bar{\partial}$-Poincaré lemma 1.3.9 and the fact that the sheaves $\mathcal{A}_X^{p,q}$ are acyclic we obtain the following generalization:

Corollary 2.6.21 *The Dolbeault cohomology of X computes the cohomology of the sheaf Ω_X^p, i.e. $H^{p,q}(X) \cong H^q(X, \Omega_X^p)$.* $\square$

The formalism can be generalized to the Dolbeault cohomology of vector bundles as follows.

Let E be a complex vector bundle over a complex manifold X.

Definition 2.6.22 By $\mathcal{A}^{p,q}(E)$ we denote the sheaf

$$U \longmapsto \mathcal{A}^{p,q}(U, E) := \Gamma(U, \textstyle\bigwedge^{p,q} X \otimes E) \,.$$

Locally a section α of $\mathcal{A}^{p,q}(E)$ can be written as $\alpha = \sum \alpha_i \otimes s_i$ with $\alpha_i \in \mathcal{A}_X^{p,q}$ and $s_i \in E$. Then one could try to define $d\alpha$ as $\sum d(\alpha_i) \otimes s_i$. This definition is clearly problematic, as $\alpha = \alpha_i g \otimes g^{-1} s_i$ for any function $g : U \to \mathbb{C}^*$, but if the function g is non constant then $d(\alpha_i g) \otimes g^{-1} s_i = d(\alpha_i) \otimes s_i + \alpha_i d(g) \otimes g^{-1} s_i \neq d(\alpha_i) \otimes s_i$.

Lemma 2.6.23 *If E is a holomorphic vector bundle then there exists a natural $\mathbb{C}$-linear operator $\bar{\partial}_E : \mathcal{A}^{p,q}(E) \to \mathcal{A}^{p,q+1}(E)$ with $\bar{\partial}_E^2 = 0$ and which satisfies the Leibniz rule $\bar{\partial}_E(f \cdot \alpha) = \bar{\partial}(f) \wedge \alpha + f \bar{\partial}_E(\alpha)$.*

Proof. Fix a local holomorphic trivialization $s = (s_1, \ldots, s_r)$ of E and write a section $\alpha \in \mathcal{A}^{p,q}(E)$ as $\alpha = \sum \alpha_i \otimes s_i$ with $\alpha \in \mathcal{A}_X^{p,q}$. Then set

$$\bar{\partial}_E \alpha := \sum \bar{\partial}(\alpha_i) \otimes s_i.$$

With respect to a different holomorphic trivialization $s' = (s'_1, \ldots, s'_r)$ we obtain an operator $\bar{\partial}'_E$. But $\bar{\partial}_E = \bar{\partial}'_E$. Indeed, if (ψ_{ij}) is the transition matrix, i.e. $s_i = \sum_j \psi_{ij} s'_j$, then $\bar{\partial}_E \alpha = \sum_i \bar{\partial}(\alpha_i) \otimes s_i$ and

$$\bar{\partial}'_E \alpha = \bar{\partial}'_E \left(\sum_{i,j} \alpha_i \otimes \psi_{ij} s'_j \right) = \sum_{i,j} \bar{\partial}(\alpha_i \psi_{ij}) \otimes s'_j$$

$$= \sum_{i,j} \bar{\partial}(\alpha_i) \psi_{ij} \otimes s'_j = \bar{\partial}_E \alpha,$$

as (ψ_{ij}) is a matrix with holomorphic entries. $\qquad\square$

By definition, the operator $\bar{\partial}_E$ on $\mathcal{A}^{p,q}(E)$ is given by $\mathrm{rk}(E)$ copies of the usual $\bar{\partial}$ operator once a holomorphic trivialization of E is chosen.

Definition 2.6.24 The *Dolbeault cohomology of a holomorphic vector bundle* E is

$$H^{p,q}(X, E) := H^q(\mathcal{A}^{p,\bullet}(X, E), \bar{\partial}_E) = \frac{\mathrm{Ker}\left(\bar{\partial}_E : \mathcal{A}^{p,q}(X, E) \to \mathcal{A}^{p,q+1}(X, E) \right)}{\mathrm{Im}\left(\bar{\partial}_E : \mathcal{A}^{p,q-1}(X, E) \to \mathcal{A}^{p,q}(X, E) \right)}.$$

As above one has

Corollary 2.6.25 $H^{p,q}(X, E) \cong H^q(X, E \otimes \Omega_X^p)$.

Proof. The complex of sheaves $\mathcal{A}^{p,0}(E) \to \mathcal{A}^{p,1}(E) \to \mathcal{A}^{p,2}(E) \to \ldots$ is a resolution of $E \otimes \Omega_X^p$ and the sheaves $\mathcal{A}^{p,q}(E)$ are acyclic. $\qquad\square$

We conclude this section by the vector bundle analogue of Theorem 2.6.19. It is in some sense a linearized version of the Newlander–Nierenberg theorem, but its proof is a more or less direct consequence of the Frobenius integrability principle. As this would lead as astray, we omit the proof.

Theorem 2.6.26 *Let E be a complex vector bundle on a complex manifold X. A holomorphic structure on E is uniquely determined by a $\mathbb{C}$-linear operator $\bar{\partial}_E : \mathcal{A}^0(E) \to \mathcal{A}^{0,1}(E)$ satisfying the Leibniz rule and the integrability condition $\bar{\partial}_E^2 = 0$.* $\qquad\square$

Exercises

2.6.1 Show that any almost complex structure is induced by at most one complex structure.

2.6.2 Show that any oriented Riemannian surface admits a natural almost complex structure (cf. Example 1.2.12). Use the result of Newlander and Nierenberg to show that any almost complex structure on a Riemann surface is induced by a complex structure.

2.6.3 Compare the almost complex structure on S^2 as given in the previous exercise with the natural complex structure on $\mathbb{P}^1$.

2.6.4 Let $f : X \to Y$ be a surjective holomorphic map between connected complex manifolds. Let $(X, f)_{\mathrm{reg}}$ be the open set of f-smooth points, i.e. the set of points where f looks locally like the projection of a product. Show that $(X, f)_{\mathrm{reg}}$ is dense in X and that its complement can be described as the zero set of a global section of a holomorphic vector bundle on X.

2.6.5 Use the Poincaré lemma to show that any hypersurface $D \subset \mathbb{C}^n$ is the defined by a global holomorphic function $f : \mathbb{C}^n \to \mathbb{C}$. (Cousin problem)

2.6.6 Show that the exterior product induces a multiplication on the full Dolbeault cohomology $\bigoplus_{p,q} H^{p,q}(X)$ and that this yields a $\mathbb{Z}^2$-graded $\mathbb{Z}/2\mathbb{Z}$-commutative algebra $\bigoplus H^{p,q}(X)$ for any complex manifold X.

2.6.7 Let X be a complex manifold. Verify that the following definition of the *Bott–Chern cohomology*

$$H_{\mathrm{BC}}^{p,q}(X) := \frac{\{\alpha \in \mathcal{A}^{p,q}(X) \mid d\alpha = 0\}}{\partial\bar\partial \mathcal{A}^{p-1,q-1}(X)}$$

makes sense. Deduce from Exercise 1.3.4 that $H_{\mathrm{BC}}^{p,q}(B) = 0$ for a polydisc $B \subset \mathbb{C}^n$ and $p, q \geq 1$. Show that there are natural maps

$$H_{\mathrm{BC}}^{p,q}(X) \longrightarrow H^{p,q}(X) \quad \text{and} \quad H_{\mathrm{BC}}^{p,q}(X) \longrightarrow H^{p+q}(X, \mathbb{C}).$$

(Bott–Chern cohomology groups are very useful when one is interested in compact complex manifolds which are not necessarily Kähler. In fact, for compact Kähler manifolds Dolbeault and Bott–Chern cohomology groups coincide via the above map. See Exercise 3.2.14.)

2.6.8 Let M be the real manifold described as a hypersurface $x_0^4 + x_1^4 + x_2^4 + x_3^4 = 0$ in $\mathbb{P}^3$. We denote the naturally induced complex structure by I. Show that (M, I) and $(M, -I)$ define isomorphic complex manifolds.

2.6.9 Let G be a real Lie group. Consider the map $\mathrm{Ad} : G \to \mathrm{Gl}(T_e G)$ with $\mathrm{Ad}(g) : T_e G \to T_e G$ defined as the differential of $G \to G$, $h \mapsto ghg^{-1}$.

   i) Show that a connected Lie group is commutative if and only if $\mathrm{Ad} \equiv \mathrm{id}$. (This requires some basic knowledge on the exponential map, etc. See [40, page 38].)

   ii) Show that for a complex Lie group the map Ad is holomorphic. (The tangent space $T_e G$ becomes complex in a natural way in this situation.) Deduce that $\mathrm{Ad} \equiv \mathrm{id}$ for any connected compact complex Lie group.

   iii) Conclude that any connected compact complex Lie group is abelian.

2.6.10   Let M be a real four-dimensional manifold and let $\sigma \in \mathcal{A}^2_{\mathbb{C}}(M)$ be a closed form such that $\sigma \wedge \sigma = 0$ and $\sigma \wedge \bar{\sigma}$ everywhere non-zero. Show that there exists a unique complex structure I on M such that σ is a holomorphic two-form on (M, I). (This observation is due to Andreotti. Hint: Define $T^{0,1}$ as the kernel of $\sigma : T_{\mathbb{C}}M \to \bigwedge_{\mathbb{C}} M$.)

2.6.11   For this exercise you need to be acquainted with the basics of hypercohomology. Consider the holomorphic de Rham complex $\mathcal{O}_X \xrightarrow{d} \Omega_X \xrightarrow{d} \Omega^2_X \to \dots$ (Note that on Ω^i one has $d = \partial$.) Show that this complex is quasi-isomorphic to $\mathbb{C}$. Conclude that $H^k(X, \mathbb{C}) = \mathbb{H}^k(X, (\Omega^*_X, d))$. Similarly, show that $H^k(X, \mathbb{C}) = \mathbb{H}^k(X, (\Omega^*_X, t \cdot d))$ for any $t \in \mathbb{C}^*$. Prove that in the limit $t = 0$ one has $\mathbb{H}^k(X, (\Omega^*_X, 0)) = \bigoplus_{p+q=k} H^{p,q}(X)$.

Comments: - It is important to be aware of which part of the geometry of a complex manifold X is linear (e.g. the bidegree decomposition $T_{\mathbb{C}} = T^{1,0} \oplus T^{0,1}$) and which is not (the differential operators $\partial, \bar{\partial}$ and the integrability condition).

- For the Newlander–Nierenberg theorem we have to refer to the original source [95].

- Andreotti's observation (Exercise 2.6.10) is crucial in the theory of K3 surfaces. It seems that it was common knowledge in the fifties, but I could not find the original source. See however Weil's report [115]

- Theorem 2.6.26 allows to study all holomorphic structures on a given complex vector bundle E by means of the set of $\bar{\partial}_E$-operators on E. Be aware that one still has to divide out by the gauge group, i.e. diffeomorphisms of E which are linear on the fibres and which cover the identity on X.

- Consider S^6 as embedded into the imaginary part $\text{Im}(\mathbb{O}) \cong \mathbb{R}^7$ of the octonions $\mathbb{O}$. Then the tangent bundle TS^6 can be identified with $\{(u, v) \mid u \perp v, u \in S^6, v \in \text{Im}(\mathbb{O})\}$ and $I : (u, v) \mapsto (u, (1/2)(u \cdot v - v \cdot v))$ (multiplication in $\mathbb{O}$) does define an almost complex structure on S^6. It can be shown that I is not integrable, but the calculation is cumbersome. It is an open problem whether S^6 admits an integrable complex structure. In contrast, an easy argument, using characteristic classes, shows that S^4 does not even admit an almost complex structure. In fact, the intersection form of a simply connected compact fourfold that admits an almost complex structure has an odd number of positive eigenvalues, see [38].

3

Kähler Manifolds

Kähler manifolds form an important class of complex manifolds. Many interesting manifolds, e.g. projective ones, are Kähler, and the notion is flexible enough for many applications. Kähler manifolds are differentiable manifolds endowed with a complex structure and a Riemannian metric satisfying a certain compatibility condition. Thus, Kähler manifolds play a central role at the cross-road of complex and Riemannian geometry. We will eventually be interested in compact Kähler manifolds, but the local study of special Kähler metrics is an highly attractive area of research too.

Section 3.1 is devoted to the definition of Kähler metrics, to the construction of a few concrete examples, and to the so-called Kähler identities, commutator relations for linear and linear differential operators. A slightly different and more algebraic point of view, close to applications in mathematical physics, will be developed in the Appendix 3.B to this chapter.

The Kähler identities together with Hodge theory for ∂ and $\bar{\partial}$ (for the analysis behind it we have to refer to the literature) are used in Section 3.2 to transfer the linear algebra for the exterior algebra of an hermitian vector space (see Section 1.2) to the cohomology of any compact Kähler manifold. This will lead us to the abstract notion of a Hodge structure explained in Appendix 3.C.

Section 3.3 presents the Hard Lefschetz theorem, also obtained by passing from $\mathcal{A}^*$ to H^*, and its application to the signature of the intersection pairing on the middle cohomology of an even dimensional Kähler manifold. For a compact Kähler manifold the set of Chern classes of all holomorphic line bundles can be described in terms of the Hodge structure on H^2. This result is called the Lefschetz theorem on $(1,1)$-classes. The Hodge conjecture proposes a generalization of it. The short paragraph presenting the Hodge conjecture might serve as a first encounter with this important open problem.

The Appendix 3.A discusses formality for compact Kähler manifolds, an algebraic property of the de Rham complex, that has far reaching consequences for the homotopy theory.

3.1 Kähler Identities

This section proves the fundamental Kähler identities expressing commutator relations between various linear and linear differential operators on any Kähler manifold. Many complex manifolds, but by far not all, possess a Kähler metric, which by definition coincides locally with the standard hermitian structure on $\mathbb{C}^n$ up to terms of order at least two. This section also contains a detailed discussion of the most important examples of compact Kähler manifolds.

Let X be a complex manifold. We denote the induced almost complex structure by I. A local version of the following definition has already been discussed in Section 1.3.

Definition 3.1.1 A Riemannian metric g on X is an *hermitian structure* on X if for any point $x \in X$ the scalar product g_x on $T_x X$ is compatible with the almost complex structure I_x. The induced real $(1,1)$-form $\omega := g(I(\ ),(\ ))$ is called the *fundamental form*.

Locally the fundamental form ω is of the form

$$\omega = \frac{i}{2} \sum_{i,j=1}^{n} h_{ij} dz_i \wedge d\bar{z}_j,$$

where for any $x \in X$ the matrix $(h_{ij}(x))$ is a positive definite hermitian matrix.

The complex manifold X endowed with an hermitian structure g is called an *hermitian manifold*. Note that the hermitian structure g is uniquely determined by the almost complex structure I and the fundamental form ω. Indeed, $g(\ ,\ ) = \omega(\ ,I(\ ))$.

One could as well define hermitian structures on almost complex manifolds. All assertions below concerning the linear structure would still be valid in this more general context. But, as soon as we use the splitting of the exterior differential $d = \partial + \bar{\partial}$, we need an integrable almost complex structure (cf. Theorem 2.6.19).

In analogy to the theory developed in Section 1.2, one defines the following vector bundle homomorphisms on any hermitian manifold of complex dimension n:

i) The *Lefschetz operator*

$$L : \bigwedge^k X \longrightarrow \bigwedge^{k+2} X, \ \alpha \longmapsto \alpha \wedge \omega$$

is an operator of degree two.

ii) The *Hodge ∗-operator*

$$* : \bigwedge^k X \longrightarrow \bigwedge^{2n-k} X$$

is induced by the metric g and the natural orientation of the complex manifold X. Here, $2n$ is the real dimension of X.

iii) The *dual Lefschetz operator*

$$\Lambda := *^{-1} \circ L \circ * : \bigwedge^k X \longrightarrow \bigwedge^{k-2} X$$

is an operator of degree -2 and depends on the Kähler form ω and the metric g (and, therefore, on the complex structure I).

All three operators can be extended $\mathbb{C}$-linearly to the complexified bundles $\bigwedge_{\mathbb{C}}^k X$. By abuse of notation, those will again be called L, $*$, and Λ, respectively. Due to the results of Section 1.2 one has the following

Corollary 3.1.2 *Let (X, g) be an hermitian manifold. Then there exists a direct sum decomposition of vector bundles*

$$\bigwedge^k X = \bigoplus_{i \geq 0} L^i (P^{k-2i} X),$$

where $P^{k-2i} X := \mathrm{Ker}(\Lambda : \bigwedge^{k-2i} X \to \bigwedge^{k-2i-2} X)$ is the bundle of primitive forms. □

The decomposition is compatible with the bidegree decomposition $\bigwedge_{\mathbb{C}}^k X = \bigoplus_{p+q=k} \bigwedge^{p,q} X$, i.e. following Remark 1.2.33 one has

$$P_{\mathbb{C}}^k X = \bigoplus_{p+q=k} P^{p,q} X,$$

where $P^{p,q} X := P_{\mathbb{C}}^{p+q} X \cap \bigwedge^{p,q} X$. Also recall the definition of the operators H and $\mathbf{I}$:

$$H = \sum_{k=0}^{2n} (k-n) \cdot \Pi^k \quad \text{and} \quad \mathbf{I} = \sum_{p,q=0}^{n} i^{p-q} \cdot \Pi^{p,q},$$

where Π^k and $\Pi^{p,q}$ are the natural projections $\mathcal{A}^*(X) \to \mathcal{A}^k(X)$ and $\mathcal{A}^*(X) \to \mathcal{A}^{p,q}(X)$, respectively. Although the definition of the operator $\mathbf{I}$ uses the bidegree decomposition, it is in fact a real operator (cf. Definition 1.2.10).

We now pass from linear operators to differential operators. On an arbitrary oriented m-dimensional Riemannian manifold (M, g) the *adjoint operator* d^* is defined as

$$d^* = (-1)^{m(k+1)+1} * \circ d \circ * : \mathcal{A}^k(M) \longrightarrow \mathcal{A}^{k-1}(M)$$

and the *Laplace operator* is given by

$$\Delta = d^* d + d d^*.$$

If the dimension of M is even, e.g. if M admits a complex structure, then $d^* = - * \circ d \circ *$. Analogously, one defines ∂^* and $\bar{\partial}^*$ as follows.

Definition 3.1.3 If (X, g) is an hermitian manifold, then

$$\partial^* := - * \circ \bar{\partial} \circ * \quad \text{and} \quad \bar{\partial}^* = - * \circ \partial \circ *.$$

Due to Lemma 1.2.24, we know that the Hodge $*$-operator maps $\mathcal{A}^{p,q}(X)$ to $\mathcal{A}^{n-q,n-p}(X)$. Thus,

$$
\begin{array}{ccc}
\mathcal{A}^{p,q}(X) & \xrightarrow{\ \partial^*\ } & \mathcal{A}^{p-1,q}(X) \\
\downarrow{\scriptstyle *} & & \uparrow{\scriptstyle -*} \\
\mathcal{A}^{n-q,n-p}(X) & \xrightarrow{\ \bar{\partial}\ } & \mathcal{A}^{n-q,n-p+1}(X)
\end{array}
$$

and, similarly, $\bar{\partial}^*(\mathcal{A}^{p,q}(X)) \subset \mathcal{A}^{p,q-1}(X)$.

The following lemma is an immediate consequence of the decomposition $d = \partial + \bar{\partial}$ which holds because the almost complex structure on a complex manifold is integrable (Proposition 2.6.15).

Lemma 3.1.4 *If (X, g) is an hermitian manifold then $d^* = \partial^* + \bar{\partial}^*$ and $\partial^{*2} = \bar{\partial}^{*2} = 0$.* $\qquad\square$

Definition 3.1.5 If (X, g) is an hermitian manifold, then the *Laplacians associated to ∂ and $\bar{\partial}$*, respectively, are defined as

$$\Delta_\partial := \partial^*\partial + \partial\partial^* \quad \text{and} \quad \Delta_{\bar{\partial}} := \bar{\partial}^*\bar{\partial} + \bar{\partial}\bar{\partial}^*.$$

Clearly, Δ_∂ and $\Delta_{\bar{\partial}}$ respect the bidegree, i.e.

$$\Delta_\partial, \Delta_{\bar{\partial}} : \mathcal{A}^{p,q}(X) \longrightarrow \mathcal{A}^{p,q}(X).$$

All these linear and differential operators behave especially well if a further compatibility condition on the Riemannian metric and the complex structure is imposed. This is the famous Kähler condition formulated for the first time by Kähler in [73].

Definition 3.1.6 A *Kähler structure* (or *Kähler metric*) is an hermitian structure g for which the fundamental form ω is closed, i.e. $d\omega = 0$. In this case, the fundamental ω form is called the *Kähler form*.

The complex manifold endowed with the Kähler structure is called a *Kähler manifold*. However, sometimes a complex manifold X is called Kähler if there exists a Kähler structure without actually fixing one. More accurately, one should speak of a complex manifold of Kähler type in this case.

Hermitian structures exist on any complex manifold (cf. Exercise 3.1.1), but, as we will see shortly, Kähler structures do not always exist.

The local version of a Kähler metric has been studied in detail in Section 1.3. There we have seen that the condition $d\omega = 0$ is equivalent to the fact that the hermitian structure g osculates in any point to order two to the standard hermitian structure (see Proposition 1.3.12 for the precise statement).

As the metric can be recovered from its fundamental form and the almost complex structure, one has:

Lemma 3.1.7 *Let ω be a closed real $(1,1)$-form on a complex manifold X. If ω is* positive definite, *i.e. ω is locally of the form $\omega = \frac{i}{2}\sum h_{ij}dz_i \wedge d\bar{z}_j$ such that $(h_{ij}(x))$ is a positive definite hermitian matrix for any $x \in X$, then there exists a Kähler metric on X such that ω is the associated fundamental form.* $\square$

Thus, the set of closed positive real $(1,1)$-forms $\omega \in \mathcal{A}^{1,1}(X)$ is the set of all Kähler forms.

Corollary 3.1.8 *The set of all Kähler forms on a compact complex manifold X is an open convex cone in the linear space $\{\omega \in \mathcal{A}^{1,1}(X) \cap \mathcal{A}^2(X) \mid d\omega = 0\}$.*

Proof. The positivity of an hermitian matrix $(h_{ij}(x))$ is an open property and, since X is compact, the set of forms $\omega \in \mathcal{A}^{1,1}(X) \cap \mathcal{A}^2(X)$ that are locally of the form $\omega = \frac{i}{2}\sum h_{ij}dz_i \wedge d\bar{z}_j$ with (h_{ij}) positive definite at every point is open. The differential equation $d\omega = 0$ ensures that the metric associated to such an ω is Kähler. (See the comment at the end of the section if you worry about the topology considered here.)

In order to see that Kähler forms form a convex cone, one has to show that for $\lambda \in \mathbb{R}_{>0}$ and Kähler forms ω_1, ω_2 also $\lambda \cdot \omega_i$ and $\omega_1 + \omega_2$ are Kähler forms. Both assertion follow from the corresponding statements for positive definite hermitian matrices. $\square$

Also multiplication with a positive function $\lambda : X \to \mathbb{R}$ yields a positive form $\lambda \cdot \omega$, which however, except in complex dimension one, is not closed and hence not Kähler.

Positivity of forms will be treated more generally and in more detail in Section 4.3.

Examples 3.1.9 i) The *Fubini–Study* metric is a canonical Kähler metric on the projective space $\mathbb{P}^n$. Let $\mathbb{P}^n = \bigcup_{i=0}^{n} U_i$ be the standard open covering and $\varphi_i : U_i \cong \mathbb{C}^n, (z_0 : \ldots : z_n) \mapsto \left(\frac{z_0}{z_i}, \ldots, \frac{\widehat{z_i}}{z_i}, \ldots, \frac{z_n}{z_i}\right)$. Then one defines

$$\omega_i := \frac{i}{2\pi}\partial\bar{\partial}\log\left(\sum_{\ell=0}^{n}|\frac{z_\ell}{z_i}|^2\right) \in \mathcal{A}^{1,1}(U_i),$$

which under φ_i corresponds to

$$\frac{i}{2\pi}\partial\bar{\partial}\log\left(\sum_{k=1}^{n}|w_k|^2 + 1\right).$$

Let us first show that $\omega_i|_{U_i \cap U_j} = \omega_j|_{U_i \cap U_j}$, i.e. that the ω_i glue to a global form $\omega_{\text{FS}} \in \mathcal{A}^{1,1}(\mathbb{P}^n)$. Indeed,

$$\log\left(\sum_{\ell=0}^{n}|\frac{z_\ell}{z_i}|^2\right) = \log\left(|\frac{z_j}{z_i}|^2\sum_{\ell=0}^{n}|\frac{z_\ell}{z_j}|^2\right) = \log\left(|\frac{z_j}{z_i}|^2\right) + \log\left(\sum_{\ell=0}^{n}|\frac{z_\ell}{z_j}|^2\right).$$

Thus, it suffices to show that $\partial\bar{\partial}\log(|\frac{z_j}{z_i}|^2) = 0$ on $U_i \cap U_j$. Since $\frac{z_j}{z_i}$ is the j-th coordinate function on U_i, this follows from

$$\partial\bar{\partial}\log|z|^2 = \partial\left(\frac{1}{z\bar{z}}\bar{\partial}(z\bar{z})\right) = \partial\left(\frac{zd\bar{z}}{z\bar{z}}\right) = \partial\left(\frac{d\bar{z}}{\bar{z}}\right) = 0$$

Next, we observe that ω_{FS} is a real $(1,1)$-form. Indeed, $\overline{\partial\bar{\partial}} = \bar{\partial}\partial = -\partial\bar{\partial}$ yields $\omega_i = \bar{\omega}_i$. Moreover, ω_{FS} is closed, as $\partial\omega_i = \frac{i}{2\pi}\partial^2\bar{\partial}\log(\ ) = 0$.

It remains to show that ω_{FS} is positive definite, i.e. that ω_{FS} really is the Kähler form associated to a metric. This can be verified on each U_i separately. A straightforward computation yields

$$\partial\bar{\partial}\log\left(1+\sum_{i=1}^{n}|w_i|^2\right) = \frac{\sum dw_i \wedge d\bar{w}_i}{1+\sum|w_i|^2} - \frac{(\sum\bar{w}_i dw_i)\wedge(\sum w_i d\bar{w}_i)}{(1+\sum|w_i|^2)^2}$$

$$= \frac{1}{(1+\sum|w_i|^2)^2}\sum h_{ij}dw_i \wedge d\bar{w}_j,$$

with $h_{ij} = (1+\sum|w_i|^2)\delta_{ij} - \bar{w}_i w_j$. The matrix (h_{ij}) is positive definite, since for $u \neq 0$ the Cauchy–Schwarz inequality for the standard hermitian product $(\ ,\ )$ on $\mathbb{C}^n$ yields

$$u^t(h_{ij})\bar{u} = (u,u) + (w,w)(u,u) - u^t\bar{w}w^t\bar{u}$$
$$= (u,u) + (w,w)(u,u) - (u,w)(w,u)$$
$$= (u,u) + (w,w)(u,u) - \overline{(w,u)}(w,u)$$
$$= (u,u) + (w,w)(u,u) - |(w,u)|^2 > 0.$$

As the Fubini–Study metric, which will come up again and again, is a very prominent example of a Kähler metric, we will dwell on it a bit a longer.

Let us consider the natural projection $\pi : \mathbb{C}^{n+1}\setminus\{0\} \to \mathbb{P}^n$. Then

$$\pi^*\omega_{\mathrm{FS}} = \frac{i}{2\pi}\partial\bar{\partial}\log(\|z\|^2).$$

Indeed, over $\pi^{-1}(U_i) = \{(z_0,\ldots,z_n) \mid z_i \neq 0\}$ one has

$$\pi^*\omega_{\mathrm{FS}} = \frac{i}{2\pi}\partial\bar{\partial}\log\left(\sum_{\ell=0}^{n}|\frac{z_\ell}{z_i}|^2\right)$$

$$= \frac{i}{2\pi}\partial\bar{\partial}\left(\log(\|z\|^2) - \log(|z_i|^2)\right),$$

but $\partial\bar{\partial}\log(|z_i|^2) = 0$, as has been shown above. (Compare this with Exercise 1.3.8.)

Notice, that if we write more abstractly a projective space as $\mathbb{P}(V)$, then the Fubini–Study metric does depend on the choice of a basis of V or, more precisely, on a chosen hermitian structure on V. See Exercise 3.1.6 for the precise statement.

We conclude this example by proving the equation

$$\int_{\mathbb{P}^1} \omega_{FS} = 1,$$

which will serve as a normalization in the definition of Chern classes (see page 197). Moreover, since $\mathbb{P}^1 \cong S^2$ and thus $H^2(\mathbb{P}^1, \mathbb{Z}) = H^2(S^2, \mathbb{Z}) \cong \mathbb{Z}$, it shows that $[\omega_{FS}] \in H^2(\mathbb{P}^1, \mathbb{Z})$ is a generator.

The integral is explicitly computed as follows:

$$\begin{aligned}
\int_{\mathbb{P}^1} \omega_{FS} &= \int_{\mathbb{C}} \frac{i}{2\pi} \frac{1}{(1+|w|^2)^2} dw \wedge d\overline{w} \\
&= \frac{1}{\pi} \int_{\mathbb{R}^2} \frac{1}{(1+\|(x,y)\|^2)^2} dx \wedge dy \\
&= 2 \int_0^{\infty} \frac{r dr}{(1+r^2)^2} = 1
\end{aligned}$$

ii) Any *complex torus* $\mathbb{C}^n/\Gamma$ can easily be endowed with a Kähler structure by taking a scalar product on the real vector space $\mathbb{C}^n$ which is compatible with the natural almost complex structure. This defines a constant metric on $\mathbb{C}^n$, which is, in particular, Γ-invariant and thus descends to an hermitian structure on $\mathbb{C}^n/\Gamma$.

iii) Any *complex curve* admits a Kähler structure. In fact, any hermitian metric is Kähler, as a two-form on a complex curve is always closed. For the existence of hermitian structures see Exercise 3.1.1.

iv) On the *unit disc* $D^n \subset \mathbb{C}^n$ one considers the form

$$\omega = \frac{i}{2} \partial \overline{\partial} \log(1 - \|z\|^2).$$

Firstly, ω is indeed a Kähler form, as might be checked by a calculation similar to i). One finds, $\omega = \frac{i}{2(1-\|z\|^2)^2} \sum \left((1-\|z\|^2)\delta_{ij} + \overline{z}_i z_j \right) dz_i \wedge d\overline{z}_j$. In particular, for $n = 1$ this says $\omega = (i/2)(1-|z|^2)^{-2} dz \wedge d\overline{z}$.

Secondly, ω is invariant under the action of $\mathrm{SU}(1,n)$ (see page 60). Thus, ω descends to a Kähler form on any ball quotient D^n/Γ.

Proposition 3.1.10 *Let g be a Kähler metric on a complex manifold X. Then the restriction $g|_Y$ to any complex submanifold $Y \subset X$ is again Kähler.*

Proof. Clearly, $g|_Y$ is again a Riemannian metric on Y. Since $T_xY \subset T_xX$ is invariant under the almost complex structure I for any $x \in Y$ and the restriction of it to T_xY is the almost complex structure I_Y on Y, the metric

$g|_Y$ is compatible with the almost complex structure on Y. Thus, $g|_Y$ defines an hermitian structure on Y. By definition, the associated Kähler form ω_Y is given by $\omega_Y = g|_Y(I_Y(\ ),(\ )) = g(I(\ ),(\ ))|_Y = \omega|_Y$. Therefore, $d_Y\omega_Y = d_Y(\omega|_Y) = (d_X\omega)|_Y = 0$. $\square$

Corollary 3.1.11 *Any projective manifold is Kähler.*

Proof. By definition a projective manifold can be realized as a submanifold of $\mathbb{P}^n$. Restricting the Fubini–Study metric yields a Kähler metric. $\square$

Note that the Kähler structure of a projective manifold X obtained by a projective embedding $X \subset \mathbb{P}^N$ depends on this embedding. Often, other Kähler structures exist that are not obtained as the restriction of the Fubini–Study metric under any projective embedding whatsoever.

We now come to a key result in the local theory of Kähler manifolds. In addition to the commutator relations for L and Λ (Proposition 1.2.26), the following Proposition calculates the mixed commutators of linear operators, e.g. L and Λ, and differential operators, e.g. ∂, $\bar{\partial}$, and Δ, explicitly. The Kähler condition $d\omega = 0$ is crucial for this.

Proposition 3.1.12 (Kähler identities) *Let X be a complex manifold endowed with a Kähler metric g. Then the following identities hold true:*
   i) $[\bar{\partial}, L] = [\partial, L] = 0$ *and* $[\bar{\partial}^*, \Lambda] = [\partial^*, \Lambda] = 0$.
   ii) $[\bar{\partial}^*, L] = i\partial$, $[\partial^*, L] = -i\bar{\partial}$ *and* $[\Lambda, \bar{\partial}] = -i\partial^*$, $[\Lambda, \partial] = i\bar{\partial}^*$.
   iii) $\Delta_\partial = \Delta_{\bar{\partial}} = \frac{1}{2}\Delta$ *and* Δ *commutes with* $*$, ∂, $\bar{\partial}$, ∂^*, $\bar{\partial}^*$, L, *and* Λ.

The theorem will be proved in terms of yet another operator d^c.

Definition 3.1.13 One defines

$$d^c := \mathbf{I}^{-1} \circ d \circ \mathbf{I} \quad \text{and} \quad d^{c*} := -*\circ d^c \circ *.$$

Since $\mathbf{I}$ is a real operator, also d^c is a real operator which is extended $\mathbb{C}$-linearly. Equivalently, one could define

$$d^c = -i(\partial - \bar{\partial}).$$

Indeed, if $\alpha \in \mathcal{A}^{p,q}(X)$ then

$$\mathbf{I}(\partial - \bar{\partial})(\alpha) = i^{p+1-q}\partial(\alpha) - i^{p-q-1}\bar{\partial}(\alpha) = i^{p+1-q}d(\alpha) = id(\mathbf{I}(\alpha)).$$

Also note that $dd^c = 2i\partial\bar{\partial}$.
    Assertion ii) implies $[\Lambda, d] = i(\bar{\partial}^* - \partial^*) = -i*(\partial - \bar{\partial})* = -d^{c*}$. In fact, using the bidegree decomposition one easily sees that $[\Lambda, d] = -d^{c*}$ is equivalent to the assertions of ii) that concern Λ.

Proof. Let us first prove i). By definition

$$[\bar{\partial}, L](\alpha) = \bar{\partial}(\omega \wedge \alpha) - \omega \wedge \bar{\partial}(\alpha) = \bar{\partial}(\omega) \wedge \alpha = 0,$$

for $\bar{\partial}(\omega)$ is the $(1, 2)$-part of $d\omega$, which is trivial by assumption. Analogously, one proves $[\partial, L](\alpha) = \partial(\omega) \wedge \alpha = 0$.

The second assertion in i) follows from the first one: For $\alpha \in \mathcal{A}^k(X)$ one finds

$$
\begin{aligned}
[\bar{\partial}^*, \Lambda](\alpha) &= - * \partial * *^{-1} L * (\alpha) - *^{-1} L * (- * \partial*)(\alpha) \\
&= - * \partial L * (\alpha) - (-1)^k *^{-1} L\partial * (\alpha) = -(*\partial L * - * L\partial*)(\alpha) \\
&= - * [\partial, L] * (\alpha) = 0.
\end{aligned}
$$

Here, we use twice that $*^2 = (-1)^\ell$ on $\mathcal{A}^\ell(X)$.

The last assertion can be proved analogously. It can also be verified by just complex conjugating: $[\partial^*, \Lambda] = \overline{[\bar{\partial}^*, \bar{\Lambda}]} = \overline{[\bar{\partial}^*, \Lambda]} = 0$, where one uses that $*$ and Λ are $\mathbb{C}$-linear extensions of real operators.

ii) Using the Lefschetz decomposition, it is enough to prove the assertion for forms of the type $L^j \alpha$ with α a primitive k-form. Then $d\alpha \in \mathcal{A}^{k+1}(X)$ can again be written according to the Lefschetz decomposition (Corollary 3.1.2) as

$$d\alpha = \alpha_0 + L\alpha_1 + L^2\alpha_2 + \dots$$

with $\alpha_j \in P^{k+1-2j}(X)$. Since L commutes with d and $L^{n-k+1}\alpha = 0$, this yields

$$0 = L^{n-k+1}\alpha_0 + L^{n-k+2}\alpha_1 + L^{n-k+3}\alpha_2 + \dots.$$

As the Lefschetz decomposition is a direct sum decomposition, this implies $L^{n-k+j+1}\alpha_j = 0$, for $j = 0, 1, \dots$. On the other hand, L^ℓ is injective on $\mathcal{A}^i(X)$ for $\ell \leq n - i$. Hence, since $\alpha_j \in \mathcal{A}^{k+1-2j}(X)$, one finds $\alpha_j = 0$ for $j \geq 2$. Thus, $d\alpha = \alpha_0 + L\alpha_1$ with $\Lambda\alpha_0 = \Lambda\alpha_1 = 0$.

Let us first compute $[\Lambda, d](L^j\alpha)$ for $\alpha \in P^k(X)$. Using $[d, L] = 0$, $\Lambda\alpha_i = 0$, and Corollary 1.2.28 one computes

$$
\begin{aligned}
\Lambda d L^j \alpha &= \Lambda L^j d\alpha = \Lambda L^j \alpha_0 + \Lambda L^{j+1}\alpha_1 \\
&= -j(k+1-n+j-1)L^{j-1}\alpha_0 - (j+1)(k-1-n+j)L^j\alpha_1
\end{aligned}
$$

and

$$
\begin{aligned}
d\Lambda L^j \alpha &= -j(k-n+j-1)L^{j-1}d\alpha \\
&= -j(k-n+j-1)(L^{j-1}\alpha_0 + L^j\alpha_1).
\end{aligned}
$$

Therefore,

$$[\Lambda, d](L^j\alpha) = -jL^{j-1}\alpha_0 - (k-n+j-1)L^j\alpha_1.$$

On the other hand, Proposition 1.2.31 applied several times yields

$$-d^{c*}L^j\alpha = *\mathbf{I}^{-1}d\,\mathbf{I}*L^j\alpha$$

$$= *\mathbf{I}^{-1}d\,\mathbf{I}\left((-1)^{\frac{k(k+1)}{2}}\frac{j!}{(n-k-j)!}\cdot L^{n-k-j}\mathbf{I}(\alpha)\right)$$

$$= (-1)^{\frac{k(k+1)}{2}+k}\frac{j!}{(n-k-j)!}\cdot\left(\mathbf{I}^{-1}*L^{n-k-j}d\alpha\right)\quad\text{using }\mathbf{I}^2|_{\bigwedge^k}=(-1)^k$$

$$= (-1)^{\frac{k(k+1)}{2}+k}\frac{j!}{(n-k-j)!}\cdot\left(\mathbf{I}^{-1}(*L^{n-k-j}\alpha_0+*L^{n-k-j+1}\alpha_1)\right)$$

$$= (-1)^{\frac{k(k+1)}{2}+k+\frac{(k+1)(k+2)}{2}}j\cdot\left(L^{j-1}\alpha_0\right)$$

$$+ (-1)^{\frac{k(k+1)}{2}+k+\frac{k(k-1)}{2}}(n-k-j+1)\cdot\left(L^j\alpha_1\right)$$

$$= -jL^{j-1}\alpha_0 - (k-n+j-1)L^j\alpha_1.$$

This yields $[\Lambda, d] = -d^{c*}$. The equalities involving L are deduced from this (cf. Exercise 3.1.9).

iii) We first show that $\partial\bar{\partial}^* + \bar{\partial}^*\partial = 0$. Indeed, assertion ii) yields $i(\partial\bar{\partial}^* + \bar{\partial}^*\partial) = \partial[\Lambda, \partial] + [\Lambda, \partial]\partial = \partial\Lambda\partial - \partial\Lambda\partial = 0$. Next,

$$\Delta_\partial = \partial^*\partial + \partial\partial^*$$

$$= i[\Lambda, \bar{\partial}]\partial + i\partial[\Lambda, \bar{\partial}]$$

$$= i(\Lambda\bar{\partial}\partial - \bar{\partial}\Lambda\partial + \partial\Lambda\bar{\partial} - \partial\bar{\partial}\Lambda)$$

$$= i(\Lambda\bar{\partial}\partial - (\bar{\partial}[\Lambda, \partial] + \bar{\partial}\partial\Lambda) + ([\partial, \Lambda]\bar{\partial} + \Lambda\partial\bar{\partial}) - \partial\bar{\partial}\Lambda)$$

$$= i(\Lambda\bar{\partial}\partial - i\partial\bar{\partial}^* - \bar{\partial}\partial\Lambda - i\bar{\partial}^*\bar{\partial} + \Lambda\partial\bar{\partial} - \partial\bar{\partial}\Lambda)$$

$$= \Delta_{\bar{\partial}}.$$

In order to compare Δ with Δ_∂, write

$$\Delta = (\partial + \bar{\partial})(\partial^* + \bar{\partial}^*) + (\partial^* + \bar{\partial}^*)(\partial + \bar{\partial})$$

$$= \Delta_\partial + \Delta_{\bar{\partial}} + (\partial\bar{\partial}^* + \bar{\partial}^*\partial) + \overline{(\partial\bar{\partial}^* + \bar{\partial}^*\partial)}$$

$$= \Delta_\partial + \Delta_{\bar{\partial}} + 0 + 0$$

$$= 2\Delta_\partial.$$

As an example that Δ commutes with all the other operators, we shall show $[\Lambda, \Delta] = 0$. The other commutativity relations are left to the reader. Using $dd^c = 2i\partial\bar{\partial} = -d^c d$, one computes $\Lambda\Delta = \Lambda dd^* + \Lambda d^*d = d\Lambda d^* - id^{c*}d^* + d^*\Lambda d = dd^*\Lambda + id^*d^{c*} + d^*d\Lambda - id^*d^{c*} = \Delta\Lambda$. $\qquad\square$

Remark 3.1.14 A different proof of ii) can be given by using Proposition 1.3.12 (cf. [35, 59]). There, one first proves the identity for $X = \mathbb{C}^n$ with the standard metric and then uses the fact that any relation for $\mathbb{C}^n$ that involves only differential operators of order ≤ 1 also holds true for an arbitrary Kähler manifold. This approach stresses more the local nature of the Kähler identities.

Exercises

3.1.1   Show that any complex manifold admits an hermitian structure.

3.1.2   Let X be a connected complex manifold of dimension $n > 1$ and let g be a Kähler metric. Show that g is the only Kähler metric in its conformal class, i.e. if $g' = e^f \cdot g$ is Kähler then f is constant (cf. Exercise 1.3.9).

3.1.3   Let (X, g) be a compact hermitian manifold of dimension n. Show that d-harmonicity equals ∂-harmonicity and $\bar{\partial}$-harmonicity for forms of degree 0 and $2n$.

3.1.4   Prove $\int_{\mathbb{P}^n} \omega_{\mathrm{FS}}^n = 1$.

3.1.5   Let $\mathbb{C}^n \subset \mathbb{C}^{n+1}$ be the standard inclusion $(z_0, \ldots, z_{n-1}) \mapsto (z_0, \ldots, z_{n-1}, 0)$ and consider the induced inclusion $\mathbb{P}^{n-1} \subset \mathbb{P}^n$. Show that restricting the Fubini–Study Kähler form $\omega_{\mathrm{FS}}(\mathbb{P}^n)$ on $\mathbb{P}^n$ yields the Fubini–Study Kähler form on $\mathbb{P}^{n-1}$.

3.1.6   Let $A \in \mathrm{Gl}(n+1, \mathbb{C})$ and denote the induced isomorphism by $F_A : \mathbb{P}^n \cong \mathbb{P}^n$. Show that $F_A^*(\omega_{\mathrm{FS}}) = \omega_{\mathrm{FS}}$ if and only if $A \in \mathrm{U}(n+1)$.

3.1.7   Show that L, d, and d^* acting on $\mathcal{A}^*(X)$ of a Kähler manifold X determine the complex structure of X.

3.1.8   Show that on a compact Kähler manifold X of dimension n the integral $\int_X \omega^n$ is $n! \cdot \mathrm{vol}(X)$ (cf. Exercise 1.2.9). Conclude from this that there exists an injective ring homomorphism $k[x]/x^{n+1} \to H^*(X, \mathbb{R})$. In particular, $b_2(X) \geq 1$. Deduce from this that S^2 is the only sphere that admits a Kähler structure.

3.1.9   Conclude the first part of assertion ii) of Proposition 3.1.12 from the second.

3.1.10   Let X and Y be two Kähler manifolds. Show that the product $X \times Y$ admits a natural Kähler structure, too.

3.1.11   Fill in the details of Remark 3.1.14.

3.1.12   Let X be a complex manifold endowed with a Kähler form and let α be a closed $(1,1)$-form which is primitive (at every point in the sense of Section 1.2). Show that α is harmonic, i.e. $\Delta(\alpha) = 0$.

3.1.13   Let M be a differentiable manifold of dimension $2n$. A closed two-form ω on M is a *symplectic structure* (or form) if ω is everywhere non-degenerate, i.e ω^n is a volume form. Show that any Kähler manifold (X, g) possesses a natural symplectic structure. Observe that symplectic structures in general do not form a convex cone. Compare this to Corollary 3.1.8.

Comments: - For some historical comments on Kähler manifolds we recommend the article by Bourguignon [18]. To some of the topics mentioned there we shall come back later.

- Not any Kähler metric is geometrically equally interesting. Usually, Kähler metrics with certain additional curvature properties are studied. Very often it happens that those are just known to exist, but cannot be explicitly constructed, at least on compact manifolds. See Appendix 4.B.

- We have been a little sloppy in Corollary 3.1.8. A priori, the openness of the set of Kähler forms depends on the topology that has been put on the infinite-dimensional space of $(1,1)$-forms. But in fact, it holds true with respect to any reasonable topology and we will not elaborate on this.

3.2 Hodge Theory on Kähler Manifolds

The main result of the previous section, the Kähler identities, is a purely local statement. In this section we will focus on compact hermitian and Kähler manifolds. The compactness allows us to apply Hodge theory, i.e. the theory of elliptic operators on compact manifolds. We assume that the reader is familiar with Hodge theory for compact Riemannian manifolds or, is at least willing to accept the Hodge decomposition in the Riemannian context (cf. Appendix A).

If X is a complex manifold with an hermitian structure g, we denote the hermitian extension of the Riemannian metric g by $g_{\mathbb{C}}$ (cf. page 30). It naturally induces hermitian products on all form bundles (cf. page 33).

Definition 3.2.1 Let (X, g) be a compact hermitian manifold. Then one defines an *hermitian product* on $\mathcal{A}_{\mathbb{C}}^*(X)$ by

$$(\alpha, \beta) := \int_X g_{\mathbb{C}}(\alpha, \beta) * 1.$$

Proposition 3.2.2 *Let (X, g) be a compact hermitian manifold. Then the following decompositions are orthogonal with respect to (,):*
  i) *The degree decomposition $\mathcal{A}_{\mathbb{C}}^*(X) = \bigoplus_k \mathcal{A}_{\mathbb{C}}^k(X)$.*
  ii) *The bidegree decomposition $\mathcal{A}_{\mathbb{C}}^*(X) = \bigoplus_{p+q=k} \mathcal{A}^{p,q}(X)$.*
  iii) *The Lefschetz decomposition $\mathcal{A}_{\mathbb{C}}^k(X) = \bigoplus_{i \geq 0} L^i P_{\mathbb{C}}^{k-2i}(X)$.*

Proof. The first assertion follows from the definition of $g_{\mathbb{C}}$. For ii) let $\alpha \in \mathcal{A}^{p,q}(X)$ and $\beta \in \mathcal{A}^{p',q'}(X)$. Then by Lemma 1.2.24 one has $g_{\mathbb{C}}(\alpha, \beta) \equiv 0$ unless $(p, q) = (p', q')$.

Assertion iii) follows again from the analogous statement for any cotangent space (see i) of Proposition 1.2.30). $\qquad\square$

Thus, each $\mathcal{A}^{p,q}(X)$ is an infinite-dimensional vector space endowed with a *scalar product* (,) and the induced *norm* $\|\alpha\|^2 = (\alpha, \alpha)$. The completion of it with respect to (,) yields the space of L^2-forms of bidegree (p, q), but we have to refer to [35, 116] for the analytical aspects.

In analogy to Lemma A.0.9 one has

Lemma 3.2.3 *Let X be a compact hermitian manifold. Then with respect to the hermitian product (,) the operators ∂^* and $\bar{\partial}^*$ are the formal adjoints of ∂ and $\bar{\partial}$, respectively.*

Proof. The proof is literally the same as for d and d^*. We recall it for ∂^*. Let $\alpha \in \mathcal{A}^{p-1,q}(X)$ and $\beta \in \mathcal{A}^{p,q}(X)$. By definition

$$(\partial\alpha, \beta) = \int_X g_{\mathbb{C}}(\partial\alpha, \beta) * 1 = \int_X \partial\alpha \wedge *\bar{\beta} \quad \text{(see page 33)}$$
$$= \int_X \partial(\alpha \wedge *\bar{\beta}) - (-1)^{p+q-1} \int_X \alpha \wedge \partial(*\bar{\beta}).$$

The first integral of the last line vanishes due to Stokes' theorem, as $\alpha \wedge *\bar{\beta}$ is a form of bidegree $(p-1,q)+(n-p,n-q)=(n-1,n)$ and, therefore, $\partial(\alpha \wedge *\bar{\beta})=d(\alpha \wedge *\bar{\beta})$.

The second integral is computed using $*^2=(-1)^k$ on $\mathcal{A}^k(X)$ as follows:

$$\int_X \alpha \wedge \partial(*\bar{\beta}) = \varepsilon \cdot \int_X g_{\mathbb{C}}(\alpha, *(\bar{\partial}(*\beta))) * 1$$

$$= \varepsilon \cdot \int_X g_{\mathbb{C}}(\alpha, -\partial^*\beta) * 1$$

$$= -\varepsilon \cdot (\alpha, \partial^*\beta)$$

with $\varepsilon = (-1)^{2n-(p+q)+1}$. Then check that $\varepsilon \cdot (-1)^{p+q-1} = 1$.   □

Recall, that $\mathcal{H}^k(X,g)$ denotes the space of $(d-)$harmonic k-forms (Definition A.0.11). Analogously, one defines $\mathcal{H}^{p,q}(X,g)$ as the space of $(d-)$harmonic (p,q)-forms. When the metric is fixed, one often drops g in the notation. For an arbitrary hermitian manifold the bidegree decomposition of Proposition 3.2.2 does not carry over to harmonic forms. In this case it is more natural to consider harmonic forms with respect to $\bar{\partial}$ or/and ∂.

Definition 3.2.4 Let (X,g) be an hermitian complex manifold. A form $\alpha \in \mathcal{A}^k(X)$ is called $\bar{\partial}$-harmonic if $\Delta_{\bar{\partial}}(\alpha)=0$. Moreover,

$$\mathcal{H}_{\bar{\partial}}^k(X,g) := \{\alpha \in \mathcal{A}_{\mathbb{C}}^k(X) \mid \Delta_{\bar{\partial}}(\alpha)=0\} \text{ and}$$
$$\mathcal{H}_{\bar{\partial}}^{p,q}(X,g) := \{\alpha \in \mathcal{A}_{\mathbb{C}}^{p,q}(X) \mid \Delta_{\bar{\partial}}(\alpha)=0\}$$

Analogously, one defines ∂-harmonic forms and the spaces $\mathcal{H}_{\partial}^k(X,g)$ and $\mathcal{H}_{\partial}^{p,q}(X,g)$.

Lemma 3.2.5 *Let (X,g) be a compact hermitian manifold (X,g). A form α is $\bar{\partial}$-harmonic (resp. ∂-harmonic) if and only if $\bar{\partial}\alpha = \bar{\partial}^*\alpha = 0$ (resp. $\partial\alpha = \partial^*\alpha = 0$).*

Proof. The assertion follows from

$$(\Delta_{\bar{\partial}}(\alpha),\alpha) = \|\bar{\partial}^*(\alpha)\|^2 + \|\bar{\partial}(\alpha)\|^2.$$

Thus, $\Delta_{\bar{\partial}}(\alpha)=0$ implies the vanishing of both terms on the right hand side, i.e. $\bar{\partial}(\alpha) = \bar{\partial}^*(\alpha) = 0$. The converse is clear.

A similar argument proves the assertion for Δ_{∂}.   □

Proposition 3.2.6 *Let (X,g) be an hermitian manifold, not necessarily compact. Then*

i) $\mathcal{H}_{\bar{\partial}}^k(X,g) = \bigoplus_{p+q=k} \mathcal{H}_{\bar{\partial}}^{p,q}(X,g)$ *and* $\mathcal{H}_{\partial}^k(X,g) = \bigoplus_{p+q=k} \mathcal{H}_{\partial}^{p,q}(X,g)$.

ii) *If (X,g) is Kähler then both decompositions coincide with $\mathcal{H}^k(X,g)_{\mathbb{C}} = \bigoplus_{p+q=k} \mathcal{H}^{p,q}(X,g)$. In particular, $\mathcal{H}^k(X,g)_{\mathbb{C}} = \mathcal{H}_{\bar{\partial}}^k(X,g) = \mathcal{H}_{\partial}^k(X,g)$.*

Proof. Let $\alpha = \sum \alpha^{p,q}$ be the bidegree decomposition of a given form α. Clearly, if $\Delta_{\bar{\partial}}(\alpha^{p,q}) = 0$ for all (p,q) then also $\Delta_{\bar{\partial}}(\alpha) = 0$. On the other hand, $\Delta_{\bar{\partial}}(\alpha) = 0$ implies $0 = \sum \Delta_{\bar{\partial}}(\alpha^{p,q})$ with $\Delta_{\bar{\partial}}(\alpha^{p,q}) \in \mathcal{A}^{p,q}(X)$. But the bidegree decomposition is direct. Thus, $\Delta_{\bar{\partial}}(\alpha^{p,q}) = 0$, i.e. the $\alpha^{p,q}$ are harmonic. The proof for the space of ∂-harmonic forms is similar.

The second assertion follows from the first and the fact that on a Kähler manifold $\Delta_{\partial} = \Delta_{\bar{\partial}} = \frac{1}{2}\Delta$ (see Proposition 3.1.12). $\qquad\square$

There are a number of useful facts about the spaces of harmonic forms, which can be easily deduced from standard results about harmonic forms on (compact) Riemannian manifolds and the linear algebra on hermitian (Kähler) manifolds. We collect some of them in the following

Remarks 3.2.7 i) Let (X, g) be an hermitian manifold, not necessarily compact. Then the Hodge $*$-operator induces $\mathbb{C}$-linear isomorphism

$$* : \mathcal{H}^{p,q}(X,g) \cong \mathcal{H}^{n-q,n-p}(X,g).$$

Indeed, $*$ induces isomorphisms $\mathcal{H}^k(X,g)_{\mathbb{C}} \cong \mathcal{H}^{2n-k}(X,g)_{\mathbb{C}}$ for any Riemannian manifold (cf. Lemma A.0.14) and $* : \mathcal{A}^{p,q}(X) \to \mathcal{A}^{n-q,n-p}(X)$ on any hermitian manifold. Similarly, the Hodge $*$-operator induces isomorphisms

$$* : \mathcal{H}_{\bar{\partial}}^{p,q}(X,g) \cong \mathcal{H}_{\partial}^{n-q,n-p}(X,g).$$

If X is Kähler, this yields automorphisms of $\mathcal{H}_{\bar{\partial}}^{*,*}(X,g)$. Also note that complex conjugation interchanges $\mathcal{H}_{\bar{\partial}}^{p,q}(X,g)$ and $\mathcal{H}_{\partial}^{q,p}(X,g)$.

ii) Let (X, g) be a compact connected hermitian manifold. Then the pairing

$$\mathcal{H}_{\bar{\partial}}^{p,q}(X,g) \times \mathcal{H}_{\bar{\partial}}^{n-p,n-q}(X,g) \longrightarrow \mathbb{C}, \quad (\alpha,\beta) \longmapsto \int_X \alpha \wedge \beta$$

is non-degenerate. Indeed, if $0 \neq \alpha \in \mathcal{H}_{\bar{\partial}}^{p,q}(X,g)$ then $*\bar{\alpha} \in \mathcal{H}_{\bar{\partial}}^{n-p,n-q}(X,g)$ and

$$\int \alpha \wedge *\bar{\alpha} = \|\alpha\|^2 > 0.$$

This yields *Serre duality* on the level of harmonic forms:

$$\mathcal{H}_{\bar{\partial}}^{p,q}(X,g) \cong \mathcal{H}_{\bar{\partial}}^{n-p,n-q}(X)^*.$$

See Section 4.1 for a more general result.

iii) If (X, g) is a (possibly non-compact) Kähler manifold of dimension n, then for any $k \leq n$ and any $0 \leq p \leq k$ the Lefschetz operator defines an isomorphism

$$L^{n-k} : \mathcal{H}^{p,k-p}(X,g) \cong \mathcal{H}^{n+p-k,n-p}(X,g).$$

Here one uses $[L, \Delta] = 0$ (see iii) of Proposition 3.1.12), which shows that L maps harmonic forms to harmonic forms. Since L^{n-k} by Proposition 1.2.30 is bijective on $\mathcal{A}^k(X)$, the induced map $\mathcal{H}^{p,k-p}(X,g) \to \mathcal{H}^{n+p-k,n-p}(X,g)$ is injective. The surjectivity is deduced from the fact that the dual Lefschetz operator Λ also commutes with Δ and, therefore, maps harmonic forms to harmonic forms. Note that here we cannot expect anything for arbitrary hermitian manifolds, as the product of an harmonic form with the Kähler form will hardly be closed (let alone harmonic) if the fundamental form is not closed.

Of course, one would very much like to have similar results for the cohomology groups. For this we need the following fundamental result which is analogous to the Hodge decomposition on compact oriented Riemannian manifolds. Unfortunately, its proof cannot be reduced to the known one, but it uses essentially the same techniques (cf. [116]).

Theorem 3.2.8 (Hodge decomposition) *Let (X, g) be a compact hermitian manifold. Then there exist two natural orthogonal decompositions*

$$\mathcal{A}^{p,q}(X) = \partial \mathcal{A}^{p-1,q}(X) \oplus \mathcal{H}_{\partial}^{p,q}(X, g) \oplus \partial^* \mathcal{A}^{p+1,q}(X).$$

and

$$\mathcal{A}^{p,q}(X) = \bar{\partial} \mathcal{A}^{p,q-1}(X) \oplus \mathcal{H}_{\bar{\partial}}^{p,q}(X, g) \oplus \bar{\partial}^* \mathcal{A}^{p,q+1}(X).$$

The spaces $\mathcal{H}^{p,q}(X, g)$ are finite-dimensional. If (X, g) is assumed to be Kähler then $\mathcal{H}_{\partial}^{p,q}(X, g) = \mathcal{H}_{\bar{\partial}}^{p,q}(X, g)$. $\qquad\square$

The orthogonality of the decomposition is easy to verify and the last assertion follows from iii) in Proposition 3.1.12. The crucial fact is the existence of the direct sum decomposition.

Corollary 3.2.9 *Let (X, g) be a compact hermitian manifold. Then the canonical projection $\mathcal{H}_{\bar{\partial}}^{p,q}(X, g) \to H^{p,q}(X)$ is an isomorphism.*

Proof. Since any $\alpha \in \mathcal{H}_{\bar{\partial}}^{p,q}(X, g)$ is $\bar{\partial}$-closed, mapping α to its Dolbeault cohomology class $[\alpha] \in H^{p,q}(X)$ defines a map $\mathcal{H}_{\bar{\partial}}^{p,q}(X, g) \to H^{p,q}(X)$.

Moreover, $\mathrm{Ker}(\bar{\partial} : \mathcal{A}^{p,q}(X) \to \mathcal{A}^{p,q+1}(X)) = \bar{\partial}(\mathcal{A}^{p,q-1}(X)) \oplus \mathcal{H}_{\bar{\partial}}^{p,q}(X, g)$, as $\bar{\partial}\bar{\partial}^* \beta = 0$ if and only if $\bar{\partial}^* \beta = 0$. Indeed, $\bar{\partial}\bar{\partial}^* \beta = 0$ implies $0 = (\bar{\partial}\bar{\partial}^* \beta, \beta) = \|\bar{\partial}^* \beta\|^2$. Thus, $\mathcal{H}_{\bar{\partial}}^{p,q}(X, g) \to H^{p,q}(X)$ is an isomorphism. $\qquad\square$

The strength of the Kähler condition is that the space of harmonic (p, q)-forms appears in two different orthogonal decompositions. The following corollary is a prominent example for the use of this fact. Although it looks like a rather innocent technical statement, it is crucial for many important results.

Corollary 3.2.10 ($\partial\bar{\partial}$-lemma) *Let X be a compact Kähler manifold. Then for a d-closed form α of type (p, q) the following conditions are equivalent:*
  i) *The form α is d-exact, i.e. $\alpha = d\beta$ for some $\beta \in \mathcal{A}_{\mathbb{C}}^{p+q-1}(X)$.*
  ii) *The form α is ∂-exact, i.e. $\alpha = \partial\beta$ for some $\beta \in \mathcal{A}^{p-1,q}(X)$.*
  iii) *The form α is $\bar{\partial}$-exact, i.e. $\alpha = \bar{\partial}\beta$ for some $\beta \in \mathcal{A}^{p,q-1}(X)$.*
  iv) *The form α is $\partial\bar{\partial}$-exact, i.e. $\alpha = \partial\bar{\partial}\beta$ for some $\beta \in \mathcal{A}^{p-1,q-1}(X)$.*

Proof. We add another equivalent condition: v) The form α is orthogonal to $\mathcal{H}^{p,q}(X,g)$ for an arbitrary Kähler metric g on X. Since X is Kähler, we don't have to specify with respect to which differential operator (d, ∂, or $\bar{\partial}$) harmonicity is considered.

Using Hodge decomposition we see that v) is implied by any of the other conditions. Moreover, iv) implies i) - iii). Thus, it suffices to show that v) implies iv).

If $\alpha \in \mathcal{A}^{p,q}(X)$ is d-closed (and thus ∂-closed) and orthogonal to the space of harmonic forms, then Hodge decomposition with respect to ∂ yields $\alpha = \partial\gamma$. Now one applies Hodge decomposition with respect to $\bar{\partial}$ to the form γ. This yields $\gamma = \bar{\partial}\beta + \bar{\partial}^*\beta' + \beta''$ for some harmonic β''. Thus, $\alpha = \partial\bar{\partial}\beta + \partial\bar{\partial}^*\beta'$. Using $\partial\bar{\partial}^* = -\bar{\partial}^*\partial$ (cf. the proof of iii) in Proposition 3.1.12) and $\bar{\partial}\alpha = 0$ one concludes $\bar{\partial}\bar{\partial}^*\partial\beta' = 0$. Since $(\bar{\partial}\bar{\partial}^*\partial\beta', \partial\beta') = \|\bar{\partial}^*\partial\beta'\|^2$, we find $\partial\bar{\partial}^*\beta' = -\bar{\partial}^*\partial\beta' = 0$. Therefore, $\alpha = \partial\bar{\partial}\beta$. $\qquad\square$

Remark 3.2.11 A local version of the $\partial\bar{\partial}$-lemma, i.e. for X a polydisc, was discussed in Exercise 1.3.3. There, no Kähler condition was needed. It is surprising to see that the Kähler condition allows to carry the local statement over to the global situation. Observe that the result holds true also for $(p,q) = (1,0)$ or $(0,1)$ (cf. Exercise 3.2.8).

Corollary 3.2.12 *Let (X,g) be a compact Kähler manifold. Then there exists a decomposition*

$$H^k(X,\mathbb{C}) = \bigoplus_{p+q=k} H^{p,q}(X).$$

This decomposition does not depend on the chosen Kähler structure.

Moreover, with respect to complex conjugation on $H^*(X,\mathbb{C}) = H^*(X,\mathbb{R}) \otimes \mathbb{C}$ one has $\overline{H^{p,q}(X)} = H^{q,p}(X)$ and Serre duality (see page 127) yields $H^{p,q}(X) \cong H^{n-p,n-q}(X)^*$.

Proof. The decomposition is induced by

$$H^k(X,\mathbb{C}) = \mathcal{H}^k(X,g)_{\mathbb{C}} = \bigoplus_{p+q=k} \mathcal{H}^{p,q}(X,g) = \bigoplus_{p+q=k} H^{p,q}(X),$$

which *a priori* might depend on the Kähler metric g. Let g' be a second Kähler metric. The two groups $\mathcal{H}^{p,q}(X,g)$ and $\mathcal{H}^{p,q}(X,g')$ are naturally identified by $\mathcal{H}^{p,q}(X,g) \cong H^{p,q}(X) \cong \mathcal{H}^{p,q}(X,g')$. Let $\alpha \in \mathcal{H}^{p,q}(X,g)$ and denote the corresponding element in $\mathcal{H}^{p,q}(X,g')$ by α'. We have to show that the associated de Rham cohomology classes $[\alpha], [\alpha'] \in H^k(X,\mathbb{C})$ coincide.

Since α and α' induce the same element in $H^{p,q}(X)$, they differ by some $\bar{\partial}\gamma$, i.e. $\alpha' = \alpha + \bar{\partial}\gamma$. But then $d\bar{\partial}\gamma = d(\alpha' - \alpha) = 0$. Moreover, $\bar{\partial}\gamma$ is orthogonal to $\mathcal{H}^k(X,g)_{\mathbb{C}}$. By Hodge decomposition with respect to the ordinary exterior differential d this yields $\bar{\partial}\gamma \in d(\mathcal{A}_{\mathbb{C}}^{k-1}(X))$. Thus, $[\alpha] = [\alpha'] \in H^k(X,\mathbb{C})$.

For the second assertion use Remarks 3.2.7 i) and iii). $\qquad\square$

Remark 3.2.13 For the above corollary the Kähler assumption is absolutely crucial, although in dimension two, i.e. for compact complex surfaces, the assertion always holds true (cf. [8]).

Definition 3.2.14 Let X be a compact Kähler manifold. The *Kähler class* associated to a Kähler structure on X is the cohomology class $[\omega] \in H^{1,1}(X)$ of its Kähler form. The *Kähler cone*

$$\mathcal{K}_X \subset H^{1,1}(X) \cap H^2(X, \mathbb{R})$$

is the set of all Kähler classes associated to any Kähler structure on X.

Hodge theory on a compact Kähler manifold X allows one to view $\mathcal{K}_X$ as an open convex cone in $H^{1,1}(X) \cap H^2(X, \mathbb{R})$ (cf. Exercise 3.2.12).

As noted above, the bidegree decomposition $H^k(X, \mathbb{C}) = \bigoplus H^{p,q}(X)$ and, of course, complex conjugation $\overline{H^{p,q}}(X) = H^{q,p}(X)$ does not depend on the chosen Kähler structure on X. In the next section we shall discuss the remaining operators $*$, L, and Λ on the level of cohomology. It will turn out that they only depend on the Kähler class $[\omega]$.

Exercises

3.2.1 Let (X, g) be a Kähler manifold. Show that the Kähler form ω is harmonic.

3.2.2 Work out the details of the proof of the second assertion of Corollary 3.2.12. In particular, show that for a compact hermitian manifold (X, g) of dimension n there exists a natural isomorphism $H^{p,q}(X) \cong H^{n-p,n-q}(X)^*$ (use ii) of Remark 3.2.7). This is a special case of Serre duality, which holds true more generally for Dolbeault cohomolgy of holomorphic vector bundles (cf. Proposition 4.1.15).

3.2.3 Let X be a compact Kähler manifold X of dimension n. Let $H^{p,q}(X) \cong H^{n-p,n-q}(X)^*$ be given by Serre duality. Observe that the direct sum decomposition of these isomorphisms yields Poincaré duality

$$H^k(X, \mathbb{C}) = \bigoplus_{p+q=k} H^{p,q}(X) \cong \bigoplus_{p+q=k} H^{n-p,n-q}(X)^* = H^{2n-k}(X, \mathbb{C})^*.$$

3.2.4 Recall Exercise 2.6.11 and show that on a compact Kähler manifold "the limit $\lim_{t \to 0}$ commutes with hypercohomology", i.e.

$$\lim_{t \to 0} \mathbb{H}^k(X, (\Omega_X^*, t \cdot d)) = \mathbb{H}^k(X, \lim_{t \to 0}(\Omega_X^*, t \cdot d)).$$

3.2.5 Show that for a complex torus of dimension one the decomposition in Corollary 3.2.12 does depend on the complex structure. It suffices to consider H^1.

3.2.6 Show that the odd Betti numbers b_{2i+1} of a compact Kähler manifold are even.

3.2.7   Are Hopf surfaces (cf. Section 2.1) Kähler manifolds?

3.2.8   Show that holomorphic forms, i.e. elements of $H^0(X, \Omega^p)$, on a compact Kähler manifold X are harmonic with respect to any Kähler metric.

3.2.9   Can you deduce Theorem 3.2.8 for compact Kähler manifolds from the Hodge decomposition for compact oriented Riemannian manifolds and the $\partial\bar{\partial}$-lemma?

3.2.10   Let (X, g) be a compact hermitian manifold. Show that any $(d\text{-})$harmonic (p, q)-form is also $\bar{\partial}$-harmonic.

3.2.11   Show that $H^{p,q}(\mathbb{P}^n) = 0$ except for $p = q \leq n$. In the latter case, the space is one-dimensional. Use this and the exponential sequence to show that $\mathrm{Pic}(\mathbb{P}^n) \cong \mathbb{Z}$.

3.2.12   Let X be a compact Kähler manifold and consider $H^{1,1}(X)$ as a subspace of $H^2(X, \mathbb{C})$. Show that the Kähler $\mathcal{K}_X$ cone is an open convex cone in $H^{1,1}(X, \mathbb{R}) := H^{1,1}(X) \cap H^2(X, \mathbb{R})$ and that $\mathcal{K}_X$ not contain any line $\{\alpha + t\beta \mid t \in \mathbb{R}\}$ for any $\alpha, \beta \in H^{1,1}(X, \mathbb{R})$ with $\beta \neq 0$.

Furthermore, show that $t\alpha + \beta$ is a Kähler class for $t \gg 0$ for any Kähler class α and any β.

3.2.13   Prove the dd^c-lemma: If $\alpha \in \mathcal{A}^k(X)$ is a d^c-exact and d-closed form on a compact Kähler manifold X then there exists a form $\beta \in \mathcal{A}^{k-2}(X)$ such that $\alpha = dd^c\beta$. (A proof will be given in Lemma 3.A.22.)

3.2.14   Let X be compact and Kähler. Show that the two natural homomorphisms $H^{p,q}_{\mathrm{BC}}(X) \to H^{p,q}(X)$ and $\bigoplus_{p+q=k} H^{p,q}_{\mathrm{BC}}(X) \to H^k(X, \mathbb{C})$ introduced in Exercise 2.6.7 are bijective.

Use this to show again that the bidegree decomposition in Corollary 3.2.12 is independent of the Kähler structure.

(Thus the main difference between Dolbeault and Bott–Chern cohomology is that a natural homomorphism from Bott–Chern cohomology to de Rham cohomology always exists, whereas Dolbeault cohomology groups only occur as the graded modules of the filtration induced by the so called Hodge–Fröhlicher spectral sequence $E_1^{p,q} = H^{p,q}(X) \Rightarrow H^{p+q}(X, \mathbb{C})$.)

3.2.15   Let (X, g) be a compact hermitian manifold and let $[\alpha] \in H^{p,q}(X)$. Show that the harmonic representative of $[\alpha]$ is the unique $\bar{\partial}$-closed form with minimal norm $\|\alpha\|$. (This is the analogue of Lemma A.0.18.)

3.2.16   Let X be a compact Kähler manifold. Show that for two cohomologous Kähler forms ω and ω', i.e. $[\omega] = [\omega'] \in H^2(X, \mathbb{R})$, there exists a real function f such that $\omega = \omega' + i\partial\bar{\partial}f$.

Comments: - For the analysis behind Hodge theory see [35] or [116].
- Be careful with the $\partial\bar{\partial}$-lemma. It is frequently stated incorrectly.

3.3 Lefschetz Theorems

We continue to transfer results valid on the level of forms to cohomology. The main ingredients are again Hodge decomposition (Theorem 3.2.8) and the Kähler identities (Proposition 3.1.12). We will see that the cohomology of a compact Kähler manifold enjoys many interesting properties. Roughly, everything we have observed for the exterior algebra of an hermitian vector space in Section 1.2 continues to hold true for the cohomology of a compact Kähler manifold.

At the end we state and briefly discuss the Hodge conjecture, one of the most famous open problems in mathematics.

Let us first explain the relation between the Picard group of a compact Kähler manifold X and the cohomology group $H^{1,1}(X)$. Recall that the exponential sequence is a natural short exact sequence of the form (see Section 2.2)

$$0 \longrightarrow \mathbb{Z} \longrightarrow \mathcal{O} \longrightarrow \mathcal{O}^* \longrightarrow 0.$$

It in particular yields a canonical boundary map $\mathrm{Pic}(X) \cong H^1(X, \mathcal{O}^*) \to H^2(X, \mathbb{Z})$ which can be composed with $H^2(X, \mathbb{Z}) \to H^2(X, \mathbb{C})$ induced by the inclusion $\mathbb{Z} \subset \mathbb{C}$. Moreover, if (X, g) is a compact Kähler manifold, the target space decomposes $H^2(X, \mathbb{C}) = H^{2,0}(X) \oplus H^{1,1}(X) \oplus H^{0,2}(X)$ (see Corollary 3.2.12).

Let $\alpha \in H^2(X, \mathbb{C})$. Then we can associate to α a class in $H^{0,2}(X, \mathbb{C}) = H^2(X, \mathcal{O}_X)$ in two, *a priori* different ways: Firstly, using the map $H^2(X, \mathbb{C}) \to H^2(X, \mathcal{O}_X)$ induced by the inclusion $\mathbb{C} \subset \mathcal{O}_X$ and, secondly, by the projection $H^2(X, \mathbb{C}) \to H^{0,2}(X)$ given by the bidegree decomposition 3.2.12. For the latter one we assume X be compact and Kähler. These two maps coincide, as is shown by the following lemma. See also Exercise 3.3.1.

Lemma 3.3.1 *Let X be a compact Kähler manifold. The two natural maps $H^k(X, \mathbb{C}) \to H^k(X, \mathcal{O}_X)$, induced by $\mathbb{C} \subset \mathcal{O}_X$, and $H^k(X, \mathbb{C}) \to H^{0,k}(X)$, given by the bidegree decomposition, coincide.*

Proof. We use the following commutative diagram that relates the standard acyclic resolutions of $\mathbb{C}$ and $\mathcal{O}_X$.

$$
\begin{array}{ccccccc}
\mathbb{C} & \longrightarrow & \mathcal{A}_{\mathbb{C}}^0(X) & \xrightarrow{\ d\ } & \mathcal{A}_{\mathbb{C}}^1(X) & \xrightarrow{\ d\ } & \mathcal{A}_{\mathbb{C}}^2(X) & \cdots \\
\downarrow & & \downarrow{\scriptstyle =} & & \downarrow{\scriptstyle \Pi^{0,1}} & & \downarrow{\scriptstyle \Pi^{0,2}} & \\
\mathcal{O}_X & \longrightarrow & \mathcal{A}_{\mathbb{C}}^0(X) & \xrightarrow{\ \bar{\partial}\ } & \mathcal{A}^{0,1}(X) & \xrightarrow{\ \bar{\partial}\ } & \mathcal{A}^{0,2}(X) & \cdots
\end{array}
$$

Here, the first vertical map is the natural inclusion $\mathbb{C} \subset \mathcal{O}$ and the other ones are given by projecting a q-form to its $(0,q)$-part. The commutativity of the diagram is obvious. Using it one can describe $H^k(X, \mathbb{C}) \to H^k(X, \mathcal{O})$ induced by the inclusion $\mathbb{C} \subset \mathcal{O}_X$ in terms of explicit representatives as follows:

For a class $[\alpha] \in H^k(X, \mathbb{C})$ we choose its harmonic representative $\alpha \in \mathcal{H}^k(X, g)$ (with respect to some Kähler metric g). Its image is represented by the $(0, k)$-part of it, which is again harmonic. Of course, this describes nothing but the projection given by the bidegree decomposition. $\qquad\square$

Warning: For $p \neq 0$ the diagram

$$
\begin{array}{ccc}
\mathcal{A}_\mathbb{C}^k(X) & \xrightarrow{\ d\ } & \mathcal{A}_\mathbb{C}^{k+1}(X) \\
\Big\downarrow{\scriptstyle \Pi^{p,k-p}} & & \Big\downarrow{\scriptstyle \Pi^{p,k-p+1}} \\
\mathcal{A}^{p,k-p}(X) & \xrightarrow{\ \bar\partial\ } & \mathcal{A}^{p,k-p+1}(X)
\end{array}
$$

does not commute anymore.

The long exact cohomology sequence of the exponential sequence shows that the composition $\mathrm{Pic}(X) \to H^2(X, \mathbb{Z}) \to H^2(X, \mathbb{C}) \to H^2(X, \mathcal{O}_X) = H^{0,2}(X)$ is trivial. Hence, the image of $\mathrm{Pic}(X) \to H^2(X, \mathbb{C})$ is contained in the image of $H^2(X, \mathbb{Z}) \to H^2(X, \mathbb{C})$ and, if X is compact and Kähler, also in the kernel of the projection $H^2(X, \mathbb{C}) = H^{2,0}(X) \oplus H^{1,1}(X) \oplus H^{0,2}(X) \to H^{0,2}(X)$. Since $H^2(X, \mathbb{R}) \subset H^2(X, \mathbb{C})$ is invariant under complex conjugation and contains the image of $H^2(X, \mathbb{Z}) \to H^2(X, \mathbb{C})$, one finds that the image of $\mathrm{Pic}(X) \to H^2(X, \mathbb{C})$ is contained in

$$
H^{1,1}(X, \mathbb{Z}) := \mathrm{Im}\left(H^2(X, \mathbb{Z}) \longrightarrow H^2(X, \mathbb{C}) \right) \cap H^{1,1}(X).
$$

As it turns out, $H^{1,1}(X, \mathbb{Z})$ describes the image completely. This is due to

Proposition 3.3.2 (Lefschetz theorem on $(1,1)$-classes) *Let X be a compact Kähler manifold. Then $\mathrm{Pic}(X) \to H^{1,1}(X, \mathbb{Z})$ is surjective.*

Proof. Let $\alpha = \rho(\tilde\alpha) \in \mathrm{Im}(H^2(X, \mathbb{Z}) \xrightarrow{\rho} H^2(X, \mathbb{C}))$. Then, with respect to the bidegree decomposition of $H^2(X, \mathbb{C})$, one can write α as $\alpha = \alpha^{2,0} + \alpha^{1,1} + \alpha^{0,2}$. Since α is real, one has $\alpha^{2,0} = \overline{\alpha^{0,2}}$. Thus, $\alpha \in H^{1,1}(X, \mathbb{Z})$ if and only if $\alpha^{0,2} = 0$.

We have seen that the projection $\alpha \mapsto \alpha^{0,2}$ is induced by the inclusion map $\mathbb{C} \subset \mathcal{O}$. Thus, $\alpha = \rho(\tilde\alpha) \in H^2(X, \mathbb{C})$ is in the image of $\mathrm{Pic}(X) \to H^2(X, \mathbb{C})$ if and only if $\tilde\alpha$ is in the kernel of the map $H^2(X, \mathbb{Z}) \to H^2(X, \mathbb{C}) \to H^2(X, \mathcal{O})$. Since the exponential sequence induces a long exact sequence of the form

$$
\cdots \longrightarrow \mathrm{Pic}(X) \longrightarrow H^2(X, \mathbb{Z}) \longrightarrow H^2(X, \mathcal{O}) \longrightarrow \cdots,
$$

this proves the assertion. $\qquad\square$

Remark 3.3.3 Often, the image of $\mathrm{Pic}(X) \to H^2(X, \mathbb{R}) \subset H^2(X, \mathbb{C})$ is called the *Neron–Severi group* $\mathrm{NS}(X)$ of the manifold X. It spans a finite-dimensional real vector space $\mathrm{NS}(X)_\mathbb{R} \subset H^2(X, \mathbb{R}) \cap H^{1,1}(X)$, where the inclusion is strict in general. The Lefschetz theorem above thus says that the natural inclusion $\mathrm{NS}(X) \subset H^{1,1}(X,\mathbb{Z})$ is an equality.

If X is projective, yet another description of the Neron–Severi group can be given. Then, $\mathrm{NS}(X)$ is the quotient of $\mathrm{Pic}(X)$ by the subgroup of numerically trivial line bundles. A line bundle L is called *numerically trivial* if L is of degree zero on any curve $C \subset X$. If X is not projective it may very well happen that there are no curves in X, but yet $\mathrm{NS}(X) \neq 0$.

Definition 3.3.4 Let X be a compact complex manifold. Then the rank of the image $\mathrm{Pic}(X) \to H^2(X,\mathbb{R})$ is called the *Picard number* $\rho(X)$.

Thus, if X is in addition Kähler the Picard number satisfies $\rho(X) = \mathrm{rk}(H^{1,1}(X,\mathbb{Z})) = \mathrm{rk}(\mathrm{NS}(X))$.

Modulo torsion in $H^2(X,\mathbb{Z})$, the image of $c_1 : \mathrm{Pic}(X) \to H^2(X,\mathbb{Z})$ can be expressed, at least if X is compact Kähler, in purely Hodge theoretic terms (see Section 3.C). Let us now turn to the continuous part of the Picard group, i.e. the kernel of $\mathrm{Pic}(X) \to H^2(X,\mathbb{Z})$. We shall see that a structure theorem for it can be proved, but we again have to assume that X is compact Kähler. Let us first give a name to the continuous part.

Definition 3.3.5 Let X be a complex manifold. Its *Jacobian* $\mathrm{Pic}^0(X)$ is the kernel of the map $\mathrm{Pic}(X) \to H^2(X,\mathbb{Z})$.

Using the exponential sequence the Jacobian can always be described as the quotient $H^1(X,\mathcal{O})/H^1(X,\mathbb{Z})$, but only for compact Kähler manifolds one should expect this to be anything nice. Note that for compact manifolds the natural map $H^1(X,\mathbb{Z}) \to H^1(X,\mathcal{O})$ is really injective. (See the argument in Section 2.2.)

Corollary 3.3.6 *If X is a compact Kähler manifold, then $\mathrm{Pic}^0(X)$ is in a natural way a complex torus of dimension $b_1(X)$.*

Proof. We use the bidegree decomposition (Corollary 3.2.12): $H^1(X,\mathbb{C}) = H^{1,0}(X) \oplus H^{0,1}(X)$, the fact that $\overline{H^{1,0}(X)} = H^{0,1}(X)$, and $H^1(X,\mathbb{C}) = H^1(X,\mathbb{R}) \otimes_{\mathbb{R}} \mathbb{C} = (H^1(X,\mathbb{Z}) \otimes_{\mathbb{Z}} \mathbb{R}) \otimes_{\mathbb{R}} \mathbb{C}$. This shows that $H^1(X,\mathbb{Z}) \to H^1(X,\mathbb{C}) \to H^{0,1}(X)$ is injective with discrete image that generates $H^{0,1}(X)$ as a real vector space. In other words, $H^1(X,\mathbb{Z}) \subset H^{0,1}(X)$ is a lattice. Thus, it suffices to show that this inclusion $H^1(X,\mathbb{Z}) \subset H^{0,1}(X)$ coincides with the one given by the exponential sequence. But this is immediate from Lemma 3.3.1. □

Clearly, the pre-image $\mathrm{Pic}^{\alpha}(X) \subset \mathrm{Pic}(X)$ of any element $\alpha \in H^2(X,\mathbb{Z})$ in the image of c_1 can be identified with $\mathrm{Pic}^0(X)$, although not canonically. Thus, for a compact Kähler manifold X such that $H^2(X,\mathbb{Z})$ is torsion-free the Picard group $\text{Pic}(X)$ is fibred by tori of complex dimension $b_1(X)$ over the discrete set $H^{1,1}(X, \mathbb{Z})$:

$$
\begin{array}{ccc}
\text{Pic}(X) & \longrightarrow\!\!\!\!\!\longrightarrow & H^{1,1}(X, \mathbb{Z}) \\
\uparrow & & \uparrow \\
\text{Pic}^\alpha(X) & \longrightarrow & \{\alpha\}
\end{array}
$$

There is another complex torus naturally associated with any compact Kähler manifold, the Albanese torus. Suppose X is compact and Kähler. Consider the dual of its space of global holomorphic one-forms $H^{1,0}(X) = H^0(X, \Omega_X)$ and the natural map $H_1(X, \mathbb{Z}) \to H^0(X, \Omega_X)^*$ given by $[\gamma] \mapsto (\alpha \mapsto \int_\gamma \alpha)$, where a homology class $[\gamma]$ is represented by a closed path γ in X. Since X is Kähler, the image forms a lattice. One way to see this, is to view $H^0(X, \Omega_X)^*$, via Serre duality, as $H^{n-1,n}(X) = H^n(X, \Omega_X^{n-1})$ and $H_1(X, \mathbb{Z}) \subset H^0(X, \Omega_X)^*$ as the image of $H^{2n-1}(X, \mathbb{Z}) \subset H^{2n-1}(X, \mathbb{R})$ under the natural projection $H^{2n-1}(X, \mathbb{C}) \to H^{n-1,n}(X)$.

Definition 3.3.7 The *Albanese torus* of a compact Kähler manifold X is the complex torus

$$\text{Alb}(X) = H^0(X, \Omega_X)^*/H_1(X, \mathbb{Z}).$$

For a fixed base point $x_0 \in X$ one defines the *Albanese map*

$$\text{alb} : X \longrightarrow \text{Alb}(X), \quad x \longmapsto \left(\alpha \longmapsto \int_{x_0}^x \alpha\right).$$

The integral $\int_{x_0}^x \alpha$ depends on the chosen path connecting x_0 and x, but for two different choices the difference is an integral over a closed path γ. Hence, $\text{alb}(x) \in \text{Alb}(X)$ is well-defined. Changing the point x_0 amounts to a translation in $\text{Alb}(X)$.

By construction, one has the following

Proposition 3.3.8 *The Albanese map* $\text{alb} : X \to \text{Alb}(X)$ *is holomorphic and the pull-back of forms induces a bijection*

$$\bigwedge\nolimits_0 \text{Alb}(X) \cong H^0(\text{Alb}(X), \Omega_{\text{Alb}(X)}) \cong H^0(X, \Omega_X).$$

The Albanese map is functorial for holomorphic maps, i.e. any holomorphic map $f : X \to X'$ *induces a commutative diagram*

$$
\begin{array}{ccc}
X & \xrightarrow{\;f\;} & X' \\
{\scriptstyle \text{alb}_X}\big\downarrow & & \big\downarrow{\scriptstyle \text{alb}_{X'}} \\
\text{Alb}(X) & \longrightarrow & \text{Alb}(X')
\end{array}
$$

Here, one has to choose the base point of X' *as* $f(x_0)$. *The map on the bottom is induced by the canonical linear map* $H^0(X', \Omega_{X'}) \to H^0(X, \Omega_X)$.

Proof. Writing the integral $\int_{x_0}^{x} \alpha$ as $\int_{x_0}^{x_1} \alpha + \int_{x_1}^{x} \alpha$, the holomorphicity becomes a local question for x near a reference point x_1. The assertion, therefore, is equivalent to the holomorphicity of $\int_{0}^{x} \alpha$ as a function of $x \in B$, where α is a holomorphic one-form on a polydisc $B \subset \mathbb{C}^n$, which can be deduced from a power series expansion.

The second assertion can similarly be proved by a local argument. The functoriality is a consequence of the formula $\int_{x_0}^{x} f^*\alpha = \int_{f(x_0)}^{f(x)} \alpha$ for any path γ connecting x_0 with x in X. $\square$

Examples 3.3.9 i) If X is of dimension one, i.e. X is a curve, then $\mathrm{Alb}(X) \cong \mathrm{Pic}^0(X)$. Indeed, Serre duality shows $H^0(X, \Omega_X)^* \cong H^1(X, \mathcal{O}_X)$. Moreover, the embeddings $H_1(X, \mathbb{Z}) \subset H^0(X, \Omega_X)^*$ and $H^1(X, \mathbb{Z}) \subset H^1(X, \mathcal{O}_X)$ are compatible with the natural isomorphism $H_1(X, \mathbb{Z}) \cong H^1(X, \mathbb{Z})$.

Furthermore, the Albanese map $\mathrm{alb} : X \to \mathrm{Alb}(X)$ coincides with the Abel–Jacobi map $X \to \mathrm{Pic}(X)$, $x \mapsto \mathcal{O}(x - x_0)$ defined in Section 2.3.

ii) Let X be a complex torus given as V/Γ with Γ a lattice inside the complex vector space V. Hence, there exist natural isomorphisms $T_0X \cong V$ and $H^0(X, \Omega_X) \cong V^*$. Moreover, $H_1(X, \mathbb{Z}) \cong \Gamma$ and the embedding $H_1(X, \mathbb{Z}) \to H^0(X, \Omega_X)^*$ is the given inclusion $\Gamma \subset V$. With $x_0 = 0$ the Albanese map $X \to \mathrm{Alb}(X) = V/\Gamma$ is the identity.

Without representing a complex torus X as a quotient V/Γ, the Albanese map enables us to write X, canonically up to the choice of a base point x_0, as $X \cong H^0(X, \Omega_X)^*/H_1(X, \mathbb{Z})$. Moreover, from the functoriality of the Albanese map one deduces that any holomorphic map $f : X \to X'$ between complex tori X and X' is described by the dual of the pull-back map $H^0(X, \Omega_X)^* \to H^0(X', \Omega_{X'})^*$ divided out by $f_* : H_1(X, \mathbb{Z}) \to H_1(X', \mathbb{Z})$. The base point of X' needs to be of the form $f(x)$.

Let us now discuss the higher cohomology groups of a compact Kähler manifold.

Corollary 3.3.10 *Let (X, g) be a compact Kähler manifold. Then the Lefschetz operator L and its dual Λ define natural operators on cohomology*

$$L : H^{p,q}(X) \longrightarrow H^{p+1,q+1}(X) \quad and \quad \Lambda : H^{p,q}(X) \longrightarrow H^{p-1,q-1}(X).$$

Moreover, L and Λ only depend on the Kähler class $[\omega] \in \mathcal{K}_X \subset H^2(X, \mathbb{R})$.

Proof. This is a consequence of iii), Remark 3.2.7 and the bidegree decomposition 3.2.12. The last assertion follows from the observation that L on cohomology is given by the exterior product with the Kähler class $[\omega]$. $\square$

Definition 3.3.11 Let (X, g) be a compact Kähler manifold. Then the *primitive cohomology* is defined as

$$H^k(X, \mathbb{R})_{\mathrm{p}} := \mathrm{Ker} \left(\Lambda : H^k(X, \mathbb{R}) \longrightarrow H^{k-2}(X, \mathbb{R}) \right)$$

and

$$H^{p,q}(X)_{\mathrm{p}} := \mathrm{Ker} \left(\Lambda : H^{p,q}(X) \longrightarrow H^{p-1,q-1}(X) \right)$$

Remark 3.3.12 It is important to note that the primitive cohomology does not really depend on the chosen Kähler structure, but only on the Kähler class $[\omega] \in H^{1,1}(X, \mathbb{R})$. Thus, one usually has an infinite-dimensional set of Kähler forms all inducing the same primitive cohomology (see Corollary 3.1.8). But once a Kähler structure realizing the given Kähler class is chosen, any class in $H^k(X, \mathbb{R})_{\mathrm{p}}$ (or, $H^{p,q}(X)_{\mathrm{p}}$) can be realized by an harmonic form that is primitive at every point (although possibly zero at some points).

Proposition 3.3.13 (Hard Lefschetz theorem) *Let (X, g) be a compact Kähler manifold of dimension n. Then for $k \leq n$*

$$L^{n-k} : H^k(X, \mathbb{R}) \cong H^{2n-k}(X, \mathbb{R}) \tag{3.1}$$

and for any k

$$H^k(X, \mathbb{R}) = \bigoplus_{i \geq 0} L^i H^{k-2i}(X, \mathbb{R})_{\mathrm{p}}. \tag{3.2}$$

Moreover, both assertions respect the bidegree decomposition. In particular, $H^k(X, \mathbb{R})_{\mathrm{p}} \otimes \mathbb{C} = \bigoplus_{p+q=k} H^{p,q}(X)_{\mathrm{p}}$.

Proof. The first assertion is iii) of Remark 3.2.7. For the second assertion one either uses the fact that L and Λ define an $\mathfrak{sl}(2)$-representation on $\mathcal{H}^*(X, g)$ and then concludes as in the proof of Proposition 1.2.30 or one applies Corollary 3.1.2. Since L and Λ respect harmonicity, Corollary 3.1.2 then proves the second assertion. The compatibility with the bidegree decomposition follows from the fact that the Kähler class is of type $(1,1)$. See also Remark 1.2.33. $\square$

Another way to phrase Proposition 3.3.13 is to say that any Kähler class on X yields a natural $\mathfrak{sl}(2)$-representation on $H^*(X, \mathbb{R})$. Thus, we have constructed, starting with the finite-dimensional $\mathfrak{sl}(2)$-representation on the exterior algebra of an hermitian vector space, via an infinite-dimensional representation on $\mathcal{A}^*(X)$, another finite-dimensional $\mathfrak{sl}(2)$-representation on the cohomology of a compact Kähler manifold.

Corollary 3.3.14 *The Hodge $*$-operator on a compact Kähler manifold (X, g) acts naturally on cohomology $H^*(X, \mathbb{C})$ inducing isomorphisms $* : H^{p,q}(X) \cong H^{n-q,n-p}(X)$. The action only depends on the Kähler class $[\omega]$.*

Proof. The first assertion follows again from Remark 3.2.7, i). By using Weil's formula 1.2.31 the Hodge ∗-operator can be described in terms of the Lefschetz operator L and the Lefschetz decomposition. In particular, everything depends only on the Kähler class (and the complex structure of X). □

The Hodge numbers $h^{p,q}(X)$ of a compact Kähler manifold X of dimension n are visualized by the *Hodge diamond*:

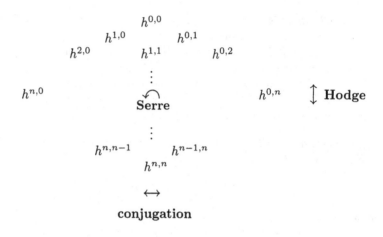

Serre duality (Corollary 3.2.12) shows that the Hodge diamond is invariant under rotation by π. Complex conjugation $\overline{H^{p,q}}(X) = H^{q,p}(X)$ provides another symmetry: Thus, the Hodge diamond is also invariant under reflection in the vertical line passing through $h^{0,0}$ and $h^{n,n}$. The Hodge ∗-operator induces a reflection in the horizontal line passing through $h^{n,0}$ and $h^{0,n}$.

Proposition 3.3.15 (Hodge–Riemann bilinear relation) *Let (X,g) be a compact Kähler manifold of dimension n with Kähler class $[\omega]$ and let $0 \neq \alpha \in H^{p,q}(X)_p$. Then*

$$i^{p-q}(-1)^{\frac{(p+q)(p+q-1)}{2}} \int_X \alpha \wedge \bar{\alpha} \wedge [\omega]^{n-(p+q)} > 0.$$

Proof. Since any $\alpha \in H^{p,q}(X)_p$ can be represented by an harmonic form $\alpha \in \mathcal{H}^{p,q}(X,g)$ which is primitive at any point $x \in X$ (see Remark 3.3.12), the assertion follows from Corollary 1.2.36. (However, since the primitive harmonic form α might be trivial at certain points, we possibly get strict inequality only on a non-empty open subset of X.) □

Before studying the intersection pairing on the middle cohomology of a general compact Kähler manifold, let us consider the surface case (cf. Exercise 1.2.7).

Corollary 3.3.16 (Hodge index theorem) *Let X be a compact Kähler surface, then the intersection pairing*

$$H^2(X,\mathbb{R}) \times H^2(X,\mathbb{R}) \longrightarrow \mathbb{R}, \quad (\alpha,\beta) \longmapsto \int_X \alpha \wedge \beta$$

has index $(2h^{2,0}(X)+1, h^{1,1}(X)-1)$. Restricted to $H^{1,1}(X)$ it is of index $(1, h^{1,1}(X)-1)$.

Proof. By the bidegree decomposition one has the orthogonal splitting

$$H^2(X,\mathbb{R}) = \left((H^{2,0}(X) \oplus H^{0,2}(X)) \cap H^2(X,\mathbb{R})\right) \oplus H^{1,1}(X,\mathbb{R}).$$

For degree reasons, any class in $H^{2,0}(X) \oplus H^{0,2}(X)$ is primitive. Thus, the intersection pairing is positive definite on the first summand. Indeed, if $\alpha = \alpha^{2,0} + \alpha^{0,2} \in H^2(X,\mathbb{R})$, then

$$\int_X \alpha^2 = 2\int_X \alpha^{2,0} \wedge \alpha^{0,2} = 2\int_X \alpha^{2,0} \wedge \overline{\alpha^{2,0}} > 0.$$

According to the Lefschetz decomposition (3.2), the second summand $H^{1,1}(X,\mathbb{R})$ decomposes further into $[\omega]\mathbb{R} \oplus H^{1,1}(X,\mathbb{R})_p$. This splitting is orthogonal, since for a primitive class α the integral $\int_X \omega \wedge \alpha$ vanishes (see also i), Proposition 1.2.30). Clearly, $\int \omega^2 > 0$ and by Hodge–Riemann $\int_X \alpha^2 < 0$ for $\alpha \in H^{1,1}(X,\mathbb{R})_p$. Thus, the intersection pairing is positive definite on $(H^{2,0}(X) \oplus H^{0,2}(X)) \cap H^2(X,\mathbb{R}) \oplus [\omega]\mathbb{R}$ and negative definite on the orthogonal complement $H^{1,1}(X,\mathbb{R})_p$. This proves the claim. $\square$

Remark 3.3.17 The corollary can be used to exclude that certain compact differentiable manifolds of dimension four admit a Kähler structure (cf. Exercise 3.3.3). Moreover, since one can show that any compact complex surface with $b_1(X)$ even admits a Kähler structure (this is a highly non-trivial statement that originally uses classification theory and a detailed investigation of K3 surfaces [101]), there are topological restriction on a simply connected compact four-dimensional manifold to carry even a complex structure. See as well the comments at the end of this section and of Section 2.6.

Let X be a compact Kähler manifold of dimension n. In particular, X is an even dimensional compact oriented differentiable manifold of real dimension $2n$. Therefore, by Poincaré duality, the intersection product defines a non-degenerate pairing on the middle cohomology $H^n(X,\mathbb{R})$. Up to sign it coincides with the Hodge–Riemann pairing, but it has the advantage of being purely topological, i.e. independent of the complex structure and the Kähler class $[\omega]$.

If n is odd then the intersection pairing on the middle cohomology $H^n(X,\mathbb{R})$ is a non-degenerate skew-symmetric pairing. If n is even, e.g. $n = 2m$, then the intersection pairing $H^{2m}(X,\mathbb{R}) \times H^{2m}(X,\mathbb{R}) \to \mathbb{R}$ is a non-degenerate symmetric bilinear form. Its signature $\mathrm{sgn}(X)$ can be computed by the following result, which generalizes the previous corollary.

Corollary 3.3.18 *If a compact Kähler manifold X is of even complex dimension $2m$, then the intersection pairing on the middle cohomology is of signature*

$$\text{sgn}(X) = \sum_{p,q=0}^{2m} (-1)^p h^{p,q}(X),$$

where $h^{p,q}(X) := \dim H^{p,q}(X)$ are the Hodge numbers.

Proof. We use that the Lefschetz decomposition (3.2) is orthogonal with respect to the intersection pairing. This follows from the corresponding statement for the Hodge–Riemann pairing on the exterior algebra of every tangent space (see Exercise 1.2.2) and the above observation that the Hodge–Riemann pairing on the middle cohomology is the intersection pairing possibly up to sign. Let us now fix a Kähler structure on X and consider the induced orthogonal decomposition:

$$
\begin{aligned}
H^{2m}(X,\mathbb{R}) &= \left(\bigoplus_{p+q=2m} H^{p,q}(X) \right)_{\mathbb{R}} \\
&= H^{m,m}(X,\mathbb{R}) \oplus \bigoplus_{\substack{p+q=2m \\ p>q}} (H^{p,q}(X) \oplus H^{q,p}(X))_{\mathbb{R}} \\
&= \bigoplus_{j\geq 0} L^j H^{m-j,m-j}(X,\mathbb{R})_{\mathrm{p}} \\
&\quad \oplus \bigoplus_{\substack{p+q=2m \\ p>q}} \bigoplus_{j\geq 0} L^j \left(H^{p-j,q-j}(X)_{\mathrm{p}} \oplus H^{q-j,p-j}(X)_{\mathrm{p}} \right)_{\mathbb{R}}.
\end{aligned}
$$

If $0 \neq \alpha \in H^{m-j,m-j}(X,\mathbb{R})_{\mathrm{p}}$, then $\int_X L^j(\alpha) \wedge L^j(\alpha) = \int_X \alpha \wedge \bar{\alpha} \wedge \omega^{2j}$. Thus, by Hodge–Riemann $(-1)^{m-j} \int_X L^j(\alpha) \wedge L^j(\alpha) > 0$, i.e. the intersection pairing on $L^j H^{m-j,m-j}(X,\mathbb{R})_{\mathrm{p}}$ is definite of sign $(-1)^{m-j}$.

Analogously, for $\alpha \in H^{p-j,q-j}(X)_{\mathrm{p}}$ one has $\int_X L^j(\alpha + \bar{\alpha}) \wedge L^j(\alpha + \bar{\alpha}) = 2\int_X \alpha \wedge \bar{\alpha} \wedge \omega^{2j}$ and thus, again by Hodge–Riemann, we find that the intersection pairing on $L^j(H^{p-j,q-j}(X)_{\mathrm{p}} \oplus H^{q-j,p-j}(X)_{\mathrm{p}})_{\mathbb{R}}$ is definite of sign $i^{p-q}(-1)^{(p+q-2j)(p+q-2j-1)/2} = i^{p-q}(-1)^{m-j}$ for $p+q=2m$.

The rest of the proof consists of the following somewhat tricky calculation:

$$
\begin{aligned}
&\sum_{j\geq 0}(-1)^{m-j} \dim H^{m-j,m-j}(X)_{\mathrm{p}} \\
&+ 2 \sum_{\substack{p+q=2m \\ p>q}} i^{p-q} \sum_{j\geq 0}(-1)^{m-j} \dim H^{p-j,q-j}(X)_{\mathrm{p}} \\
&= \sum_{p+q=2m} i^{p-q} \sum_{j\geq 0}(-1)^{m-j} \dim H^{p-j,q-j}(X)_{\mathrm{p}}
\end{aligned}
$$

$$= \sum_{p+q=2m} (-1)^p \sum_{j\geq 0} (-1)^j \dim H^{p-j,q-j}(X)_{\mathrm{p}}$$

$$\overset{(*)}{=} \sum_{p+q=2m} (-1)^p \left(h^{p,q}(X) + 2\sum_{j>0} (-1)^j h^{p-j,q-j}(X) \right)$$

$$= \sum_{p+q=2m} (-1)^p \left(h^{p,q}(X) + \sum_{j} (-1)^j h^{p-j,q-j}(X) \right)$$

(since $h^{p-j,q-j} = h^{q-j,p-j} = h^{p+j,q+j}$ for $p + q = 2m$)

$$= \sum_{p+q\equiv 0(2)} (-1)^p h^{p,q}(X) = \sum_{p,q} (-1)^p h^{p,q}(X)$$

(since $(-1)^p h^{p,q} + (-1)^q h^{q,p} = 0$ for $p + q \equiv 1(2)$).

Here, $(*)$ uses $h^{p,q}(X) = \sum_{j\geq 0} \dim H^{p-j,q-j}(X)_{\mathrm{p}}$ which yields

$$h^{p,q}(X) + 2\sum_{j>0} (-1)^j h^{p-j,q-j}(X)$$

$$= \sum_{k\geq 0} \dim H^{p-k,q-k}(X)_{\mathrm{p}} + 2\sum_{j>0} (-1)^j \sum_{k\geq 0} \dim H^{p-j-k,q-j-k}(X)_{\mathrm{p}}$$

$$= \sum_{j\geq 0} (-1)^j \dim H^{p-j,q-j}(X)_{\mathrm{p}}.$$

$\square$

It is interesting to note that the signature of the intersection pairing, which is purely topological, can be computed using the Hodge numbers associated to a complex structure on the manifold. In fact, each single Hodge number might really depend on the complex structure, but the above alternating sum does not.

Moreover, in the proof we had to choose a Kähler structure on the complex manifold in order to decompose the cohomology into its primitive parts. In Section 5.1 we will state the Hirzebruch signature formula which expresses $\mathrm{sgn}(X)$ in terms of characteristic classes.

The Lefschetz theorem on $(1,1)$-classes (Proposition 3.3.2) yields a surjective homomorphism $\mathrm{Pic}(X) \to H^{1,1}(X,\mathbb{Z}) = \mathrm{NS}(X)$. To any hypersurface $D \subset X$ one associates via this homomorphism the image of $\mathcal{O}(D)$. It is not hard to see, and we will actually prove this in Section 5.3, that $H^{1,1}(X,\mathbb{Z})$ is generated by classes induced by hypersurfaces provided the manifold X is projective. This basic fact is generalized by the Hodge conjecture. In order to state it, we will need the following

Definition 3.3.19 The *fundamental class* $[Z] \in H^{p,p}(X)$ of a complex submanifold $Z \subset X$ of codimension p in X is defined by the condition

$$\int_X \alpha \wedge [Z] = \int_Z \alpha|_Z$$

for all $\alpha \in H^{2n-2p}(X)$.

It can be shown that $[Z]$ is integral, i.e. contained in the image of $H^{2p}(X, \mathbb{Z}) \to H^{2p}(X, \mathbb{C})$ (cf. Appendix B). Thus, $[Z]$ is an element in $H^{p,p}(X, \mathbb{Z})$, which is by definition the intersection of $H^{p,p}(X)$ and the image of $H^{2p}(X, \mathbb{Z}) \to H^{2p}(X, \mathbb{C})$. Moreover, using Exercise 1.1.17 one sees that $[Z] \in H^{p,p}(X, \mathbb{Z})$ is also defined for singular analytic subvarieties $Z \subset X$.

Definition 3.3.20 A class in $H^{c,c}(X, \mathbb{Q}) := H^{c,c}(X) \cap H^*(X, \mathbb{Q})$ is called *analytic* if it is contained in the $\mathbb{Q}$-vector space generated by all fundamental classes $[Z] \in H^{c,c}(X, \mathbb{Z})$.

The following conjecture is one of the most prominent open question in mathematics. It has been included in the Clay Millennium prize problems [27].

Hodge Conjecture. Let X be a projective complex manifold. Then any class in $H^{c,c}(X, \mathbb{Q})$ is analytic.

Remarks 3.3.21 i) It is known that the Hodge conjecture is false when formulated over $\mathbb{Z}$, i.e. not every class in $H^{c,c}(X, \mathbb{Z})$ is contained in the group generated by fundamental classes. Also note that for the more general class of compact Kähler manifolds the assertion is known to fail. In fact, this can be seen already for $c = 1$, but this is no exception.

ii) The Lefschetz theorem on $(1,1)$-classes provides evidence for the Hodge conjecture. First of all, one verifies that the image of $\mathcal{O}(D)$ under $\mathrm{Pic}(X) \to H^2(X, \mathbb{R})$ is in fact $[D]$. This will be proved in Proposition 4.4.13. Then, we will see that the Kodaira vanishing theorem (cf. Proposition 5.2.2) implies that on a projective manifold any line bundle is associated to a divisor. Thus, any class in $H^{1,1}(X, \mathbb{Z})$ can be written as linear combination of fundamental classes of hypersurfaces. In fact, those can even be chosen to be smooth (see comments to Section 5.3).

iii) If one wants to avoid singular subvarieties and integration over those one might just consider cycles associated to smooth subvarieties only. Or, slightly more general, one could define a fundamental class for any generically finite holomorphic map $f : Z \to X$ of a compact complex manifold Z to X. By definition $[Z, f]$ is then determined by $\int_X [Z, f] \wedge \alpha = \int_Z f^*(\alpha)$. Then Hodge conjecture would assert that any class in $H^{c,c}(X, \mathbb{Q})$ can be written as a linear combination of fundamental classes $[Z, f]$. In fact both versions of the Hodge conjecture are equivalent. This follows from the resolution of singularities (cf. the comments to Section 2.5), which was also one of the main open problems of Grothendieck's list [62] in 1969 and which was solved shortly afterwards by Hironaka [67].

Exercises

3.3.1 Go back to the proof of Lemma 3.3.1 and convince yourself that we have not actually used that X is Kähler. This comes in when considering the bidegree decomposition of H^2.

3.3.2 Let X be a Hopf surface. Show that the Jacobian, i.e. $H^1(X, \mathcal{O}_X)/H^1(X, \mathbb{Z})$, is not a compact torus in a natural way. In fact, $H^1(X, \mathbb{Z}) = \mathbb{Z}$ and $H^1(X, \mathcal{O}_X) = \mathbb{C}$.

3.3.3 Show that the oriented differentiable manifold given by the connected sum $\mathbb{P}^2 \sharp \mathbb{P}^2$ does not underly any Kähler surface. In fact, it cannot even be a complex surface, but in order to see this one would have to use the fact that any complex surface with even b_1 is in fact Kähler. (The four-sphere can be treated analogously. See also the comments to Section 2.6.)

3.3.4 Let X be a Kähler surface with $\mathrm{kod}(X) = -\infty$. Show that its signature is $(1, h^{1,1}(X) - 1)$.

3.3.5 Let X be a compact Kähler manifold of dimension n and let $Y \subset X$ be a smooth hypersurface such that $[Y] \in H^2(X, \mathbb{R})$ is a Kähler class. Show that the canonical restriction map $H^k(X, \mathbb{R}) \to H^k(Y, \mathbb{R})$ is injective for $k \leq n - 1$. (This is one half of the so-called weak Lefschetz theorem (cf. Proposition 5.2.6). It can be proven by using Poincaré duality on X and the Hard Lefschetz theorem 3.3.13).

3.3.6 Construct a complex torus $X = \mathbb{C}^2/\Gamma$ such that $\mathrm{NS}(X) = 0$. Conclude that such a torus cannot be projective and, moreover, that $K(X) = \mathbb{C}$. Observe that in any case $\mathrm{Pic}(X) \neq 0$.

3.3.7 Show that any complex line bundle on $\mathbb{P}^n$ can be endowed with a unique holomorphic structure. Find an example of a compact (Kähler, projective) manifold and a complex line bundle that does not admit a holomorphic complex structure.

3.3.8 Show that on a complex torus $\mathbb{C}^n/\Gamma$ the trivial complex line bundle admits many (how many?) holomorphic non-trivial structures.

3.3.9 Let X be a compact Kähler manifold. Show that the fundamental class $[Y] \in H^{p,p}(X)$ of any compact complex submanifold $Y \subset X$ of codimension p is non-trivial. Is this true for Hopf manifolds?

3.3.10 Let X be a complex torus $\mathbb{C}^n/\Gamma$. Show that $\mathrm{Pic}^0(\mathrm{Pic}^0(X))$ is naturally isomorphic to X.

3.3.11 Show that $\mathrm{Alb}(X)$ and $\mathrm{Pic}^0(X)$ of a compact Kähler manifold X are dual to each other, i.e. $\mathrm{Pic}^0(\mathrm{Alb}(X)) \cong \mathrm{Pic}^0(X)$.

Comments: - Fujiki's article [52] discusses the Lefschetz decomposition from the view point of holonomy and also in other situations.

- We alluded to the fact that any compact complex surface X with $b_1(X)$ even is in fact of Kähler type. The first proof was given by using classification theory and reducing to K3 surfaces. For K3 surfaces it was eventually proved by Todorov and Siu. A more direct proof was recently found by Lamari [82].

- Kähler manifolds are special instances of symplectic manifolds. To any symplectic manifold one can find a compatible almost complex structure. So, one might wonder whether every symplectic manifold is in fact Kähler. Even in dimension four, this does not hold. The first example of a symplectic fourfold which is not a Kähler surface was constructed by Thurston. The reader interested in real fourfolds might ask whether complex surfaces play a role in their classification. In fact, for some time one had tried to prove that irreducible fourfolds are always complex. This turned out to be false due to counterexamples of Gompf and Mrowka. Recently, one has realized that symplectic fourfolds can, to a large extent, be investigated by similar methods as complex surfaces. This direction has been initiated by pioneering work of Gromov and Donaldson.

- Of course, one has tried to prove the Hodge conjecture in special cases. But even for abelian varieties, i.e. projective complex tori, the question is still open. See [56] for a survey.

- Picard and Albanese torus are instances of a whole series of complex tori associated to any compact Kähler manifolds, so called *intermediate Jacobians*. In analogy to $\mathrm{Pic}^0(X) = H^{0,1}(X)/H^1(X,\mathbb{Z})$ and $\mathrm{Alb}(X) = H^{n-1,n}(X)/H^{2n-1}(X,\mathbb{Z})$ one defines the k-the intermediate Jacobian as $H^{k-1,k}(X)/H^{2k-1}(X,\mathbb{Z})$. They are a useful tool in the investigation of cycles of codimension k.

Appendix to Chapter 3

3.A Formality of Compact Kähler Manifolds

In this section we shall first define the notion of a differentiable graded algebra, which captures the algebraic aspects of the de Rham complex of a differentiable manifold, and then move on to discuss how the $\partial\bar{\partial}$-lemma 3.2.10 implies that the differential graded algebra associated with a compact Kähler manifold is particularly simple, is "formal".

Definition 3.A.1 A *differential graded algebra* (or, *dga* for short) over a field k is a graded k-algebra $\mathcal{A} = \bigoplus_{i\geq 0} \mathcal{A}^i$ together with a k-linear map $d : \mathcal{A} \to \mathcal{A}$ such that
   i) The k-algebra structure of $\mathcal{A}$ is given by an inclusion $k \subset \mathcal{A}^0$.
   ii) The multiplication is graded commutative, i.e. for $\alpha \in \mathcal{A}^i$ and $\beta \in \mathcal{A}^j$ one has $\alpha \cdot \beta = (-1)^{i \cdot j} \beta \cdot \alpha \in \mathcal{A}^{i+j}$.
   iii) The Leibniz rule holds: $d(\alpha \cdot \beta) = d(\alpha) \cdot \beta + (-1)^i \alpha \cdot d(\beta)$ for $\alpha \in \mathcal{A}^i$.
   iv) The map d is a differential, i.e. $d^2 = 0$.

Any graded k-algebra becomes a dga by taking $d = 0$. If the context is clear, the field k will usually not be mentioned explicitly. What has just been defined is actually a commutative differential graded algebra, but as we will not encounter any non-commutative ones, we can drop the adjective.
   Any dga $(\mathcal{A} = \bigoplus \mathcal{A}^i, d)$ gives rise to a complex of k-vector spaces

$$\mathcal{A}^0 \xrightarrow{\ d\ } \mathcal{A}^1 \xrightarrow{\ d\ } \mathcal{A}^2 \xrightarrow{\ d\ } \cdots .$$

Definition 3.A.2 The *i-th cohomology* of a dga $(\mathcal{A} = \bigoplus \mathcal{A}^i, d)$ is

$$H^i(\mathcal{A}, d) := \frac{\mathrm{Ker}(d : \mathcal{A}^i \to \mathcal{A}^{i+1})}{\mathrm{Im}(d : \mathcal{A}^{i-1} \to \mathcal{A}^i)}.$$

Lemma 3.A.3 *If $(\mathcal{A} = \bigoplus \mathcal{A}^i, d)$ is a dga, then its cohomology $(H^*(\mathcal{A}, d) = \bigoplus H^i(\mathcal{A}, d), d_{H^*} = 0)$ has the structure of a dga.*

Proof. Indeed, due to the Leibniz rule the product on $\mathcal{A}$ induces a natural product on $H^*(\mathcal{A}, d)$. $\qquad\square$

Definition 3.A.4 Let $(\mathcal{A}, d_\mathcal{A})$ and $(\mathcal{B}, d_\mathcal{B})$ be two dga's. A *dga-homomorphism* between $\mathcal{A}$ and $\mathcal{B}$ is a k-linear map $f : \mathcal{A} \to \mathcal{B}$ such that i) $f(\mathcal{A}^i) \subset \mathcal{B}^i$, ii) $f(\alpha \cdot \beta) = f(\alpha) \cdot f(\beta)$, and iii) $d_\mathcal{B} \circ f = f \circ d_\mathcal{A}$.

Lemma 3.A.5 *Any dga-homomorphism $f : (\mathcal{A}, d_\mathcal{A}) \to (\mathcal{B}, d_\mathcal{B})$ induces a dga-homomorphism $f^* : H^*(\mathcal{A}, d_\mathcal{A}) \to H^*(\mathcal{B}, d_\mathcal{B})$.*

Proof. This is straightforward to check.   □

Definition 3.A.6 A dga-homomorphism $f : (\mathcal{A}, d_\mathcal{A}) \to (\mathcal{B}, d_\mathcal{B})$ is a *quasi-isomorphism* if f^* is an isomorphism.

Example 3.A.7 Let M be a differentiable manifold. Then $(\mathcal{A}^*(M), d)$, where d is the exterior differential on forms, is a dga. Its cohomology equals the de Rham cohomology of M, i.e. $H^*(M, \mathbb{R}) = H^*(\mathcal{A}^*(M), d)$.

Definition 3.A.8 A dga $(\mathcal{A}, d)$ is called *connected* if the inclusion $k \subset \mathcal{A}^0$ induces an isomorphism $k \cong H^0(\mathcal{A}, d)$. It is called *simply connected* if in addition $H^1(\mathcal{A}, d) = 0$.

Remark 3.A.9 Clearly, a differentiable manifold M is connected if and only if its de Rham algebra $(\mathcal{A}^*(M), d)$ is connected. If M is simply connected, then also $(\mathcal{A}^*(M), d)$ is simply connected, but the converse does not hold. Any differentiable manifold with finite fundamental group is a counterexample.

As the above example already shows, a general dga can be very complicated. Thus, one tries to simplify a given dga $(\mathcal{A}, d)$ as far as possible. It turns out that the concept of a "minimal dga" is the right one in this context. Moreover, at least every simply connected dga admits minimal model, which still encodes most of the properties of the original dga.

Definition 3.A.10 A dga $(\mathcal{M} = \bigoplus \mathcal{M}^i, d)$ is called *minimal* if the following conditions are satisfied:

i) $\mathcal{M}^0 = k$.

ii) $\mathcal{M}^+ := \bigoplus_{i>0} \mathcal{M}^i$ is free, i.e. there exist homogeneous elements $x_1, x_2, \ldots$ of positive degree $d_1, d_2, \ldots$, such that $\mathcal{M}^+ = S^+\langle x_1, x_2, \ldots \rangle$ (the positive part of the (super)symmetric algebra of the vector space generated by $x_1, x_2, \ldots$, cf. Section 3.B).

iii) The x_i's in ii) can be chosen such that $d_1 \leq d_2 \leq \ldots$ and $dx_i \in S^+\langle x_1, \ldots, x_{i-1} \rangle$.

Remark 3.A.11 More explicitly, condition ii) says that any $\alpha \in \mathcal{M}^+$ is a linear combination of terms of the form $x_1^{i_1} \cdot x_2^{i_2} \cdot \ldots$ and that all relations are generated by the commutativity relations $x_i \cdot x_j = (-1)^{d_i d_j} x_j \cdot x_i$.

Lemma 3.A.12 *A minimal dga $(\mathcal{M} = \bigoplus \mathcal{M}^i, d)$ is simply connected if and only if $\mathcal{M}^1 = 0$.*

Proof. Clearly, $\mathcal{M}^1 = 0$ implies that $\mathcal{M}$ is simply connected. Thus, we have to show that conversely $H^1(\mathcal{M}, d) = 0$ implies $\mathcal{M}^1 = 0$. So, let us assume $H^1(\mathcal{M}, d) = 0$.

Since $\mathcal{M}$ is minimal, we may choose non-trivial elements $x_1, x_2, \ldots$ satisfying ii) and iii). In particular, $d_1 \leq d_2 \leq \ldots$ and $dx_1 = 0$. If $\mathcal{M}^1 \neq 0$ then $d_1 = 1$. Hence, $x_1 \in \text{Ker}(d : \mathcal{M}^1 \to \mathcal{M}^2)$. On the other hand, $\mathcal{M}^0 = k$ and, therefore, $d\mathcal{M}^0 = 0$. Thus, x_1 would induce a non-trivial element $[x_1] \in H^1(\mathcal{M}, d)$. Contradiction.   □

Remark 3.A.13 Let $(\mathcal{M}, d)$ be a minimal dga. Then d is decomposable, i.e. $d(\mathcal{M}) \subset \mathcal{M}^+ \cdot \mathcal{M}^+$. Indeed, if $x_1, x_2, \ldots$ are chosen as in ii) and iii), then $dx_j \in \mathcal{M}^{j+1}$ is a linear combination of monomials $x_1^{i_1} \cdot \ldots \cdot x_{j-1}^{i_{j-1}}$. Since $d_1 \leq \ldots \leq d_{j-1} < d_j + 1$, we must have $i_1 + \ldots + i_{j-1} \geq 2$.

Moreover, if $(\mathcal{M}, d)$ is a simply connected dga satisfying i), ii) and such that $d(\mathcal{M}) \subset \mathcal{M}^+ \cdot \mathcal{M}^+$, then $\mathcal{M}$ is minimal. In fact, i) and ii) suffice to deduce $\mathcal{M}^1 = 0$ as in the proof of the previous lemma. If we order the elements $x_1, x_2, \ldots$ in ii) such that $d_1 \leq d_2 \leq \ldots$, then dx_i can be written in terms of $x_1, \ldots x_j$ with $d_1 \leq d_2 \leq \ldots \leq d_j \leq d_i + 1$. Since $dx_i \in \mathcal{M}^+ \cdot \mathcal{M}^+$, we may assume that $d_j \leq d_i$. Eventually, using $\mathcal{M}^1 = 0$, we obtain $d_1 \leq \ldots \leq d_j < d_i$. Hence, $j < i$.

Let us study a few minimal dga's in detail.

Examples 3.A.14 i) For a given $n \geq 0$ let $\mathcal{M} = \bigoplus \mathcal{M}^i$ such that

$$\mathcal{M}^i = \begin{cases} k & i = 0, 2n+1 \\ 0 & \text{else} \end{cases}$$

We let d be the trivial differential. Moreover, apart from the k-linear structure, the multiplication is the trivial one. (In fact, there is no other choice in this case.) Clearly, $\mathcal{M}$ is minimal with $\mathcal{M}^+ = S^+\langle x \rangle$, where x is of degree $2n+1$, and the cohomology of $\mathcal{M}$ is concentrated in degree 0 and $2n+1$.

ii) Let $\mathcal{M} := S^*\langle x, y \rangle$ with x and y homogeneous elements of degree $2n$ and $4n-1$, respectively. We set $dx = 0$ and $dy = x^2$. This determines the structure of $\mathcal{M}$ completely. Then $(\mathcal{M}, d)$ is a minimal dga with cohomology concentrated in degree 0 and $2n$.

iii) Let $\mathcal{M} := S^*\langle x, y \rangle$ with x and y homogeneous elements of degree 2 and $2n+1$ ($n > 0$), respectively. The differential d is determined by the two conditions $dx = 0$ and $dy = x^{n+1}$. Clearly, $\mathcal{M}$ is minimal with $H^i(\mathcal{M}, d) = k$ for $i = 0, 2, \ldots, 2n$ and $H^i(\mathcal{M}, d) = 0$ for i odd or $i > 2n$.

Definition 3.A.15 The *minimal model* of a dga $(\mathcal{A}, d_\mathcal{A})$ is a minimal dga $(\mathcal{M}, d_\mathcal{M})$ together with a dga-quasi-isomorphism $f : (\mathcal{M}, d_\mathcal{M}) \to (\mathcal{A}, d_\mathcal{A})$.

If $(\mathcal{M}, d_\mathcal{M})$ is a minimal model of $(\mathcal{A}, d_\mathcal{A})$ and $(\mathcal{A}, d_\mathcal{A}) \to (\mathcal{B}, d_\mathcal{B})$ is a dga-quasi-isomorphism, then via the composition $\mathcal{M} \to \mathcal{A} \to \mathcal{B}$ the dga $(\mathcal{M}, d_\mathcal{M})$ is also a minimal model of $(\mathcal{B}, d_\mathcal{B})$.

The existence of a minimal model is a result due to Sullivan:

Proposition 3.A.16 *Every simply connected dga $(\mathcal{A}, d_\mathcal{A})$ admits a minimal model, which is unique up to isomorphism.*

Proof. We indicate the main ideas. The reader will easily fill in the missing arguments.

Firstly, there are two techniques that allow one to create and annihilate cohomology of a dga.

i) Let $(\mathcal{M}, d_{\mathcal{M}})$ be a dga and $i \in \mathbb{N}$. Then $\mathcal{M}[x]$ admits naturally the structure of a dga, such that $\mathcal{M} \subset \mathcal{M}[x]$ is a dga homomorphism, $\deg(x) = i$, and

$$H^j(\mathcal{M}[x], d) = \begin{cases} H^i(\mathcal{M}, d_{\mathcal{M}}) \oplus [x] \cdot \mathbb{R} & j = i \\ H^j(\mathcal{M}, d_{\mathcal{M}}) & j < i \end{cases}$$

In particular, $dx = 0$. Note that $x^2 = 0$ if i is odd.

ii) Let $\mathcal{M}$ be a dga and $c \in \mathcal{M}^{i+1}$ a closed element. Then $\mathcal{M}[x]$ admits naturally the structure of a dga containing $\mathcal{M}$, such that $\deg(x) = i$, $dx = c$, and

$$H^j(\mathcal{M}[x], d) = \begin{cases} H^i(\mathcal{M}, d_{\mathcal{M}})/[c] & j = i + 1 \\ H^j(\mathcal{M}, d_{\mathcal{M}}) & j \leq i \end{cases}$$

In fact, in order to make this work we need the additional assumption that the dga is simply connected. Indeed, introducing x of degree i we might a priori also create new cohomology of degree $i + 1$ given by products $x \cdot y$, with y of degree one. An element $x \cdot y$ is closed if and only if $d(x) \cdot y = 0$ and $d(y) = 0$. If, however, the dga is simply connected, a closed element y of degree one is of the form $y = d(z)$. Hence, $\pm[x \cdot y] = [d(x)z - d(xz)] = [cz]$ is contained in the cohomology of $\mathcal{M}$ and, in particular, is not newly created by adjoining x.

Next, one constructs inductively minimal dga $(\mathcal{M}_i, d_i)$ and dga homomorphisms $f_i : \mathcal{M}_i \to \mathcal{A}$ such that $f_i^* : H^j(\mathcal{M}_i, d_i) \to H^j(\mathcal{A}, d_{\mathcal{A}})$ is bijective for $j \leq i - 1$ and injective for $j = i$. If $(\mathcal{M}_i, d_i)$ is already constructed one uses i) to create additional cohomology classes and ii) to ensure that $H^{i+1}(\mathcal{M}_{i+1}, d_{i+1}) \to H^{i+1}(\mathcal{A}, d_{\mathcal{A}})$ is injective. $\square$

Examples 3.A.17 i) and ii) of Examples 3.A.14 provide us with the minimal models of the de Rham complexes of the spheres. Indeed, for the odd dimensional sphere S^{2n+1} let $\mathcal{M}$ be as in i). Then $\mathcal{M} \to \mathcal{A}^*(S^{2n+1})$, $x \mapsto$ vol is a dga-quasi-isomorphism. For the even dimensional sphere S^{2n} we choose $\mathcal{M}$ as in ii) and define $\mathcal{M} \to \mathcal{A}^*(S^{2n})$ by $x \mapsto$ vol and $y \mapsto 0$.

Definition 3.A.18 Two dga $(\mathcal{A}, d_{\mathcal{A}})$ and $(\mathcal{B}, d_{\mathcal{B}})$ are *equivalent* if there exists a sequence of dga-quasi-isomorphisms

$$
\begin{array}{ccccccc}
& (\mathcal{C}_1, d_{\mathcal{C}_1}) & & \cdots & & (\mathcal{C}_n, d_{\mathcal{C}_n}) & \\
\swarrow & & \searrow & & \swarrow & & \searrow \\
(\mathcal{A}, d_{\mathcal{A}}) & & (\mathcal{C}_2, d_{\mathcal{C}_2}) & & \cdots & & (\mathcal{B}, d_{\mathcal{B}}).
\end{array}
$$

This is analogous to what happens if one passes from the category of complexes to the derived category. There, quasi-isomorphic complexes yield isomorphic objects in the derived category.

Remark 3.A.19 Using Proposition 3.A.16 one finds that a simply connected dga $(\mathcal{A}, d_{\mathcal{A}})$ is equivalent to a dga $(\mathcal{B}, d_{\mathcal{B}})$ if and only if there exists a dga-quasi-isomorphism $(\mathcal{M}, d_{\mathcal{M}}) \to (\mathcal{B}, d_{\mathcal{B}})$, where $(\mathcal{M}, d_{\mathcal{M}})$ is the minimal model of $(\mathcal{A}, d_{\mathcal{A}})$. Indeed, one can lift recursively $(\mathcal{M}, d_{\mathcal{M}}) \to (\mathcal{C}_{2i}, d_{\mathcal{C}_{2i}})$ to $(\mathcal{M}, d_{\mathcal{M}}) \to (\mathcal{C}_{2i+1}, d_{\mathcal{C}_{2i+1}})$.

Definition 3.A.20 A dga $(\mathcal{A}, d_\mathcal{A})$ is called *formal* if $(\mathcal{A}, d_\mathcal{A})$ is equivalent to a dga $(\mathcal{B}, d_\mathcal{B})$ with $d_\mathcal{B} = 0$.

Clearly, $(\mathcal{A}, d_\mathcal{A})$ is formal if and only if $(\mathcal{A}, d_\mathcal{A})$ is equivalent to its cohomology dga $(H^*(\mathcal{A}, d_\mathcal{A}), d = 0)$.

Definition 3.A.21 A differentiable manifold M is *formal* if its de Rham algebra $(\mathcal{A}^*(M), d)$ is a formal dga.

A general differentiable manifold is not formal, but some interesting classes of manifolds are. In order to show that any compact Kähler manifold is formal (Proposition 3.A.28), we will first prove yet another version of the $\partial\bar{\partial}$-lemma 3.2.10.

Let X be a complex manifold. Recall that $d^c : \mathcal{A}^k(X) \to \mathcal{A}^{k+1}(X)$ is the real differential operator $d^c = \mathbf{I}^{-1}d\,\mathbf{I} = -i(\partial - \bar{\partial})$ (Definition 3.1.13). We have seen that $dd^c = -d^c d = 2i\partial\bar{\partial}$ and one easily checks $(d^c)^2 = 0$.

The following is Exercise 3.2.13, so you might prefer your own proof.

Lemma 3.A.22 (dd^c-lemma) *Let X be a compact Kähler manifold and let $\alpha \in \mathcal{A}^k(X)$ be a d^c-exact and d-closed form. Then there exists a form $\beta \in \mathcal{A}^{k-2}(X)$ with $\alpha = dd^c\beta$. (We do not assume α be pure.)*

Proof. Write $\alpha = d^c\gamma$ and consider the Hodge decomposition $\gamma = d\beta + \mathcal{H}(\gamma) + d^*\delta$. Since X is Kähler, the harmonic part $\mathcal{H}(\gamma)$ of γ is also ∂- and $\bar{\partial}$-closed. Hence, $d^c\gamma = d^c d\beta + d^c d^*\delta$.

It suffices to show that $d^c d^*\delta = 0$. In order to see this, we use $0 = d\alpha = dd^c d^*\delta$ and $d^c d^* = -d^* d^c$ (as in the proof of the $\partial\bar{\partial}$-lemma). Hence, $0 = (dd^* d^c\delta, d^c\delta) = \|d^* d^c\delta\|^2$ and thus $d^c d^*\delta = -d^* d^c\delta = 0$. $\square$

Remark 3.A.23 The dd^c-lemma above immediately yields a $d^c d$-lemma: If $\alpha \in \mathcal{A}^k(X)$ is a d^c-closed and d-exact form on a compact Kähler manifold X, then $\alpha = d^c d\beta$ for some $\beta \in \mathcal{A}^{k-2}(X)$. Indeed, α is d^c-closed and d-exact if and only if $\mathbf{I}(\alpha)$ is d-closed and d^c-exact. Thus, using the lemma one writes $\mathbf{I}(\alpha) = dd^c\beta$ and hence $\alpha = (\mathbf{I}^{-1}d\mathbf{I}^{-1}d)(\mathbf{I}(\beta)) = (-1)^{k-1}(d^c d)(\mathbf{I}(\beta)) = d^c d\tilde{\beta}$ with $\tilde{\beta} = (-1)^{k-1}\mathbf{I}(\beta)$.

Next consider the sub-dga $(\mathcal{A}^*(X)^c, d) \subset (\mathcal{A}^*(X), d)$ of all d^c-closed forms. Since $dd^c = -d^c d$, one has $d(\mathcal{A}^k(X)^c) \subset \mathcal{A}^{k+1}(X)^c$.

Corollary 3.A.24 *Let X be a compact Kähler manifold. Then the inclusion $i : (\mathcal{A}^*(X)^c, d) \subset (\mathcal{A}^*(X), d)$ is a dga-quasi-isomorphism.*

Proof. Let $\alpha \in \mathcal{A}^k(X)^c$ be d-exact. Then by Lemma 3.A.22 one has $\alpha = dd^c\beta$ for some $\beta \in \mathcal{A}^{k-2}(X)$. Since $d^c\beta$ is clearly d^c-closed, this shows that i^* is injective.

Any cohomology class in $H^k(X, \mathbb{R})$ can be represented by a harmonic (with respect to a fixed Kähler metric) form $\alpha \in \mathcal{A}^k(X)$. By Proposition 3.2.6 any harmonic form is also ∂- and $\bar{\partial}$-harmonic. Thus, α is d^c-closed and, therefore, α is in the image of i. This proves the surjectivity of i^*. $\qquad\square$

Definition 3.A.25 Let X be a complex manifold. Then

$$H^k_{d^c}(X) := \frac{\mathrm{Ker}(d^c : \mathcal{A}^k(X) \to \mathcal{A}^{k+1}(X))}{\mathrm{Im}(d^c : \mathcal{A}^{k-1}(X) \to \mathcal{A}^k(X))}.$$

Using $dd^c = -d^c d$, one sees that d induces a natural differential $d :$ $H^k_{d^c}(X) \to H^{k+1}_{d^c}(X)$.

Corollary 3.A.26 *If X is a compact Kähler manifold, then the natural projection $p : (\mathcal{A}^*(X)^c, d) \to (H^*_{d^c}(X), d)$ is a dga-quasi-isomorphism.*

Proof. Let $\alpha \in \mathcal{A}^k(X)$ be d-closed and d^c-exact. Then Lemma 3.A.22 asserts that $\alpha = dd^c\beta$ for some β. In particular, α is in the image of $d : \mathcal{A}^{k-1}(X)^c \to \mathcal{A}^k(X)^c$. Hence, p^* is injective.

Let an element in the cohomology of $(H^*_{d^c}(X), d)$ be represented by the d^c-closed form α. Then $d\alpha$ is d-exact and d^c-closed. Thus, $d\alpha = dd^c\beta$ by Lemma 3.A.22. Hence, $\alpha - d^c\beta$ is d^c- and d-closed and represents the same class as α in $H^k_{d^c}(X)$. This shows that p^* is surjective. $\qquad\square$

Corollary 3.A.27 *Let X be a compact Kähler manifold. Then the exterior differential d is trivial on $H^*_{d^c}(X)$.*

Proof. Indeed, if α is d^c-closed, then $d\alpha$ is d-exact and d^c-closed and thus of the form $d\alpha = d^c d\beta$ for some β. Hence, $0 = [d\alpha] \in H^{k+1}_{d^c}(X)$. $\qquad\square$

Proposition 3.A.28 *Any compact Kähler manifold is formal.*

Proof. If X is compact Kähler then $i : (\mathcal{A}^*(X)^c, d) \subset (\mathcal{A}^*(X), d)$ and $p : (\mathcal{A}^*(X)^c, d) \to (H^*_{d^c}(X), d) = (H^*_{d^c}(X), 0)$ are dga-quasi-isomorphisms. Hence, via the diagram

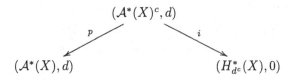

the de Rham complex $(\mathcal{A}^*(X), d)$ is equivalent to a dga with trivial differential. $\qquad\square$

Remark 3.A.29 In particular, we have proven that for a compact Kähler manifold $H^*(X, \mathbb{R}) = H^*_{d^c}(X)$. This does not hold for arbitrary complex manifolds.

Remark 3.A.30 Merkulov has recently shown in [90] a weak version (forgetting the multiplicative structure) of formality for any symplectic manifold whose cohomology satisfies the Hard Lefschetz theorem. In this sense, it seems that in the case of a Kähler manifold it is not the complex structure or the metric but the Kähler form that matters.

There are certainly manifolds that are formal without being Kähler, but sometimes the formality property of Kähler manifolds can indeed be used to exclude a given compact complex manifold from being Kähler. In order to use the formality property for this purpose, we shall make a detour by introducing Massey triple-products.

Let M be a differentiable manifold and let $\alpha \in H^p(M, \mathbb{R})$, $\beta \in H^q(M, \mathbb{R})$, and $\gamma \in H^r(M, \mathbb{R})$ be cohomology classes satisfying $0 = \alpha\beta \in H^{p+q}(M, \mathbb{R})$ and $0 = \beta\gamma \in H^{q+r}(M, \mathbb{R})$. Thus, if $\tilde{\alpha}, \tilde{\beta}, \tilde{\gamma}$ represent α, β, and γ, respectively, then $\tilde{\alpha} \wedge \tilde{\beta} = df$ and $\tilde{\beta} \wedge \tilde{\gamma} = dg$ for certain forms $f \in \mathcal{A}^{p+q-1}(M)$ and $g \in \mathcal{A}^{q+r-1}(M)$.

Definition 3.A.31 The *Massey triple-product*

$$\langle \alpha, \beta, \gamma \rangle \in H^{p+q+r-1}(M, \mathbb{R})/(H^{p+q-1} \cdot H^r + H^p \cdot H^{q+r-1})(M, \mathbb{R})$$

is the cohomology class of $f \wedge \tilde{\gamma} - (-1)^p \tilde{\alpha} \wedge g$.

We have to verify that $\langle \alpha, \beta, \gamma \rangle$ is well-defined.

Firstly, $f \wedge \tilde{\gamma} - (-1)^p \tilde{\alpha} \wedge g$ is closed, for $d(f \wedge \tilde{\gamma} - (-1)^p \tilde{\alpha} \wedge g) = d(f) \wedge \tilde{\gamma} - \tilde{\alpha} \wedge d(g) = (\tilde{\alpha} \wedge \tilde{\beta}) \wedge \tilde{\gamma} - \tilde{\alpha} \wedge (\tilde{\beta} \wedge \tilde{\gamma}) = 0$.

Secondly, if we represent α by $\tilde{\alpha}' = \tilde{\alpha} + d\eta$ for some $\eta \in \mathcal{A}^{p-1}(M)$, then $\tilde{\alpha}' \wedge \tilde{\beta} = \tilde{\alpha} \wedge \tilde{\beta} + d(\eta \wedge \tilde{\beta})$ and $\tilde{\alpha}' \wedge g = \tilde{\alpha} \wedge g + d(\eta) \wedge g$. Thus, with this new choice for a representative of α we obtain for the Massey triple-product:

$$\begin{aligned}
&(f + \eta \wedge \tilde{\beta}) \wedge \tilde{\gamma} - (-1)^p (\tilde{\alpha} \wedge g + d(\eta) \wedge g) \\
&= (f \wedge \tilde{\gamma} - (-1)^p \tilde{\alpha} \wedge g) + (\eta \wedge \tilde{\beta} \wedge \tilde{\gamma} - (-1)^p d(\eta) \wedge g) \\
&= (f \wedge \tilde{\gamma} - (-1)^p \tilde{\alpha} \wedge g) - (-1)^p d(\eta \wedge g).
\end{aligned}$$

This shows that the corresponding cohomology classes in $H^{p+q+r-1}(M, \mathbb{R})$ coincide for both choices $\tilde{\alpha}$ and $\tilde{\alpha}'$. Of course, the same argument applies to the other two classes β and γ.

Thirdly, if we replace f by $f + f_0$ with $df_0 = 0$, then the Massey triple-product becomes

$$f \wedge \tilde{\gamma} - (-1)^p \tilde{\alpha} \wedge g + f_0 \wedge \gamma,$$

i.e. modulo the class $[f_0 \wedge \gamma] \in H^{p+q-1}(M, \mathbb{R}) \cdot H^r(M, \mathbb{R})$ it remains unchanged. A similar argument applies to any other choice of g. Thus, the Massey triple-product is well-defined.

Remark 3.A.32 If one in addition fixes a Riemannian metric on M, the definition of the Massey triple-product can be refined. Then, instead of choosing f and g arbitrary, we let

$$f := Gd^*(\tilde{\alpha} \wedge \tilde{\beta}) \quad \text{and} \quad g := Gd^*(\tilde{\beta} \wedge \tilde{\gamma}),$$

where G is the Green operator, i.e. G is the inverse of the Laplace operator Δ on $\text{Im}(D) \oplus \text{Im}(d^*)$ and $G = 0$ on the space of harmonic forms $\mathcal{H}(M, g)$. Since G commutes with d, one has $dGd^*(\tilde{\alpha} \wedge \tilde{\beta}) = Gdd^*(\tilde{\alpha} \wedge \tilde{\beta}) = G\Delta(\tilde{\alpha} \wedge \tilde{\beta}) = \tilde{\alpha} \wedge \tilde{\beta}$ and, similarly, $dGd^*(\tilde{\beta} \wedge \tilde{\gamma}) = \tilde{\beta} \wedge \tilde{\gamma}$. Thus, we obtain a class

$$\left[Gd^*(\tilde{\alpha} \wedge \tilde{\beta}) \wedge \tilde{\gamma} - (-1)^p \tilde{\alpha} \wedge Gd^*(\tilde{\beta} \wedge \tilde{\gamma}) \right] \in H^{p+q+r-1}(M, \mathbb{R})$$

which depends on the chosen metric and maps to the Massey triple-product under the natural projection onto $H^{p+q+r-1}/(H^{p+q-1} \cdot H^r + H^p \cdot H^{q+r-1})$.

In fact, a slight modification of this definition allows one to define a triple-product even for cohomology classes α, β, γ which do not satisfy the assumption $\alpha\beta = 0 = \beta\gamma$. As the above triple product involving G might not be closed, one replaces it by its harmonic part

$$\left[\mathcal{H} \left(Gd^*(\tilde{\alpha} \wedge \tilde{\beta}) \wedge \tilde{\gamma} - (-1)^p \tilde{\alpha} \wedge Gd^*(\tilde{\beta} \wedge \tilde{\gamma}) \right) \right].$$

Clearly, if the product of two arbitrary harmonic forms is again harmonic, then this triple-product is trivial. The usual product in $H^*(M, \mathbb{R})$ together with the new triple-product depending on a Riemannian metric are part of an even larger structure. It turns out that the de Rham complex of a compact Riemannian manifold is endowed with the structure of an A_∞-algebra (cf. [43]).

Before stating the next proposition, let us point out that the definition of the original Massey triple-product $\langle \alpha, \beta, \gamma \rangle$ makes perfect sense for any dga $(\mathcal{A}, d_{\mathcal{A}})$. Thus, if $(\mathcal{A}, d_{\mathcal{A}})$ is a dga and $\alpha \in H^p(\mathcal{A}, d_{\mathcal{A}})$, $\beta \in H^q(\mathcal{A}, d_{\mathcal{A}})$, and $\gamma \in H^r(\mathcal{A}, d_{\mathcal{A}})$ such that $\alpha\beta = 0 = \beta\gamma$, then there is a natural class $\langle \alpha, \beta, \gamma \rangle \in H^{p+q+r-1}(\mathcal{A}, d_{\mathcal{A}})/(H^{p+q-1} \cdot H^r + H^p \cdot H^{q+r-1})(\mathcal{A}, d_{\mathcal{A}})$.

Proposition 3.A.33 *If $(\mathcal{A}, d_{\mathcal{A}})$ is a formal dga, then all Massey triple-products are trivial.*

Proof. The assertion follows easily from the observation that for any dga homomorphism $F : (\mathcal{C}, d_{\mathcal{C}}) \to (\mathcal{B}, d_{\mathcal{B}})$ the Massey product is compatible with F^*, i.e. $F^*\langle \alpha, \beta, \gamma \rangle = \langle F^*\alpha, F^*\beta, F^*\gamma \rangle$. For a formal dga $(\mathcal{A}, d_{\mathcal{A}})$ this shows that we can compute its Massey products in its cohomology dga $(H^*(\mathcal{A}, d_{\mathcal{A}}), d = 0)$. But for a dga with trivial differential the Massey products clearly vanish. $\square$

Example 3.A.34 It is easy to find examples of manifolds which are not formal. The following is a real version of the complex Iwasawa manifold of dimension three discussed in Section 2.1. Consider the Heisenberg group G of all real invertible matrices of the form

$$\begin{pmatrix} 1 & x & y \\ 0 & 1 & z \\ 0 & 0 & 1 \end{pmatrix}$$

and let $\Gamma \subset G$ be the discrete subgroup $G \cap \mathrm{Gl}(3, \mathbb{Z})$. Then $(a, b, c) \in \Gamma$ acts on G by $(x, y, z) \mapsto (x + a, y + az + b, z + c)$. We will write elements of G and Γ as vectors (x, y, z). Then $M = G/\Gamma$ is a compact three-dimensional manifold.

The differentials $\alpha := dx$, $\beta := dz$, and $\delta = x\,dz - dy$ are Γ-invariant one-forms on G and, therefore, descend to one-forms on M. Observe, that $[dx], [dy]$ generate $H^1(M, \mathbb{R})$. One way to see this is to compute that $(\Gamma/[\Gamma, \Gamma])_{\mathbb{R}}$ is generated by these classes. In fact, $(0, 1, 0) = [(1, 0, 0), (0, 0, 1)]$.

The following relations are easy to verify: $d\alpha = d\beta = 0$, $d\delta = \alpha \wedge \beta$, and $\alpha \wedge \beta \wedge \delta = dx \wedge dy \wedge dz$. Hence, we can compute the Massey triple-product $\langle \alpha, \beta, \beta \rangle = \delta \wedge \beta - \alpha \wedge 0$. The latter is a non-trivial cohomology class in $H^2(M, \mathbb{R})$, as $\int_M \alpha \wedge \delta \wedge \beta \neq 0$.

In order to show that $\langle \alpha, \beta, \beta \rangle$ is non-trivial in the quotient $H^2(M, \mathbb{R})/H^1 \cdot H^1$ one has to make sure that $\langle \alpha, \beta, \beta \rangle$ is not contained in $H^1 \cdot H^1$. Since $H^1(M, \mathbb{R})$ is generated by $[\alpha]$ and $[\beta]$, any class γ in $H^1 \cdot H^1$ satisfies $\gamma \wedge [\alpha] = \gamma \wedge [\beta] = 0$, but $[\delta \wedge \beta] \wedge [\alpha] \neq 0$. Thus, there are non-trivial Massey triple-products on M and, hence, M is not formal.

The argument generalizes easily to the complex Iwasawa manifold $X = G/\Gamma$, where $G \subset \mathrm{Gl}(3, \mathbb{C})$ is the complex Heisenberg and $\Gamma = G \cap \mathrm{Gl}(3, \mathbb{Z} + i\mathbb{Z})$. Thus, one obtains a compact complex manifold X of dimension three with $b_1(X) = 4$, which does not admit any Kähler structure. Note that the necessary condition $b_1(X) \equiv 0(2)$ for a complex manifold to be Kähler, is satisfied in this example. Thus, only the finer information about the non-formality shows that X is non-Kähler.

Non-formality of X does not only show that the complex manifold X is not Kähler, but that there is no complex structure on the underlying differentiable manifold that is Kähler.

Remark 3.A.35 To conclude, it might be worth pointing out the importance of the concepts touched upon in this section for the algebraic topology of compact Kähler manifolds. If M is a compact manifold, its cohomology $H^*(M, \mathbb{R})$ in general reflects only a small part of its topology. Surprisingly, the de Rham complex $\mathcal{A}^*(M)$ encodes much more of it, in fact the real homotopy of M is determined by it. This approach to homotopy theory using differential forms has been developed by Sullivan [103]. The formality of compact Kähler manifolds can thus be used to conclude that the real homotopy type of a compact Kähler manifold is determined by its cohomology.

Exercises

3.A.1 Work out the details of the examples 3.A.14. E.g. compute the dimensions of all graded parts $\mathcal{M}^i$.

3.A.2 Show that iii) of Examples 3.A.14 can be used as a minimal model of $\mathbb{P}^n$.

3.A.3 Show that all spheres are formal.

3.A.4 Show hat the generalized Massey triple-product in Remark 3.A.32 is trivial as soon as the product of any harmonic forms is again harmonic. (The converse does not hold.)

3.A.5 Prove that for the complex Iwasawa manifold of dimension three X one has $b_1(X) = 4$, $b_2(X) = 8$, and $b_3(X) = 10$. Thus, X satifies all numerical conditions that can be deduced from the results in Sections 3.2 and 3.3 for compact Kähler manifolds.

Comment: There are quite a few excellent references for this. We recommend [1, 43, 60].

3.B SUSY for Kähler Manifolds

This section shows that, using super Lie algebras, some of the commutator calculations encountered before can be put in a language that is more algebraic in spirit and, hopefully, more conceptual. This approach has also the advantage to present the results in a form that is more suitable for applications to mathematical physics, more precisely to quantum mechanics. I hope that presenting material from the previous sections once more in this language will be helpful for the reader interested in non-linear sigma models with target a Kähler manifold and string theory. The reader must not be deceived, however. This section is just a first step on the long road from mathematics to conformal field theory, e.g. we do not discuss the superconformal algebra at all.

The two basic objects needed for a reasonable field theory in physics are a Hilbert space and a Hamiltonian. In a geometric context, the first proposal for the Hilbert space would be the space of L^2-functions on the given Riemannian manifold with the Laplacian as the Hamiltonian. The symmetries of the theory are then encoded by linear operators on the Hilbert space that commute with the Hamiltonian, i.e. the Laplacian. Already from a purely mathematical standpoint it does not seem very natural to consider only functions and one soon passes on to the space of all differential forms. The space of differential forms can still be completed to a Hilbert space by introducing L^2-forms, but the new space shows one additional feature: Every element can be decomposed into its even and its odd part, where even and odd is meant with respect to the usual degree of a differential form. In physicists language, differential forms of even degree, e.g. functions, are *bosonic* and odd forms, e.g. one-forms, are *fermionic*.

Depending on the geometry of the specific Riemannian manifold, one introduces a natural super Lie algebra of symmetry operators, e.g. certain differential operators for differential forms which commute with the Laplacian. Usually, this Lie algebra is given in terms of generators and commutator relations. One could require the operators to generate the Lie algebra as a Lie algebra, but often the stronger condition that they span the underlying vector space is imposed. (Otherwise the Lie algebra might become too big.) Roughly, the size of the Lie algebra in physicists jargon is measured by the amount of *supersymmetry*.

The aim of this section is to introduce some basic concepts and, by going through a list of examples, to gain some feeling for the physicists point of view on Kähler and Riemannian geometry.

Let us begin with the following basic definition.

Definition 3.B.1 A $\mathbb{Z}/2\mathbb{Z}$-vector space (or *super vector space*) is a vector space V endowed with a direct sum decomposition $V = V_0 \oplus V_1$. Elements in V_0 are called *even* and those in V_1 are called *odd*.

If an element $a \in V$ is homogeneous, i.e. $a \in V_0$ or $a \in V_1$, then $\tilde{a}$ denotes its degree. Thus, for a homogeneous element one has $a \in V_{\tilde{a}}$. Many standard constructions in algebra can be carried over to the the super world by introducing an extra sign $(-1)^{\tilde{a} \cdot \tilde{b}}$ whenever the order of two homogeneous elements a and b is changed.

Example 3.B.2 The elementary constructions in linear algebra work as well for super vector spaces. Let us illustrate this in a few cases. We recommend the first sections of [34] where different ways are proposed to deal with all possible sign questions. The most elegant way hides all signs in a commutativity isomorphism in an appropriate tensor category.

i) Let $V = V_0 \oplus V_1$ and $W = W_0 \oplus W_1$ be two super vector spaces. Then $V \otimes W$ is a super vector space by posing $(V \otimes W)_0 = V_0 \otimes W_0 \oplus V_1 \otimes W_1$ and $(V \otimes W)_1 = V_0 \otimes W_1 \oplus V_1 \otimes W_0$.

ii) The second symmetric power $S^2(V)$ of an ordinary vector space V is the quotient of $V \otimes V$ by the relation $x \otimes y = y \otimes x$. If $V = V_0 \oplus V_1$ is a super vector space then, one defines $S^2(V)$ in the super world, as the quotient of the super vector space $V \otimes V$ by the relation $x \otimes y = (-1)^{\tilde{x} \cdot \tilde{y}} y \otimes x$ for any two homogeneous elements $x \in V_{\tilde{x}}$ and $y \in V_{\tilde{y}}$. The quotient is again a super vector space. Of course, one can similarly define the symmetric algebra $S^*(V)$ of a super vector space V.

A purely odd super vector space $V = V_1$ can also be regarded as an ordinary vector space by just forgetting the grading. Let us denote it by $A(V)$. Then, slightly surprising at first sight, $S^2(V) = \bigwedge^2 A(V)$.

iii) If $V = V_0 \oplus V_1$ is a super vector space then also the algebra of $\mathbb{C}$-linear endomorphisms $\mathrm{End}(V)$ is a super vector space in a natural way. Indeed, $\mathrm{End}(V) = \mathrm{End}_0(V) \oplus \mathrm{End}_1(V)$ where $F \in \mathrm{End}_0(V)$ if and only if $F(V_i) = V_i$, $i = 0, 1$, and F is odd if and only if $F(V_i) = F(V_{i+1})$.

The endomorphism algebra $\mathrm{End}(V)$ of a super vector space provides the most basic example of a super Lie algebra:

Definition 3.B.3 A *super Lie algebra* is a super vector space $\mathfrak{g} = \mathfrak{g}_0 \oplus \mathfrak{g}_1$ with a $\mathbb{C}$-linear even homomorphism $[\ ,\ ] : \mathfrak{g} \otimes \mathfrak{g} \to \mathfrak{g}$, i.e. $[\mathfrak{g}_i, \mathfrak{g}_j] \subset \mathfrak{g}_{i+j}$, such that for homogeneous elements a, b, c one has:

$$[a, b] = -(-1)^{\tilde{a} \cdot \tilde{b}}[b, a] \quad \text{and} \quad [a, [b, c]] = [[a, b], c] + (-1)^{\tilde{a} \cdot \tilde{b}}[b, [a, c]].$$

The first sign should be clear from the meta-rule for passing from ordinary to super constructions. For the second one, the usual Jacobi identity needs first to be written in the suitable form.

If V is a super vector space, then $\mathrm{End}(V)$ is super Lie algebra with $[A, B] = A \circ B - (-1)^{\tilde{A} \cdot \tilde{B}} B \circ A$ for homogeneous endomorphisms $A, B \in \mathrm{End}(V)$. One has to check that the Jacobi identity holds true, but we leave this straightforward argument to the reader.

Let us see how these algebraic notions work for the space of differential forms on a manifold M. The space of complex differential forms $\mathcal{A}_{\mathbb{C}}^*(M)$ becomes a super vector space by writing it as

$$\mathcal{A}_{\mathbb{C}}^*(M) = \underbrace{\left(\bigoplus_k \mathcal{A}_{\mathbb{C}}^{2k}(M) \right)}_{\text{even}} \oplus \underbrace{\left(\bigoplus_k \mathcal{A}_{\mathbb{C}}^{2k+1}(M) \right)}_{\text{odd}}.$$

Usually, the vector spaces of interest in quantum physics will be infinite-dimensional and endowed with a scalar product. The scalar product naturally appears when M is compact and endowed with a Riemannian metric. Often one requires the space to be complete (in order to have an honest Hilbert space). In general, any non-complete space can be completed and for algebraic aspects of the theory this procedure is harmless. So, we will ignore this point here and work with the spaces of differentiable functions and forms and not with their L^2-completions.

The next step is to define the Hamiltonian and the symmetry operators of the theory in question. In the following list we exhibit a number of relevant situations where the geometry of M gives rise to interesting super Lie algebra contained in the endomorphism algebra $\mathrm{End}(\mathcal{A}_{\mathbb{C}}^*(M))$.

Riemannian geometry. If (M, g) is just a compact oriented Riemannian manifold, then in addition to the usual exterior differential operator d one has d^* and the Laplacian Δ. Clearly, d, d^* are odd operators and Δ is an even one. As mentioned before, the Laplacian Δ plays the role of the Hamiltonian.

Instead of these operators, one might also work with the odd operators $Q_1 := d + d^*$ and $Q_2 := i(d - d^*)$. These are the usual symmetry operators considered in physics (see Witten's fundamental article [117]). They satisfy

i) $[Q_1, Q_1] = [Q_2, Q_2](= 2\Delta)$ and ii) $[Q_1, Q_2] = 0$.

Conversely, from Q_1 and Q_2 one easily recovers d and d^* by $d = (1/2)(Q_1 - iQ_2)$ and $d^* = (1/2)(Q_1 + iQ_2)$. Then the conditions i) and ii) together are equivalent to

iii) $d^2 = d^{*2} = 0$ and iv) $Q_1^2 = Q_2^2$,

considered in [46].

It easy to see that both operators d and d^* (or, equivalently, Q_1 and Q_2) commute with the Laplacian Δ, i.e. $[d, \Delta] = [d^*, \Delta] = 0$. Since $\Delta = [d, d^*]$, this can be seen as an immediate consequence of the Jacobi identity for the super Lie algebra $\mathrm{End}(\mathcal{A}_{\mathbb{C}}^*(M))$.

Let $\mathfrak{g}$ be the super Lie subalgebra of $\mathrm{End}(\mathcal{A}_{\mathbb{C}}^*(M))$ generated by d, d^*, Δ. Then $\mathfrak{g}$ is of dimension three with a one-dimensional even part $\mathfrak{g}_0$ spanned by Δ.

In physics jargon, this algebra is sometimes called the $N = (1, 1)$ *supersymmetry algebra*. See [46] for comments on the (sometimes confusing) naming.

Complex geometry. Suppose that the differentiable manifold M is in addition endowed with a complex structure I. This leads to the decomposition of the exterior differential $d = \partial + \bar{\partial}$. Physicists sometimes prefer the notation $G = \partial$ and $\bar{G} = \bar{\partial}$. Thus, $d = G + \bar{G}$ and the following commutator relations hold true

$$[G, d] = 0 \text{ and } [G, \bar{G}] = 0.$$

The integrability of the complex structure I is needed at this point.

So far, we have not imposed any compatibility condition on the complex and the Riemannian structure. But in any case, the Riemannian structure can be used to define the supplementary operators ∂^* and $\bar{\partial}^*$. One might then try to generate a super Lie algebra by the operators $\Delta, \partial, \bar{\partial}, \partial^*, \bar{\partial}^*$, but this, in general, gives something messy. They are not 'closed', i.e. the set of generators is not preserved by taking brackets or, in other words, these operators do not, in general, span the vector space of the super Lie algebra they generate, i.e. for the smallest super Lie subalgebra of $\text{End}(V)$ containing them all. Even worse, the super Lie algebra might be of infinite dimension. First of all, other Laplacians $\Delta_\partial = [\partial, \partial^*]$ and $\Delta_{\bar{\partial}} = [\bar{\partial}, \bar{\partial}]$ come up. But, more seriously, in general $[\partial, \bar{\partial}^*] \neq 0$.

Kähler geometry. The Kähler condition has two effects. Firstly, due to the Kähler identitities the super Lie algebra spanned by $\Delta, \partial, \bar{\partial}, \partial^*, \bar{\partial}^*$ is closed. Its even part is one-dimensional and the odd part is of dimension four. The Lie algebra generators form in fact a basis of the underlying vector space.

Secondly, we can enrich the situation by taking the Lefschetz operators L and Λ into account. Since the commutator of these two involves the counting operator H (which should, despite the fact that H is the standard notation for the Hamiltonian, not be confused with it), we necessarily add H to our super Lie algebra.

Thus, besides H there are two sets of generators $\{\partial, \bar{\partial}, L\}$ and $\{\partial^*, \bar{\partial}^*, \Lambda\}$. Two operators of one of these two sets always commute and the commutator between two operators from different sets are expressed by the Kähler identities 3.1.12.

Let us illustrate the advantage of considering these operators as elements in the super Lie algebra $\text{End}(\mathcal{A}_{\mathbb{C}}^*(M))$. Indeed, the fact that the Laplacian commutes with the other operators follows from the Jacobi identity. Let us check this for the dual Lefschetz operator Λ. One computes

$$\begin{aligned}[\Lambda, \Delta] &= [\Lambda, dd^* + d^*d] = [\Lambda, [d, d^*]] \\ &= [[\Lambda, d], d^*] + [d, [\Lambda, d^*]] \\ &= [-d^{c*}, d^*] + 0 = 0.\end{aligned}$$

The enriched super Lie algebra $\mathfrak{g}$ generated by $\partial, \partial^*, \bar{\partial}, \bar{\partial}^*, L, \Lambda, H$ has a four-dimensional even part $\mathfrak{g}_0$, spanned by $\Delta, L, \Lambda,$ and H, and a (unchanged) four-dimensional odd part $\mathfrak{g}_1$. This is the $N = (2, 2)$ supersymmetry algebra. (Warning: There are conformal versions of these super Lie algebras which are always infinite-dimensional and which incorporate copies of the Virasoro algebra.)

Hermitian geometry. Before treating Kähler manifolds, we could have looked at the hermitian case. Let us do this now. Thus, we impose that I be compatible with the Riemannian metric, but not that the fundamental form is closed. The problems encountered in the complex case are still up, they are indeed only cured by imposing the Kähler condition. But L and Λ might be already introduced at this level. The commutator relations between these linear differential operators are still valid, but together with the differential operators $\partial, \bar{\partial}$, etc., they make the super Lie algebra even bigger. (E.g. there is a non-trivial commutator relation even between the linear operator L and the differential operator d in the non-Kähler case.)

Beyond. For so called hyperkähler manifolds one obtains $N = (4, 4)$ supersymmetry algebras. Roughly, a hyperkähler metric is a Kähler metric which is Kähler with respect to a whole two-dimensional sphere worth of complex structures. For the linear version of this see Exercise 1.2.5. All the induced Lefschetz and differential operators are used to form this bigger algebra. See [44] for a discussion of the hyperkähler case from a physical point of view.

It seems that no other physically relevant finite-dimensional supersymmetry algebra is expected to appear naturally in geometry. This should maybe be seen as an analogue of the fact that Berger's list of irreducible holonomy groups is finite (see Theorem 4.A.16).

In [23] a uniform abstract algebraic construction of the supersymmetry algebras appearing in Riemannian, Kähler, and hyperkähler geometry is proposed. There they are naturally associated to the Lie algebras $\mathfrak{so}(1, \mathbb{C})$, $\mathfrak{so}(3, \mathbb{C}) = \mathfrak{sl}(2, \mathbb{C})$, and $\mathfrak{so}(5, \mathbb{C})$. The $\mathfrak{sl}(2)$ is, not surprisingly, the $\mathfrak{sl}(2)$ spanned by L, Λ, and H.

The advantage of working with these structures in this algebraic fashion is that the computations can be carried over to less geometric situations where only the Hilbert space and the operators are given. This is the view point of [46], which aims at non-commutative generalizations of classical geometric structures.

3.C Hodge Structures

Some of the structures described in Sections 3.2 and 3.3 are special instances of a more general notion, so called Hodge structures. This section introduces the reader to some basic aspects of this abstract notion, which is perfectly suited to study compact Kähler and projective manifolds (even over other base fields) and their deformations. Although we will not pursue this approach very far, the section should help to digest what has been explained before.

Definition 3.C.1 A *rational Hodge structure of weight* k consists of a rational vector space H and a direct sum decomposition

$$H \otimes_{\mathbb{Q}} \mathbb{C} = \bigoplus_{p+q=k} H^{p,q} \tag{3.3}$$

satisfying $\overline{H^{p,q}} = H^{q,p}$.

Any rational Hodge structure of weight k gives rise to a real representation of $\mathbb{C}^*$, i.e. a group homomorphism $\rho : \mathbb{C}^* \to \mathrm{Gl}(H_{\mathbb{R}})$. Indeed, one defines $\rho(z)(\alpha) = (z^p \bar{z}^q) \cdot \alpha$ for $\alpha \in H^{p,q}$. In order to check that this representation is real, take $\alpha \in H_{\mathbb{R}} := H \otimes_{\mathbb{Q}} \mathbb{R}$ and consider its decomposition $\alpha = \sum \alpha^{p,q}$ according to (3.3). Then $\overline{\alpha^{p,q}} = \alpha^{q,p}$ and $\rho(z)(\alpha) = \sum (z^p \bar{z}^q) \cdot \alpha^{p,q}$ is still real, as $\overline{(z^p \bar{z}^q) \cdot \alpha^{p,q}} = \bar{z}^p z^q \alpha^{q,p}$. Note that the induced representation of $\mathbb{R}^*$ is given by $\rho(t)(\alpha) = t^k \cdot \alpha$.

Proposition 3.C.2 *There is a natural bijection between rational Hodge structures of weight* k *on a rational vector space* H *and algebraic representations* $\rho : \mathbb{C}^* \to \mathrm{Gl}(H_{\mathbb{R}})$ *with* $\mathbb{R}^*$ *acting by* $\rho(t)(\alpha) = t^k \cdot \alpha$.

Proof. We only have to give an inverse construction that associates to an algebraic representation $\rho : \mathbb{C}^* \to \mathrm{Gl}(H_{\mathbb{R}})$ a Hodge structure.

Let us denote the $\mathbb{C}$-linear extension of ρ by $\rho_{\mathbb{C}} : \mathbb{C}^* \to \mathrm{Gl}(H_{\mathbb{C}})$ and let

$$H^{p,q} := \{ v \in H_{\mathbb{C}} \mid \rho_{\mathbb{C}}(z)(v) = (z^p \bar{z}^q) \cdot v \text{ for all } z \in \mathbb{C}^* \}.$$

Since $\mathbb{C}^*$ is abelian, the representation $\rho_{\mathbb{C}}$ splits into a direct sum of one-dimensional representations $\lambda_i : \mathbb{C}^* \to \mathbb{C}^*$. In order to show that $H_{\mathbb{C}} = \bigoplus H^{p,q}$, one has to argue that every one-dimensional representation λ_i that might occur is of the form $\lambda_i(z) = z^p \bar{z}^q$ with $p + q = k$. At this point the assumption that ρ is algebraic comes in.

By writing $z \in \mathbb{C}^*$ as $z = x + iy$, one can identify $\mathbb{C}^*$ with the subgroup of $\mathrm{Gl}(2, \mathbb{R})$ of all matrices of the form $\begin{pmatrix} x & y \\ -y & x \end{pmatrix}$. A representation $\rho : \mathbb{C}^* \to \mathrm{Gl}(H_{\mathbb{R}}) \cong \mathrm{Gl}(\dim(H), \mathbb{R})$ is algebraic if $\rho \begin{pmatrix} x & y \\ -y & x \end{pmatrix}$ is a matrix whose entries are polynomials in x, y, and the inverse of the determinant $(x^2 + y^2)^{-1}$. Hence, $\lambda_i(z)$ must be a polynomial in z, $\bar{z}$, and $(z\bar{z})$ and, henceforth, of the form $z^p \bar{z}^q$ for some p, q with $p + q = k$. □

Examples 3.C.3 i) By $\mathbb{Q}(k)$ one denotes the unique one-dimensional weight $-2k$ Hodge structure given by $z \mapsto z^{-k}\bar{z}^{-k}$.

ii) Let V be a rational vector space such that $V_{\mathbb{R}}$ is endowed with an almost complex structure. Then, $\bigwedge^k V$ has a natural weight k Hodge structure given by the bidegree decomposition $\bigwedge_{\mathbb{C}}^k V = \bigoplus V^{p,q}$ (see Section 1.2).

iii) The cohomology $H^k(X, \mathbb{Q})$ of any compact Kähler manifold X comes with a natural Hodge structure of weight k. But let us first consider the case of an hermitian manifold (X, g). Then then the Hodge structure (over $\mathbb{R}$) of example ii) exists on the space of k-forms $\bigwedge_x^k X$ at any point $x \in X$ and one might wonder whether one can pass to cohomology. But, as the exterior differential is not well-behaved with respect to the operator $\mathbf{I}$, this is not obvious and in general not possible. Note that, on the other hand, the Dolbeault cohomology $H^{p,q}(X)$ is naturally endowed with the representation $z \mapsto z^p \bar{z}^q$.

If (X, g) is compact and Kähler then Hodge decomposition implies that there exists a natural bidegree decomposition of the space of harmonic forms $\mathcal{H}^k(X, g)_{\mathbb{C}}$ and thus on $H^k(X, \mathbb{C})$. Thus, $H^k(X, \mathbb{Q})$ has a natural Hodge structure of weight k which is independent of the Kähler structure, but does depend on the complex structure defining the complex manifold X. To have a concrete example in mind, check that $H^{2n}(\mathbb{P}^n, \mathbb{Q}) \cong \mathbb{Q}(-n)$.

iv) Suppose that X is a compact Kähler manifold such that the chosen Kähler class $[\omega]$ is rational. Then the primitive cohomology is defined over $\mathbb{Q}$ and we obtain a rational Hodge struture on $H^k(X, \mathbb{Q})_{\mathrm{p}}$.

Definition 3.C.4 Let H be a rational Hodge structure of weight k. The induced *Hodge filtration* $\dots F^{i+1} H_{\mathbb{C}} \subset F^i H_{\mathbb{C}} \subset \dots \subset H_{\mathbb{C}}$ is given by

$$F^i H_{\mathbb{C}} := \bigoplus_{p \geq i} H^{p,q}.$$

Clearly, $F^p H_{\mathbb{C}} \cap \overline{F^{k-p} H_{\mathbb{C}}} = H^{p,k-p}$. Thus, from the Hodge filtration one recovers the Hodge stucture itself. The reason why one nevertheless often works with the Hodge filtration instead of the Hodge structure itself is that the Hodge filtration behaves better in families. More precisely, if $\{X_t\}$ is a smooth family of compact Kähler manifolds over the base $S = \{t\}$, then the $H^k(X_t, \mathbb{C})$ form a (locally constant) holomorphic vector bundle on S and similarly the vector spaces $F^i H^k(X_t, \mathbb{C})$. This is no longer true for the single components $H^{p,q}(X_t)$.

Let H and H' be two Hodge structures of weight k and k', respectively. The tensor product $H \otimes_{\mathbb{Q}} H'$ has a natural Hodge structure of weight $k + k'$, which can be described in the following two equivalent ways:

Firstly, if ρ and ρ' are the representations defining the Hodge structures on H respectively H', then $\rho \otimes \rho'$ defines a representation of $\mathbb{C}^*$ on $(H \otimes_{\mathbb{Q}} H')_{\mathbb{R}}$.

Secondly and equivalently, one sets $(H \otimes_{\mathbb{Q}} H')_{\mathbb{C}} = H_{\mathbb{C}} \otimes_{\mathbb{C}} H'_{\mathbb{C}} = \bigoplus (H \otimes H')^{r,s}$ with

$$(H \otimes H')^{r,s} = \bigoplus_{\substack{p+p'=r \\ q+q'=s}} H^{p,q} \otimes H'^{p',q'}.$$

The tensor product $H \otimes \mathbb{Q}(k)$ is called the kth *Tate twist* of H and is abbreviated by $H(k)$.

If X and X' are two compact Kähler manifolds, then $H^k(X \times X', \mathbb{Q})$ is a sum of weight k Hodge structures $H^\ell(X, \mathbb{Q}) \otimes H^{\ell'}(X', \mathbb{Q})$ for $\ell + \ell' = k$. E.g. $H^k(X \times \mathbb{P}^1, \mathbb{Q}) = H^k(X, \mathbb{Q}) \oplus H^{k-2}(X, \mathbb{Q})(-1)$.

Definition 3.C.5 Let H be a Hodge structure of weight k. A *polarization* of H is a bilinear form

$$(\, , \,) : H \times H \longrightarrow \mathbb{Q}$$

satisfying the two conditions:

i) $(\rho(z)\alpha, \rho(z)(\beta)) = (z\bar{z})^k (\alpha, \beta)$ and ii) $(\, , \rho(i) \,)$ is symmetric and positive definite.

Here are a few easy consequences. We leave the verification to the reader.

Lemma 3.C.6 *Let* $(\, , \,)$ *be a polarization of a weight k Hodge structure H. Then*

i) *The pairing* $(\, , \,)$ *is symmetric if k is even and alternating otherwise.*

ii) *With respect to the $\mathbb{C}$-linear extension of the pairing the direct sum decomposition $H_{\mathbb{C}} = \bigoplus_{p \geq q} (H^{p,q} \oplus H^{q,p})$ is orthogonal.*

iii) *On the real part of $H^{p,q} \oplus H^{q,p}$ the pairing $i^{p-q}(\, , \,)$ is positive definite.*   □

Example 3.C.7 Let $(\, , \,)$ be a polarization of a weight one Hodge structure H. Then $(\, , \,)$ is alternating and can be considered as an element in $H^{1,0^*} \otimes H^{0,1^*} \subset \bigwedge^2 H_{\mathbb{C}}^*$. Since the polarization is rational, this two-form is in fact contained in the $\mathbb{Q}$-vector space $\bigwedge^2 H^*$.

Remark 3.C.8 If one defines a strict homomorphism $\varphi : H \to H'$ between two Hodge structures of weight k as a $\mathbb{Q}$-linear map with $\varphi(H^{p,q}) \subset H'^{p,q}$ (or, equivalently, with $\varphi(\rho(z)(\alpha)) = \rho'(z)(\varphi(\alpha))$), then a polarization is in fact a strict homomorphism of Hodge structures $(\, , \,) : V \otimes V \to \mathbb{Q}(-k)$ satisfying the positivity condition ii).

Example 3.C.9 The motivating example is, of course, the Hodge–Riemann pairing $(\alpha, \beta) = (-1)^{k(k-1)/2} \int_X \alpha \wedge \beta \wedge \omega^{n-k}$ on the primitive cohomology $H^k(X)_{\mathrm{p}}$ of a compact Kähler manifold X of dimension n. Of course, only if ω is a rational Kähler class this pairing is rational. But for the positivity condition ii) the rationality of ω is of no importance. We leave this straightforward calculation to the reader.

Starting with a free $\mathbb{Z}$-free module H instead of a $\mathbb{Q}$-vector space, one can as well introduce the notion of an integral Hodge structure. Two integral Hodge structures H and H' are isomorphic if and only if there exists a $\mathbb{Z}$-module isomorphism $\varphi : H \cong H'$ such that its $\mathbb{C}$-linear extension satisfies $\varphi_{\mathbb{C}}(H^{p,q}) = H'^{p,q}$.

Proposition 3.C.10 *There is a natural bijection between the set of isomorphism classes of integral Hodge structures of weight one and the set of isomorphism classes of complex tori.*

Proof. Let H be an integral Hodge structure of weight one. Then $H \subset H_{\mathbb{C}}$ can be projected injectively into $H^{1,0}$. This yields a lattice $H \subset H^{1,0}$ and $H^{1,0}/H$ is a complex torus. Clearly, if H and H' are isomorphic weight one integral Hodge structures then $H^{1,0}/H$ and $H'^{1,0}/H'$ are isomorphic complex tori.

Conversely, if $\mathbb{C}^n/\Gamma$ is a complex torus then $\mathbb{C}^n$ can be regarded as $\Gamma_{\mathbb{R}}$ endowed with an almost complex structure. This yields a decomposition $(\Gamma_{\mathbb{R}})_{\mathbb{C}} = (\Gamma_{\mathbb{R}})^{1,0} \oplus (\Gamma_{\mathbb{R}})^{0,1}$, i.e. an integral Hodge structure of weight one. (This has been explained in detail in Section 1.2.) The two constructions are inverse to each other, due to the existence of a $\mathbb{C}$-linear isomorphism $\mathbb{C}^n \cong \Gamma_{\mathbb{R}} \cong (\Gamma_{\mathbb{R}})^{1,0}$. To conclude, one has to verify that any isomorphism between two complex tori $\mathbb{C}^n/\Gamma$ and $\mathbb{C}^n/\Gamma'$ is induced by a $\mathbb{C}$-linear isomorphism $\varphi : \mathbb{C}^n \cong \mathbb{C}^n$ with $\varphi(\Gamma) = \Gamma'$. For this see page 136. $\qquad\square$

Proposition 3.C.11 *Let H be an integral Hodge structure of weight one equipped with a polarization (for the induced rational Hodge structure). Then the induced complex torus is abelian, i.e. it is projective.*

The proposition can e.g. be proved by applying the Kodaira embedding theorem that will be explained in Section 5.3 and the interpretation of $(\ ,\ )$ as a rational Kähler form. See Corollary 5.3.5. By multiplication with an integer we may assume that the polarization is in fact integral

Comments: - Since Hodge structures are so important, there are many excellent introductions to the subject. E.g. [53]. Note that various conventions for the action of $\mathbb{C}^*$ are used in the literature, e.g. one could define $H^{p,q}$ as the subspace on which $\mathbb{C}^*$ acts as $z^{-p}\bar{z}^{-q}$. This leads to different conventions for the definition of a polarization.

- For variations of Hodge structures and period domains we recommend the original sources [21] and the recent book [22].

- If an algebraic manifold is not compact, then the cohomology carries what is called a mixed Hodge structure. See [41, 113].

- For the theory of abelian varieties see [28, 76].

4

Vector Bundles

This chapter provides indispensable tools in the study of complex manifolds: connections, curvature, and Chern classes. In contrast to previous sections, we will not focus on the holomorphic tangent bundle of a complex manifold but allow arbitrary holomorphic vector bundles. However, we will not be in the position to undertake an indepth analysis of certain fundamental questions. E.g. the question whether there exist non-trivial bundles on a given complex manifold (or holomorphic structures on a given complex bundle) will not be addressed. This is partially due to the limitations of the book, but also to the state of art. The situation is fairly well understood only for curves and projective surfaces.

In Sections 5.1 to 5.3 the reader can find a number of central results in complex algebraic geometry. Except for the Hirzebruch–Riemann–Roch theorem, complete proofs, in particular of Kodaira's vanishing and embedding theorems, are provided. These three results are of fundamental importance in the global theory of complex manifolds. Roughly, in conjunction they allow to determine the size of linear systems on a manifold X and, if X is projective, how it can be embedded into a projective space.

In the appendices we discuss the interplay between the complex geometry of holomorphic vector bundles and related structures: Appendix 4.A tries to clarify the relation between Riemannian and Kähler geometry. In particular, we will show that for Kähler manifolds the Levi-Civita connection coincides with the Chern connection. The concept of holonomy, well-known in classical Riemannian geometry, allows to view certain features in complex geometry from a different angle. Appendix 4.B outlines fundamental results about Kähler–Einstein and Hermite–Einstein metrics. Before, the hermitian structure on a holomorphic vector bundle was used as an additional datum in order to apply Hodge theory, etc. One might wonder, whether natural hermitian structures, satisfying certain compatibility conditions, can be found. This leads to the concept of Hermite–Einstein metrics, which exist on certain privileged holomorphic bundles. If the holomorphic bundle happens to be the tangent bundle, this is related to Kähler–Einstein metrics.

4.1 Hermitian Vector Bundles and Serre Duality

In Chapter 3 we studied complex manifolds together with a compatible Riemannian metric, so called hermitian manifolds or, more restrictive, Kähler manifolds. The Riemannian metric gives rise to an hermitian metric on the (holomorphic) tangent bundle. More generally, one could and should be interested in hermitian metrics on arbitrary holomorphic and complex vector bundles. This twisted version will be discussed now. Many of the arguments will be familiar to the reader. Repeating Hodge theory on compact hermitian manifolds, this time for vector bundles, might help to get used to this important technique.

Let E be a complex vector bundle over a real manifold M.

Definition 4.1.1 An *hermitian structure* h on $E \to M$ is an hermitian scalar product h_x on each fibre $E(x)$ which depends differentiably on x. The pair (E, h) is called an *hermitian vector bundle*.

The latter condition can be made more precise in terms of local trivializations. Let $\psi : E|_U \cong U \times \mathbb{C}^r$ be a trivialization over some open subset U. Then, for any $x \in U$ the form $h_x(\psi_x^{-1}(\ ), \psi_x^{-1}(\ ))$ defines an hermitian scalar product on $\mathbb{C}^r$. In other words, h_x is given by a positive-definite hermitian matrix $(h_{ij}(x))$ (which depends on ψ) and we require the map $(h_{ij}) : U \to \mathrm{Gl}(r, \mathbb{C})$ to be differentiable.

Examples 4.1.2 i) Let L be a (holomorphic) line bundle and let $s_1, \ldots, s_k$ be global (holomorphic) sections generating L everywhere, i.e. at every point at least one of them is non-trivial. Then one defines an hermitian structure on L by

$$h(t) = \frac{|\psi(t)|^2}{\sum |\psi(s_i)|^2},$$

where t is a point in the fibre $L(x)$ and ψ is a local trivialization of L around the point x. The definition does not depend on the chosen trivialization, as two of them only differ by a scalar factor. Observe that h is not holomorphic, i.e. even if a trivialization of L over an open subset is chosen holomorphic, the induced map $h : U \to \mathbb{C}^*$ is usually not holomorphic. By abuse of language, one sometimes says that h is given by $(\sum |s_i|^2)^{-1}$.

The standard example is $L = \mathcal{O}(1)$ over the projective space $\mathbb{P}^n$ and the standard globally generating sections $z_0, \ldots, z_n \in H^0(\mathbb{P}^n, \mathcal{O}(1))$.

ii) If (X, g) is an hermitian manifold then the tangent, the cotangent, and all form bundles $\bigwedge^{p,q} X$ have natural hermitian structures.

iii) If E and F are endowed with hermitian structures, then the associated bundles $E \oplus F$, $E \otimes F$, $\mathrm{Hom}(E, F)$, etc., inherit natural hermitian structures.

iv) If (E, h) is an hermitian vector bundle and $F \subset E$ is a subbundle, then the restriction of h to F endows F with an hermitian structure. One can define the orthogonal complement, $F^\perp \subset E$ of F with respect to h. It is easy to see, that the pointwise condition indeed yields a complex vector bundle. Moreover, the bundle E can be decomposed as $E = F \oplus F^\perp$ and $F^\perp$ is canonically (as a complex vector bundle) isomorphic to the quotient E/F. In particular, h also induces an hermitian structure on the quotient E/F.

v) If (E, h) is an hermitian vector bundle over an hermitian manifold (X, g), then the twisted form bundles $\bigwedge^{p,q} X \otimes E$ have natural hermitian structures.

Example 4.1.3 Let us consider the projective space $\mathbb{P}^n$ and the Euler sequence twisted by $\mathcal{O}(-1)$:

$$0 \longrightarrow \mathcal{O}(-1) \longrightarrow \mathcal{O}^{\oplus n+1} \longrightarrow T_{\mathbb{P}^n}(-1) \longrightarrow 0 .$$

The constant standard hermitian structure on $\mathcal{O}^{\oplus n+1}$ induces canonical hermitian structures h_1 on $\mathcal{O}(-1)$ and h_2 on $T_{\mathbb{P}^n}(-1)$ (see the previous example).

The hermitian structure h_1 on $\mathcal{O}(-1)$ is nothing but the dual of the canonical hermitian structure on $\mathcal{O}(1)$ determined by the choice of the basis $z_0, \ldots, z_n \in H^0(\mathbb{P}^n, \mathcal{O}(-1))$ as in i) of Examples 4.1.2. This is straightforward to verify and we leave the general version of this assertion as Exercise 4.1.1.

We next wish to identify h_2 as the tensor product of the Fubini–Study metric on $T_{\mathbb{P}^n}$ and h_1 (up to the constant factor 2π).

The verification is done on the open subset $U_0 = \{(z_0 : \ldots : z_n) \mid z_0 \neq 0\}$ with coordinates $w_i = \frac{z_i}{z_0}$, $i = 1, \ldots, n$. Hence, with respect to the bases $\frac{\partial}{\partial w_i}$ the Fubini–Study metric on $T_{\mathbb{P}^n}|_{U_i}$ is given (up to the factor 2π) by the matrix

$$H := \left(1 + \sum |w_i|^2\right)^{-2} \cdot \left(\left(1 + \sum |w_i|^2\right)\delta_{ij} - \bar{w}_i w_j\right)_{ij}$$

(see i), Examples 3.1.9). The induced hermitian structure on the dual $\Omega_{\mathbb{P}^n}|_{U_0}$ with respect to the dual basis $dw_1, \ldots, dw_n$ corresponds thus to the matrix $\bar{H}^{-1} = \left(1 + \sum |w_i|^2\right) \cdot (\delta_{ij} + \bar{w}_i w_j)_{ij}$

On the other hand, the inclusion $\Omega_{\mathbb{P}^n} \subset \mathcal{O}(-1)^{\oplus n+1}$ given by the Euler sequence is on U_0 explicitly given by $dw_i \mapsto e_i - w_i \cdot e_0$ (see the proof of Proposition 2.4.4). Since h_1^* on $\mathcal{O}(-1)|_{U_0}$ is the scalar function $(1 + \sum |w_i|^2)$, one finds that the hermitian structure on $\Omega_{\mathbb{P}^n}|_{U_0}$ induced by this inclusion is

$$\left(1 + \sum |w_i|^2\right) \cdot \left((e_i - w_i \cdot e_0, e_j - w_j \cdot e_0)\right)_{ij} = \bar{H}^{-1}.$$

Note that choosing another basis of $H^0(\mathbb{P}^n, \mathcal{O}(1))$, which in general results in a different hermitian structure on $\mathcal{O}(1)$, amounts to choosing a different, though still constant, hermitian structure on the trivial bundle $\mathcal{O}^{\oplus n+1}$ on the middle term of the Euler sequence. So, more invariantly, one could work with the Euler sequence on $\mathbb{P}(V)$, where the middle term is $V \otimes \mathcal{O}$, and an hermitian structure on V.

An hermitian structure h on a vector bundle E defines a $\mathbb{C}$-antilinear isomorphism (of real bundles) $E \cong E^*$. Here, E^* is the dual complex bundle of E. Generalizing Exercise 3.1.1 we observe the following

Proposition 4.1.4 *Every complex vector bundle admits an hermitian metric.*

Proof. Choose an open covering $X = \bigcup U_i$ trivializing a given vector bundle E. Then one might glue the constant hermitian structures on the trivial vector bundles $U_i \times \mathbb{C}^r$ over U_i by means of a partition of unity.

Here, we use that any positive linear combination of positive definite hermitian products on $\mathbb{C}^n$ is again positive definite and hermitian. $\square$

Let $f : M \to N$ be a differentiable map and let E be a vector bundle on N endowed with an hermitian structure h. Then the pull-back vector bundle f^*E gets a natural hermitian structure f^*h by $(f^*h)_x = h_{f(x)}$ on $(f^*E)(x) = E(f(x))$.

Example 4.1.5 Let X be a complex manifold and let $s_0, \ldots, s_k$ be globally generating holomorphic sections of a holomorphic line bundle L (see Remark 2.3.27, ii)). By Proposition 2.3.26 there exists an induced morphism $\varphi : X \to \mathbb{P}^k$, $x \mapsto (s_0(x) : \ldots : s_k(x))$ with $\varphi^*\mathcal{O}(1) = L$ and $\varphi^*(z_i) = s_i$. The natural hermitian structures h on $\mathcal{O}(1)$ and h' on L induced by $z_0, \ldots, z_k$ and $s_0, \ldots, s_k$, respectively, (see Example 4.1.2, i)) are compatible under φ, i.e. $\varphi^*h = h'$.

Let (X, g) be an hermitian manifold and let (E, h) be an hermitian vector bundle on X. Then the induced hermitian structures on $\bigwedge^{p,q} X \otimes E$ will be denoted $(\ , \ )$.

Definition 4.1.6 Let E be a complex vector bundle over an hermitian manifold (X, g) of complex dimension n. An hermitian structure h on E is interpreted as a $\mathbb{C}$-antilinear isomorphism $h : E \cong E^*$. Then

$$\bar{*}_E : \bigwedge^{p,q} X \otimes E \longrightarrow \bigwedge^{n-p,n-q} X \otimes E^*$$

is defined by $\bar{*}_E(\varphi \otimes s) = \bar{*}(\varphi) \otimes h(s) = \overline{*(\varphi)} \otimes h(s) = *(\bar\varphi) \otimes h(s)$. (Recall that $*$ is $\mathbb{C}$-linear on $\bigwedge^{p,q} X$.)

Clearly, $\bar{*}_E$ is a $\mathbb{C}$-antilinear isomorphism that depends on g and h. Note that with this definition we have

$$(\alpha, \beta) * 1 = \alpha \wedge \bar{*}_E(\beta)$$

for α, β sections of $\bigwedge^{p,q} X \otimes E$, where "$\wedge$" is the exterior product in the form part and the evaluation map $E \otimes E^* \to \mathbb{C}$ in the bundle part. It is not difficult to verify that, as for the usual Hodge $*$-operator, one has $\bar{*}_{E^*} \circ \bar{*}_E = (-1)^{p+q}$ on $\bigwedge^{p,q} X \otimes E$.

The aim of this section is to generalize Poincaré duality for compact manifolds (or rather Serre duality (cf. Remark 3.2.7, ii) and Exercise 3.2.2) to a duality for the cohomology groups of holomorphic vector bundles. In order to do this we need to discuss Hodge theory in analogy to the discussion in Section 3.2 for hermitian vector bundles. Let us begin with the definition of the adjoint operator of $\bar{\partial}_E$.

Definition 4.1.7 Let (E, h) be a holomorpic vector bundle together with an hermitian structure h on an hermitian manifold (X, g). The operator $\bar{\partial}_E^* : \mathcal{A}^{p,q}(E) \to \mathcal{A}^{p,q-1}(E)$ is defined as

$$\bar{\partial}_E^* := -\bar{*}_{E^*} \circ \bar{\partial}_{E^*} \circ \bar{*}_E.$$

Remark 4.1.8 For $E = \mathcal{O}_X$ with a constant hermitian structure one recovers the adjoint operator $\bar{\partial}^* = - * \circ \partial \circ *$. Indeed, $-\bar{*}(\bar{\partial}(\bar{*}\varphi)) = -\bar{*}(\bar{\partial}(\overline{*\varphi})) = -\bar{*}(\overline{\partial * \varphi}) = - * (\partial * \varphi)$.

Definition 4.1.9 Let E be a holomorphic vector bundle endowed with an hermitian structure h on an hermitian manifold (X, g), then the *Laplace operator* on $\mathcal{A}^{p,q}(E)$ is defined by

$$\Delta_E := \bar{\partial}_E^* \bar{\partial}_E + \bar{\partial}_E \bar{\partial}_E^*.$$

Definition 4.1.10 Let (E, h) be an hermitian holomorphic vector bundle over an hermitian manifold (X, g). A section α of $\bigwedge^{p,q} X \otimes E$ is called *harmonic* if $\Delta_E(\alpha) = 0$. The space of all harmonic forms is denoted $\mathcal{H}^{p,q}(X, E)$, where we omit g and h in the notation.

Observe that $\bar{*}_E$ induces a $\mathbb{C}$-antilinear isomorphism

$$\bar{*}_E : \mathcal{H}^{p,q}(X, E) \cong \mathcal{H}^{n-p,n-q}(X, E^*).$$

Definition 4.1.11 Let (E, h) be an hermitian vector bundle on a compact hermitian manifold (X, g). Then a natural hermitian scalar product on $\mathcal{A}^{p,q}(X, E)$ is defined by

$$(\alpha, \beta) := \int_X (\alpha, \beta) * 1,$$

where $(,)$ is the hermitian product on $\bigwedge^{p,q} X \otimes E$ depending on h and g (cf. Example 4.1.2).

Lemma 4.1.12 *Let (E, h) be an hermitian holomorphic vector bundle on a compact hermitian manifold (X, g). Then, with respect to $(,)$, the operator $\bar{\partial}_E^*$ on $\mathcal{A}^{p,q}(X, E)$ is adjoint to $\bar{\partial}_E$ and Δ_E is self-adjoint.*

Proof. By definition, the second assertion follows from the first one which in turn is proved by the following purely formal calculation:

For $\alpha \in \mathcal{A}^{p,q}(X, E)$ and $\beta \in \mathcal{A}^{p,q+1}(X, E)$ one has

$$(\alpha, \bar{\partial}_E^* \beta) = -(\alpha, \bar{*}_{E^*} \circ \bar{\partial}_{E^*} \circ \bar{*}_E \beta)$$

$$= -\int_X \alpha \wedge \bar{*}_E \bar{*}_{E^*} \bar{\partial}_{E^*} \bar{*}_E \beta$$

$$= (-1)^{n-p+n-q-1} \int_X \alpha \wedge \bar{\partial}_{E^*} \bar{*}_E \beta$$

$$= \int_X \bar{\partial}_E(\alpha) \wedge \bar{*}_E \beta = (\alpha, \beta).$$

Here we use the Leibniz rule $\bar{\partial}(\alpha \wedge \bar{*}_E \beta) = \bar{\partial}_E(\alpha) \wedge \bar{*}_E \beta + (-1)^{p+q} \alpha \wedge \bar{\partial}_{E^*} \bar{*}_E \beta$ and Stokes' theorem $\int_X \bar{\partial}(\alpha \wedge \bar{*}_E \beta) = \int_X d(\alpha \wedge \bar{*}_E \beta) = 0$. □

Using the lemma the reader may check that a form $\alpha \in \mathcal{A}^{p,q}(X, E)$ over a compact manifold X is harmonic if and only if α is $\bar{\partial}_E$- and $\bar{\partial}_E^*$-closed (cf. Lemma 3.2.5).

Theorem 4.1.13 (Hodge decomposition) *Let E be a holomorphic vector bundle together with an hermitian structure h on a compact hermitian manifold (X, g). Then*

$$\mathcal{A}^{p,q}(X, E) = \bar{\partial}_E \mathcal{A}^{p,q-1}(X, E) \oplus \mathcal{H}^{p,q}(X, E) \oplus \bar{\partial}_E^* \mathcal{A}^{p,q+1}(X, E) \qquad (4.1)$$

and $\mathcal{H}^{p,q}(X, E)$ is finite-dimensional. □

The case of the trivial vector bundle $E \cong \mathcal{O}_X$ with a constant hermitian structure corresponds to Hodge decomposition of compact hermitian manifolds (cf. Theorem 3.2.8). As in this case, we obtain

Corollary 4.1.14 *The natural projection $\mathcal{H}^{p,q}(X, E) \to H^{p,q}(X, E)$ is bijective. In particular, $H^{p,q}(X, E) \cong H^q(X, E \otimes \Omega_X^p)$ is finite-dimensional.*

Proof. Indeed, as any harmonic section of $\bigwedge^{p,q} X \otimes E$ is $\bar{\partial}_E$-closed, the projection is well-defined. Moreover, the space of $\bar{\partial}_E$-closed forms in $\mathcal{A}^{p,q}(X, E)$ is $\bar{\partial}_E \mathcal{A}^{p,q-1}(X, E) \oplus \mathcal{H}^{p,q}(X, E)$, as $(\bar{\partial}_E \bar{\partial}_E^* \alpha, \alpha) = \|\bar{\partial}_E^* \alpha\|^2 \neq 0$ for $\bar{\partial}_E^* \alpha \neq 0$.

Thus, the projection is surjective and its kernel is the space of forms, which are $\bar{\partial}_E$-exact and harmonic. But since the decomposition (4.1) in theorem 4.1.13 is direct, this space is trivial. □

Let E be a holomorphic vector bundle over a compact manifold X of dimension n and consider the natural pairing

$$H^{p,q}(X, E) \times H^{n-p,n-q}(X, E^*) \longrightarrow \mathbb{C}, \quad (\alpha, \beta) \longmapsto \int_X \alpha \wedge \beta,$$

where as before $\alpha \wedge \beta$ is the exterior product in the form part and the evaluation map in the bundle part. The pairing is well-defined, i.e. does not depend on the $\bar{\partial}$-closed representatives $\alpha \in \mathcal{A}^{p,q}(E)$ and $\beta \in \mathcal{A}^{n-p,n-q}(E)$.

Proposition 4.1.15 (Serre duality) *Let X be a compact complex manifold. For any holomorphic vector bundle E on X the natural pairing*

$$H^{p,q}(X, E) \times H^{n-p,n-q}(X, E^*) \longrightarrow \mathbb{C}$$

is non-degenerate.

Proof. Fix hermitian structures h and g on E and X, respectively. Then consider the pairing $\mathcal{H}^{p,q}(X, E) \times \mathcal{H}^{n-p,n-q}(X, E^*) \to \mathbb{C}$. In order to show that this pairing is non-degenerate, we have to show that for any $0 \neq \alpha \in \mathcal{H}^{p,q}(X, E)$ there exists an element $\beta \in \mathcal{H}^{n-p,n-q}(X, E^*)$ with $\int_X \alpha \wedge \beta \neq 0$. Now choose $\beta := \bar{*}_E \alpha$, then $\int \alpha \wedge \beta = \int \alpha \wedge \bar{*}_E \alpha = \int (\alpha, \alpha) * 1 = \|\alpha\|^2 \neq 0$. $\square$

Serre duality (together with the Hirzebruch–Riemann–Roch theorem 5.1.1 and Kodaira vanishing theorem 5.2.2) is one of the most useful tools to control the cohomology of holomorphic vector bundles.

Let us mention a few special cases and reformulations.

Corollary 4.1.16 *For any holomorphic vector bundle E over a compact complex manifold X there exist natural $\mathbb{C}$-linear isomorphisms* (Serre duality):

$$H^{p,q}(X, E) \cong H^{n-p,n-q}(X, E^*)^*$$
$$H^q(X, \Omega^p \otimes E) \cong H^{n-q}(X, \Omega^{n-p} \otimes E^*)^*$$
$$H^q(X, E) \cong H^{n-q}(X, K_X \otimes E^*)^*$$

$\square$

For the trivial bundle this yields $H^{p,q}(X) \cong H^{n-p,n-q}(X)^*$ (cf. Exercise 3.2.2). Moreover, if X is Kähler these isomorphisms are compatible with the bidegree decomposition $H^k(X, \mathbb{C}) = \bigoplus H^{p,q}(X)$ and Poincaré duality (cf. Exercise 3.2.3).

Remark 4.1.17 The isomorphism $\bar{*}_E : \mathcal{H}^{p,q}(X, E) \cong \mathcal{H}^{n-p,n-q}(X, E^*)$ induces an isomorphism $H^{p,q}(X, E) \cong H^{n-p,n-q}(X, E^*)$. But this isomorphism is only $\mathbb{C}$-antilinear and depends on the chosen hermitian structures g and h. Thus, Serre duality $H^{p,q}(X, E) \cong H^{n-p,n-q}(X, E^*)^*$ is better behaved in both respects.

Exercises

4.1.1 Let L be a holomorphic line bundle which is globally generated by sections $s_1, \ldots, s_k \in H^0(X, L)$. Then L admits a canonical hermitian structure h defined in Example 4.1.2. The dual bundle L^* obtains a natural hermitian structure h' via the inclusion $L^* \subset \mathcal{O}^{\oplus k}$. Describe h' and and show that $h' = h^*$.

4.1.2 Let L be a holomorphic line bundle of degree $d > 2g(C) - 2$ on a compact curve C. Show that $H^1(C, L) = 0$. Here, for our purpose we define the *genus* $g(C)$ of C by the formula $\deg(K_X) = 2g(C) - 2$.

In other words, $H^1(C, K_C \otimes L) = 0$ for any holomorphic line bundle L with $\deg(L) > 0$. In this form, it will later be generalized to the Kodaira vanishing theorem for arbitrary compact Kähler manifolds.

4.1.3 Show, e.g. by writing down an explicit basis, that

$$h^n(\mathbb{P}^n, \mathcal{O}(k)) = \begin{cases} 0 & k > -n - 1 \\ \binom{-k-1}{-n-1-k} & k \le -n - 1 \end{cases}$$

4.1.4 Let E be an hermitian holomorphic vector bundle on a compact Kähler manifold X. Show that any section $s \in H^0(X, \Omega^p \otimes E)$ is harmonic.

4.1.5 Compare this section with the discussion in Sections 3.2 and 3.3. In particular, check whether the Lefschetz operator L is defined on $H^{p,q}(X, E)$ and whether it defines isomorphisms $H^{p,k-p}(X, E) \to H^{n+p-k,n-p}(X, E)$ (cf. Remark 3.2.7, iii)).

Comments: Serre duality does in fact hold, in an appropriately modified form, for arbitrary coherent sheaves. Moreover, it is a special case of the so called Grothendieck–Verdier duality which is a duality statement for the direct image of coherent sheaves under proper morphisms. An algebraic proof, i.e. without using any metrics, can be given in case the manifold is projective

4.2 Connections

Let M be a real manifold and let $\pi : E \to M$ be a complex vector bundle on M. As before, we denote by $\mathcal{A}^i(E)$ the sheaf of i-forms with values in E. In particular, $\mathcal{A}^0(E)$ is just the sheaf of sections of E. Sections of E cannot be differentiated canonically, i.e. the exterior differential is in general not defined (see the discussion in Section 2.6). A substitute for the exterior differential is provided by a connection on E, which is not canonical, but always available.

This section introduces the reader to the fundamental notion of a connection and studies various compatibility conditions with additional data, like holomorphic or hermitian structures. At the end of this section, a short discussion of a purely holomorphic and more rigid analogue is introduced.

We will focus on complex vector bundles, but for almost everything a real version exists. We leave it to the reader to work out the precise formulation in each case.

Definition 4.2.1 A *connection* on a vector bundle E is a $\mathbb{C}$-linear sheaf homomorphism $\nabla : \mathcal{A}^0(E) \to \mathcal{A}^1(E)$ which satisfies the *Leibniz rule*

$$\nabla(f \cdot s) = d(f) \otimes s + f \cdot \nabla(s) \tag{4.2}$$

for any local function f on M and any local section s of E.

Definition 4.2.2 A section s of a vector bundle E is called *parallel* (or *flat* or *constant*) with respect to a connection ∇ on E if $\nabla(s) = 0$.

Proposition 4.2.3 *If ∇ and ∇' are two connections on a vector bundle E, then $\nabla - \nabla'$ is $\mathcal{A}_M^0$-linear and can, therefore, be considered as an element in $\mathcal{A}^1(M, \operatorname{End}(E))$. If ∇ is a connection on E and $a \in \mathcal{A}^1(M, \operatorname{End}(E))$, then $\nabla + a$ is again a connection on E.*

Proof. We have to show that $(\nabla - \nabla')(f \cdot s) = f \cdot (\nabla - \nabla')(s)$, which is an immediate consequence of the Leibniz rule (4.2).

An element $a \in \mathcal{A}^1(M, \operatorname{End}(E))$ acts on $\mathcal{A}^0(E)$ by multiplication in the form part and evaluation $\operatorname{End}(E) \times E \to E$ on the bundle component. In order to prove the second assertion, one checks $(\nabla + a)(f \cdot s) = \nabla(f \cdot s) + a(f \cdot s) = d(f) \otimes s + f \cdot \nabla(s) + fa(s) = d(f) \otimes s + f \cdot (\nabla + a)(s)$. Thus, $\nabla + a$ satisfies the Leibniz rule and is, therefore, a connection. $\qquad\square$

As a consequence of this proposition and Exercise 4.2.1 one obtains

Corollary 4.2.4 *The set of all connections on a vector bundle E is in a natural way an affine space over the (infinite-dimensional) complex vector space $\mathcal{A}^1(M, \operatorname{End}(E))$.* $\qquad\square$

Remark 4.2.5 Often, local calculations are performed by using the following statement: Any connection ∇ on a vector bundle E can locally be written as $d + A$, where A is a matrix valued one-form.

Indeed, if E is the trivial vector bundle $E = M \times \mathbb{C}^r$, then $\mathcal{A}^k(E) = \bigoplus_{i=1}^r \mathcal{A}_M^k$ and one defines the trivial connection $d : \mathcal{A}^0(E) \to \mathcal{A}^1(E)$ on E by applying the usual exterior differential to each component. Any other connection ∇ is then of the form $\nabla = d + A$, where $A \in \mathcal{A}^1(M, \mathrm{End}(E))$. For the trivial vector bundle, the latter is just a matrix valued one-form.

Let E be an arbitrary vector bundle on M endowed with a connection ∇. With respect to a trivialization $\psi : E|_U \cong U \times \mathbb{C}^r$ we may write $\nabla = d + A$ or, more precisely, $\nabla = \psi^{-1} \circ (d + A) \circ \psi$. If the trivialization is changed by $\phi : U \to \mathrm{Gl}(r, \mathbb{C})$, i.e. one considers $\psi'_x = \phi(x) \circ \psi_x$, then $\nabla = \psi'^{-1} \circ (d + A') \circ \psi'$ with $A' = \phi^{-1} d(\phi) + \phi^{-1} A \phi$.

For a given point $x_0 \in M$ one can always choose the local trivialization such that $A(x_0) = 0$. Indeed, a given trivialization can be changed by a local $\phi : U \to \mathrm{Gl}(r, \mathbb{C})$, whose Taylor expansion is of the form

$$\phi(x) = \mathrm{Id} - \sum x_i A_i(0) + \text{higher order terms.}$$

Here, $x_1, \ldots, x_n$ are local coordinates with x_0 as the origin and the connection matrix A is written as $A = \sum A_i dx_i$.

Given connections induce new connections on associated vector bundles. Here is a list of the most important examples of this principle:

Examples 4.2.6 i) Let E_1 and E_2 be vector bundles on M endowed with connections ∇_1 and ∇_2, respectively. If s_1, s_2 are local section of E_1 and E_2, respectively, we set

$$\nabla(s_1 \oplus s_2) = \nabla_1(s_1) \oplus \nabla_2(s_2).$$

This defines a natural connection on the direct sum $E_1 \oplus E_2$.

ii) In order to define a connection on the tensor product $E_1 \otimes E_2$ one defines

$$\nabla(s_1 \otimes s_2) = \nabla_1(s_1) \otimes s_2 + s_1 \otimes \nabla_2(s_2).$$

iii) A natural connection on $\mathrm{Hom}(E_1, E_2)$ can be defined by:

$$\nabla(f)(s_1) = \nabla_2(f(s_1)) - f(\nabla_1(s_1)).$$

Here, f is a local homomorphism $E_1 \to E_2$. Then $f(s_1)$ is a local section of E_2 and ∇_2 can be applied. In the second term the homomorphism f is applied to the one-form $\nabla_1(s_1)$ with values in E_1 according to $f(\alpha \otimes t) = \alpha \otimes f(t)$, for $\alpha \in \mathcal{A}^1$ and $t \in \mathcal{A}^0(E)$.

iv) If we endow the trivial bundle with the natural connection given by the exterior differential, then the last construction yields as a special case a connection ∇^* on the dual E^* of any bundle E equipped with a connection ∇. Explicitly, one has

$$\nabla^*(f)(s) = d(f(s)) - f(\nabla(s)).$$

Clearly, changing any of the given connections by a one-form as in Proposition 4.2.3 induces new connections on the associated bundles. E.g. if $\nabla' = \nabla + a$, then $\nabla'^* = \nabla^* - a^*$. The other cases are left to the reader (cf. Exercise 4.2.2).

v) Let $f : M \to N$ be a differentiable map and let ∇ be a connection on a vector bundle E over N. Let ∇ over an open subset $U_i \subset N$ be of the form $d + A_i$ (after trivializing $E|_{U_i}$). Then the pull-back connection $f^*\nabla$ on the pull-back vector bundle f^*E over M is locally defined by $f^*\nabla|_{f^{-1}(U_i)} = d + f^*A_i$. It is straightforward to see that the locally given connections glue to a global one on f^*E.

Next we shall describe how conversely a connection ∇ on the direct sum $E = E_1 \oplus E_2$ induces connections ∇_1 and ∇_2 on E_1 and E_2, respectively. Denote by p_1 and p_2 the two projections $E_1 \oplus E_2 \to E_i$. Clearly, any section s_i of E_i can also be regarded as a section of E and thus ∇ can be applied. Then we set $\nabla_i(s_i) := p_i(\nabla(s_i))$. The verification of the Leibniz rule for ∇_i is straightforward. Thus we obtain

Lemma 4.2.7 *The connection ∇ on $E = E_1 \oplus E_2$ induces natural connections ∇_1, ∇_2 on E_1 and E_2, respectively.* $\square$

The difference between the direct sum $\nabla_1 \oplus \nabla_2$ of the two induced connections and the connection ∇ on E we started with is measured by the second fundamental form. Let E_1 be a subbundle of a vector bundle E and assume that a connection on the latter is given.

Definition 4.2.8 The *second fundamental form* of $E_1 \subset E$ with respect to the connection ∇ on E is the section $b \in \mathcal{A}^1(M, \mathrm{Hom}(E_1, E/E_1))$ defined for any local section s of E_1 by

$$b(s) = \mathrm{pr}_{E/E_1}(\nabla(s)).$$

If we choose a splitting of $E \twoheadrightarrow E/E_1$, i.e. we write $E = E_1 \oplus E_2$ with $E_2 \cong E/E_1$ via the projection, then $b(s) = \nabla(s) - \nabla_1(s)$. Using the Leibniz rule (4.2) for ∇ and ∇_1 one proves $b(f \cdot s) = f \cdot b(s)$. Thus, b really defines an element in $\mathcal{A}^1(X, \mathrm{Hom}(E_1, E_2))$. Often we will consider situations where E is the trivial vector bundle together with the trivial connection

If E is endowed with an additional datum, e.g. an hermitian or a holomorphic structure, then one can formulate compatibility conditions for connections on E. Let us first discuss the hermitian case.

Definition 4.2.9 Let (E, h) be an hermitian vector bundle. A connection ∇ on E is an *hermitian connection* with respect to h if for arbitrary local sections s_1, s_2 one has

$$d(h(s_1, s_2)) = h(\nabla(s_1), s_2) + h(s_1, \nabla(s_2)). \tag{4.3}$$

Here, the exterior differential is applied to the function $h(s_1, s_2)$ and by definition $h(\alpha \otimes s, s') := \alpha h(s, s')$ for a (complex) one-form α and sections s and s'. Analogously, $h(s, \alpha \otimes s') = \bar{\alpha} h(s, s')$. If the bundle E is real and the hermitian product is a real hermitian product, then one speaks of metric connections. See also Exercise 4.2.8 for an alternative description of (4.3).

Let ∇ be an hermitian connection and let $a \in \mathcal{A}^1(M, \mathrm{End}(E))$. By Proposition 4.2.3 one knows that $\nabla' = \nabla + a$ is again a connection. Then, ∇' is hermitian if and only if $h(a(s_1), s_2) + h(s_1, a(s_2)) = 0$ for all section s_1, s_2. Thus, ∇' is hermitian if and only if a can locally be written as $a = \alpha \otimes A$ with $a \in \mathcal{A}_M^1$ and where A at each point is contained in the Lie algebra $\mathfrak{u}(E(x), h(x))$, which, after diagonalization of h, is the Lie algebra of all skew-hermitian matrices.

Definition 4.2.10 Let (E, h) be an hermitian vector bundle. By $\mathrm{End}(E, h)$ we denote the subsheaf of sections a of $\mathrm{End}(E)$ satisfying

$$h(a(s_1), s_2) + h(s_1, a(s_2)) = 0$$

for all local sections s_1, s_2.

Note that $\mathrm{End}(E, h)$ has the structure of a real vector bundle. For line bundles, i.e. $\mathrm{rk}(E) = 1$, the vector bundle $\mathrm{End}(E)$ is the trivial complex vector bundle $\mathbb{C} \times M$ and $\mathrm{End}(E, h)$ is the imaginary part $i \cdot \mathbb{R} \times M$ of it. Then using Corollary 4.2.4 and Exercise 4.2.1 we find

Corollary 4.2.11 *The set of all hermitian connections on an hermitian vector bundle (E, h) is an affine space over the (infinite-dimensional) real vector space $\mathcal{A}^1(M, \mathrm{End}(E, h))$.* □

So far, the underlying manifold M was just a real manifold and the vector bundle E was a differentiable complex (or real) vector bundle. In what follows, we consider a holomorphic vector bundle E over a complex manifold X. Recall from Section 2.6 that in this case there exists the $\bar{\partial}$-operator $\bar{\partial} : \mathcal{A}^0(E) \to \mathcal{A}^{0,1}(E)$.

Using the decomposition $\mathcal{A}^1(E) = \mathcal{A}^{1,0}(E) \oplus \mathcal{A}^{0,1}(E)$ we can decompose any connection ∇ on E in its two components $\nabla^{1,0}$ and $\nabla^{0,1}$, i.e. $\nabla = \nabla^{1,0} \oplus \nabla^{0,1}$ with

$$\nabla^{1,0} : \mathcal{A}^0(E) \longrightarrow \mathcal{A}^{1,0}(E) \quad \text{and} \quad \nabla^{0,1} : \mathcal{A}^0(E) \longrightarrow \mathcal{A}^{0,1}(E).$$

Note that $\nabla^{0,1}$ satisfies $\nabla^{0,1}(f \cdot s) = \bar{\partial}(f) \otimes s + f \cdot \nabla^{0,1}(s)$, i.e. it behaves similarly to $\bar{\partial}$. (Of course, the decomposition $\nabla = \nabla^{1,0} \oplus \nabla^{0,1}$ makes sense also when E is not holomorphic.)

Definition 4.2.12 A connection ∇ on a holomorphic vector bundle E is *compatible with the holomorphic structure* if $\nabla^{0,1} = \bar{\partial}$.

Similarly to Corollaries 4.2.4 and 4.2.11 one proves

Corollary 4.2.13 *The space of connections ∇ on a holomorphic vector bundle E compatible with the holomorphic structure forms an affine space over the (infinite-dimensional) complex vector space $\mathcal{A}^{1,0}(X, \mathrm{End}(E))$.* $\quad\square$

The existence of at least one such connection (which is needed for the corollary) can be proved directly or it can be seen as a consequence of the following existence result.

Proposition 4.2.14 *Let (E, h) be a holomorphic vector bundle together with an hermitian structure. Then there exists a unique hermitian connection ∇ that is compatible with the holomorphic structure. This connection is called the* Chern connection *on (E, h).*

Proof. Let us first show the uniqueness. This is is a purely local problem. Thus, we may assume that E is the trivial holomorphic vector bundle, i.e. $E = X \times \mathbb{C}^r$. According to Remark 4.2.5 the connection ∇ is of the form $\nabla = d + A$, where $A = (a_{ij})$ is a matrix valued one-form on X. The hermitian structure on E is given by a function H on X that associates to any $x \in X$ a positive-definite hermitian matrix $H(x) = (h_{ij}(x))$.

Let e_i be the constant i-th unit vector considered as a section of E. Then the assumption that ∇ be compatible with the hermitian structure yields $dh(e_i, e_j) = h(\sum a_{ki} e_k, e_j) + h(e_i, \sum a_{\ell j} e_\ell)$ or, equivalently,

$$dH = A^{\mathrm{t}} \cdot H + H \cdot \bar{A}.$$

Since ∇ is compatible with $\bar{\partial}$, the matrix A is of type $(1, 0)$. A comparison of types of both sides yields $\partial H = H \cdot \bar{A}$ and, after complex conjugation

$$A = \bar{H}^{-1} \partial(\bar{H}).$$

Thus, A is uniquely determined by H.

Equivalently, by using Corollaries 4.2.11 and 4.2.13 one could argue that $\mathcal{A}^1(X, \mathrm{End}(E, h)) \cap \mathcal{A}^{1,0}(X, \mathrm{End}(E)) = 0$. Indeed, any endomorphism a in this intersection satisfies $h(a(s_1), s_2) + h(s_1, a(s_2)) = 0$, where the first summand is a $(1, 0)$-form and the second is of type $(0, 1)$. Thus, both have to be trivial and hence $a = 0$.

In any case, describing the connection form A explicitly in terms of the hermitian structure H turns out to be helpful for the existence result as well. On argues as follows: Going the argument backwards, we find that locally one can find connections which are compatible with both structures. Due to the uniqueness, the locally defined connections glue. $\quad\square$

Example 4.2.15 Let E be a holomorphic line bundle. Then an hermitian structure H on E is given by a positive real function and the Chern connection E is locally given as $\nabla = d + \partial \log H$.

The proposition can be applied to the geometric situation. Let (X, g) be an hermitian manifold and let E be the holomorphic (co)tangent bundle. Then the Proposition asserts the existence of a natural hermitian connection $\nabla = \nabla^{1,0} + \bar{\partial}$ on the (co)tangent bundle T_X (respectively Ω_X). Let us study this in two easy cases.

Examples 4.2.16 i) If we endow the complex torus $\mathbb{C}^n / \Gamma$ with a constant hermitian structure, then the Chern connection on the trivial tangent bundle is the exterior differential.

ii) The second example is slightly more interesting. We study the Fubini–Study metric on $\mathbb{P}^n$ introduced in Examples 3.1.9, i). Recall that on the standard open subset $U_i \subset \mathbb{P}^n$ with coordinates $w_1, \ldots, w_n$ it is given by

$$H = \frac{1}{2\pi} \left(\frac{\delta_{ij}}{1 + \sum |w_k|^2} - \frac{\bar{w}_i w_j}{(1 + \sum |w_k|^2)^2} \right)_{i,j=1,\ldots,n}.$$

Then the distinguished connection we are looking for is locally on U_i given by $\nabla = d + \bar{H}^{-1}(\partial \bar{H})$.

Let us now study the second fundamental form for connections compatible with a given hermitian and/or holomorphic structure.

• Let

$$0 \longrightarrow E_1 \longrightarrow E \longrightarrow E_2 \longrightarrow 0$$

be a short exact sequence of holomorphic vector bundles. In general, a sequence like this need not split. However, the sequence of the underlying differentiable complex bundles can always be split and hence $E = E_1 \oplus E_2$ as complex bundles. (See Appendix A.)

If $\nabla^{0,1} = \bar{\partial}_E$, then also $\nabla_1^{0,1} = \bar{\partial}_{E_1}$ for the induced connection ∇_1 on E_1, because E_1 is a holomorphic subbundle of E. Thus, the second fundamental form b_1 is of type $(1,0)$, i.e. $b_1 \in \mathcal{A}^{1,0}(X, \mathrm{Hom}(E_1, E_2))$. The analogous statement holds true for ∇_2 on E_2 and the second fundamental form b_2 if and only if the chosen split is holomorphic.

• Let (E, h) be an hermitian vector bundle and assume that $E = E_1 \oplus E_2$ is an orthogonal decomposition, i.e. E_1, E_2 are both endowed with hermitian structures h_1 and h_2, respectively, such that $h = h_1 \oplus h_2$.

Let ∇ be a hermitian connection on (E, h). Then the induced connections ∇_1, ∇_2 are again hermitian (cf. Exercise 4.2.5) and for the fundamental forms b_1 and b_2 one has

$$h_1(s_1, b_2(t_2)) + h_2(b_1(s_1), t_2) = 0$$
$$h_1(b_2(s_2), t_1) + h_2(s_2, b_1(t_1)) = 0$$

for any local sections s_i and t_i of E_1 respectively E_2. This follows easily from

$$dh(s_1 \oplus s_2, t_1 \oplus t_2)$$
$$= h((\nabla_1(s_1) + b_2(s_2)) \oplus (\nabla_2(s_2) + b_1(s_1)), t_1 \oplus t_2)$$
$$+ h(s_1 \oplus s_2, (\nabla_1(t_1) + b_2(t_2)) \oplus (\nabla_2(t_2) + b_1(t_1)))$$

Usually, one combines both situations. A short exact sequence of holomorphic vector bundles can be splitted as above by choosing the orthogonal complement $E^\perp \cong E_2$ of $E_1 \subset E$ with respect to a chosen hermitian structure on E.

We conclude this section with a brief discussion of the notion of a holomorphic connection, which should not be confused with the notion of a connection compatible with the holomorphic structure. In fact, the notion of a holomorphic connection is much more restrictive, but has the advantage to generalize to the purely algebraic setting.

Definition 4.2.17 Let E be a holomorphic vector bundle on a complex manifold X. A *holomorphic connection* on E is a $\mathbb{C}$-linear map (of sheaves) $D : E \to \Omega_X \otimes E$ with

$$D(f \cdot s) = \partial(f) \otimes s + f \cdot D(s)$$

for any local holomorphic function f on X and any local holomorphic section s of E.

Here, E denotes both, the vector bundle and the sheaf of holomorphic sections of this bundle. Clearly, if f is a holomorphic function, then $\partial(f)$ is a holomorphic section of $\bigwedge^{1,0} X$, i.e. a section of Ω_X (use $\bar{\partial}\partial(f) = -\partial\bar{\partial}(f)$). (See Proposition 2.6.11.)

Most of what has been said about ordinary connections holds true for holomorphic connections with suitable modifications. E.g. if D and D' are holomorphic connections on E, then $D - D'$ is a holomorphic section of $\Omega_X \otimes \mathrm{End}(E)$. Locally, any holomorphic connection D is of the form $\partial + A$ where A is a holomorphic section of $\Omega_X \otimes \mathrm{End}(E)$.

Writing a holomorphic connection D locally as $\partial + A$ shows that D also induces a $\mathbb{C}$-linear map $D : \mathcal{A}^0(E) \to \mathcal{A}^{1,0}(E)$ which satisfies $D(f \cdot s) = \partial(f) \otimes s + f \cdot D(s)$. Thus, D looks like the $(1,0)$-part of an ordinary connection and, indeed, $\nabla := D + \bar{\partial}$ defines an ordinary connection on E.

However, the $(1,0)$-part of an arbitrary connection need not be a holomorphic connection in general. It might send holomorphic sections of E to those of $\mathcal{A}^{1,0}(E)$ that are not holomorphic, i.e. not contained in $\Omega_X \otimes E$. In fact, holomorphic connections exist only on very special bundles (see Remark 4.2.20 and Exercise 4.4.12).

We want to introduce a natural cohomology class whose vanishing decides whether a holomorphic connection on a given holomorphic bundle can be found. Let E be a holomorphic vector bundle and let $X = \bigcup U_i$ be an open covering such that there exist holomorphic trivializations $\psi_i : E|_{U_i} \cong U_i \times \mathbb{C}^r$.

Definition 4.2.18 The *Atiyah class*

$$A(E) \in H^1(X, \Omega_X \otimes \operatorname{End}(E))$$

of the holomorphic vector bundle E is given by the Čech cocycle

$$A(E) = \{U_{ij}, \psi_j^{-1} \circ (\psi_{ij}^{-1} d\psi_{ij}) \circ \psi_j\}.$$

Due to the cocycle condition $\psi_{ij} \psi_{jk} \psi_{ki} = 1$, the collection $\{U_{ij}, \psi_j^{-1} \circ (\psi_{ij}^{-1} d\psi_{ij}) \circ \psi_j\}$ really defines a cocycle. The definition of $A(E)$ is indeed independent of the cocycle $\{\psi_{ij}\}$. We leave the straightforward proof to the reader (see Exercise 4.2.9).

Proposition 4.2.19 *A holomorphic vector bundle E admits a holomorphic connection if and only if its Atiyah class $A(E) \in H^1(X, \Omega_X \otimes \operatorname{End}(E))$ is trivial.*

Proof. First note that $d\psi_{ij} = \partial \psi_{ij}$, as the ψ_{ij} are holomorphic.

Local holomorphic connections on $U_i \times \mathbb{C}^r$ are of the form $\partial + A_i$. Those can be glued to a connection on the bundle E if and only if

$$\psi_i^{-1} \circ (\partial + A_i) \circ \psi_i = \psi_j^{-1} \circ (\partial + A_j) \circ \psi_j$$

on U_{ij} or, equivalently,

$$\psi_i^{-1} \circ \partial \circ \psi_i - \psi_j^{-1} \circ \partial \circ \psi_j = \psi_j^{-1} A_j \psi_j - \psi_i^{-1} A_i \psi_i. \qquad (4.4)$$

The left hand side of (4.4) can be written as

$$\psi_j^{-1} \circ (\psi_{ij}^{-1} \circ \partial \circ \psi_i \circ \psi_j^{-1}) \circ \psi_j - \psi_j^{-1} \circ \partial \circ \psi_j$$
$$= \psi_j^{-1} \circ (\psi_{ij}^{-1} \circ \partial \circ \psi_{ij} - \partial) \circ \psi_j = \psi_j^{-1} \circ (\psi_{ij}^{-1} \partial(\psi_{ij})) \circ \psi_j$$

The right hand side of (4.4) is the boundary of $\{B_i \in \Gamma(U_i, \Omega \otimes \operatorname{End}(E))\}$ with $B_i = \psi_i^{-1} A_i \psi_i$. Thus, $A(E) = 0$ if and only if there exist local connections on E that can be glued to a global one. $\qquad \square$

Remarks 4.2.20 i) Later we will see that $A(E)$ is related to the curvature of E. Roughly, a holomorphic connection on a vector bundle E over a compact manifold exists, i.e. $A(E) = 0$, if and only if the bundle is flat. Moreover, we will see that $A(E)$ encodes all characteristic classes of E.

ii) Note that for vector bundles of rank one, i.e. line bundles, one has $A(L) = \{\partial \log(\psi_{ij})\}$. This gives yet another way of defining a first Chern class of a holomorphic line bundle as $A(E) \in H^1(X, \Omega_X) = H^1(X, \Omega_X \otimes \operatorname{End}(E))$. A comparison of the various possible definitions of the first Chern class encountered so far will be provided in Section 4.4.

Exercises

4.2.1 i) Show that any (hermitian) vector bundle admits a(n hermitian) connection.

ii) Show that a connection ∇ is given by its action on the space of global sections $\mathcal{A}^0(X, E)$.

4.2.2 Let ∇_i be connections on vector bundles E_i, $i = 1, 2$. Change both connections by one-forms $a_i \in \mathcal{A}^1(X, \mathrm{End}(E_i))$ and compute the new connections on the associated bundles $E_1 \oplus E_2$, $E_1 \otimes E_2$, and $\mathrm{Hom}(E_1, E_2)$.

4.2.3 Prove that connections on bundles E_i, $i = 1, 2$ which are compatible with given hermitian or holomorphic structures induce compatible connections on the associated bundles $E_1 \oplus E_2$, $E_1 \otimes E_2$, and $\mathrm{Hom}(E_1, E_2)$.

4.2.4 Study connections on an hermitian holomorphic vector bundle (E, h) that admits local holomorphic trivialization which are at the same time orthogonal with respect to the hermitian structure.

4.2.5 Let (E, h) be an hermitian vector bundle. If $E = E_1 \oplus E_2$, then E_1 and E_2 inherit natural hermitian structures h_1 and h_2. Are the induced connections ∇_i on E_i again hermitian with respect to these hermitian structures? What can you say about the second fundamental form?

4.2.6 Let ∇ be a connection on E. Describe the induced connections on $\bigwedge^2 E$ and $\det(E)$.

4.2.7 Show that the pull-back of an hermitian connection is hermitian with respect to the pull-back hermitian structure. Analogously, the pull-back of a connection compatible with the holomorphic structure on a holomorphic vector bundle under a holomorphic map is again compatible with the holomorphic structure on the pull-back bundle.

4.2.8 Show that a connection ∇ on an hermitian bundle (E, h) is hermitian if and only if $\nabla(h) = 0$, where by ∇ we also denote the naturally induced connection on the bundle $(E \otimes \bar{E})^*$.

4.2.9 Show that the definition of the Atiyah class does not depend on the chosen trivialization. Proposition 4.3.10 will provide an alternative proof of this fact.

Comments: - To a large extent, it is a matter of taste whether one prefers to work with connections globally or in terms of their local realizations as $d + A$ with the connections matrix A. However, both approaches are useful. Often, an assertion is first established locally or even fibrewise and afterwards, and sometimes more elegantly, put in a global language.

- For this and the next section we recommend Kobayashi's excellent textbook [78].

- The Atiyah class was introduced by Atiyah in [3]. There is a way to define the Atiyah class via the jet-sequence or, equivalently, using the first infinitesimal neighbourhood of the diagonal. See [31].

4.3 Curvature

In the previous paragraph we introduced the notion of a connection generalizing the exterior differential to sections of vector bundles. Indeed, by definition a connection satisfies the Leibniz rule. However, in general a connection ∇ need not satisfy $\nabla^2 = 0$, i.e. ∇ is not a differential. The obstruction for a connection to define a differential is measured by its curvature. This concept will be explained now.

The reader familiar with the basic concepts in Riemannian geometry will find the discussion similar to the treatment of the curvature of a Riemannian manifold. As before, we shall also study what happens in the presence of additional structures, e.g. hermitian and holomorphic ones.

Let E be a vector bundle on a manifold M endowed with a connection $\nabla : \mathcal{A}^0(E) \to \mathcal{A}^1(E)$. Then a natural extension

$$\nabla : \mathcal{A}^k(E) \longrightarrow \mathcal{A}^{k+1}(E)$$

is defined as follows: If α is a local k-form on M and s is a local section of E then

$$\nabla(\alpha \otimes s) = d(\alpha) \otimes s + (-1)^k \alpha \wedge \nabla(s).$$

Observe that for $k = 0$ this is just the Leibniz formula (4.2) which also ensures that the extension is well-defined, i.e. $\nabla(\alpha \otimes (f \cdot s)) = \nabla(f\alpha \otimes s)$ for any local function f. Moreover, a generalized Leibniz rule also holds for this extension, i.e. for any section t of $\mathcal{A}^\ell(E)$ and any k-form β one has

$$\nabla(\beta \wedge t) = d(\beta) \wedge t + (-1)^k \beta \wedge \nabla(t).$$

Indeed, if $t = \alpha \otimes s$ then

$$\begin{aligned}
\nabla(\beta \wedge t) &= \nabla((\beta \wedge \alpha) \otimes s) = d(\beta \wedge \alpha) \otimes s + (-1)^{k+\ell}(\beta \wedge \alpha) \otimes \nabla(s) \\
&= d(\beta) \wedge t + (-1)^k((\beta \wedge d(\alpha)) \otimes s + (-1)^\ell(\beta \wedge \alpha) \otimes \nabla(s)) \\
&= d(\beta) \wedge t + (-1)^k \beta \wedge \nabla(t)
\end{aligned}$$

Definition 4.3.1 The *curvature* F_∇ of a connection ∇ on a vector bundle E is the composition

$$F_\nabla := \nabla \circ \nabla : \mathcal{A}^0(E) \longrightarrow \mathcal{A}^1(E) \longrightarrow \mathcal{A}^2(E).$$

Usually, the curvature F_∇ will be considered as a global section of $\mathcal{A}^2(\text{End}(E))$, i.e. $F_\nabla \in \mathcal{A}^2(M, \text{End}(E))$. This is justified by the following result.

Lemma 4.3.2 *The curvature homomorphism* $F_\nabla : \mathcal{A}^0(E) \to \mathcal{A}^2(E)$ *is* $\mathcal{A}^0$-*linear.*

Proof. For a local section s of E and a local function f on M one computes

$$
\begin{aligned}
F_\nabla(f \cdot s) &= \nabla(\nabla(f \cdot s)) = \nabla\left(df \otimes s + f \cdot \nabla(s)\right) \\
&= \underbrace{d^2(f)}_{0} \otimes s \underbrace{-df \wedge \nabla(s) + df \wedge \nabla(s)}_{0} + f \cdot \nabla(\nabla(s)) \\
&= f \cdot F_\nabla(s)
\end{aligned}
$$

$\square$

Examples 4.3.3 i) Let us compute the curvature of a connection on the trivial bundle $M \times \mathbb{C}^r$. If ∇ is the trivial connection, i.e. $\nabla = d$, then $F_\nabla = 0$.

Any other connection is of the form $\nabla = d + A$, where A is a matrix of one-forms. For a section s one obtains $F_\nabla(s) = (d + A)(d + A)(s) = (d + A)(d(s) + A \cdot s) = d^2(s) + d(A \cdot s) + A \cdot d(s) + A(A(s)) = d(A)(s) + (A \wedge A)(s)$, i.e.

$$
F_\nabla = d(A) + A \wedge A.
$$

ii) For a line bundle this calculation yields $F_\nabla = d(A)$. In this case, the curvature is an ordinary two-form.

For $a \in \mathcal{A}^1(M, \mathrm{End}(E))$ the two-form $a \wedge a \in \mathcal{A}^2(M, \mathrm{End}(E))$ is given by exterior product in the form part and composition in $\mathrm{End}(E)$. Using this, the example is easily generalized to

Lemma 4.3.4 *Let ∇ be a connection on a vector bundle E and let $a \in \mathcal{A}^1(M, \mathrm{End}(E))$. Then $F_{\nabla + a} = F_\nabla + \nabla(a) + a \wedge a$.* $\square$

Any connection ∇ on a vector bundle E induces a natural connection on $\mathrm{End}(E)$ which will also be called ∇. In particular, this connection can be applied to the curvature $F_\nabla \in \mathcal{A}^2(M, \mathrm{End}(E))$ of the original connection on E. Here is the next remarkable property of the curvature:

Lemma 4.3.5 (Bianchi identity) *If $F_\nabla \in \mathcal{A}^2(M, \mathrm{End}(E))$ is the curvature of a connection ∇ on a vector bundle E, then*

$$
0 = \nabla(F_\nabla) \in \mathcal{A}^3(M, \mathrm{End}(E)).
$$

Proof. This follows from $\nabla(F_\nabla)(s) = \nabla(F_\nabla(s)) - F_\nabla(\nabla(s)) = \nabla(\nabla^2(s)) - \nabla^2(\nabla(s)) = 0$. Here we use Exercise 4.3.1. $\square$

Example 4.3.6 For the connection $\nabla = d + A$ on the trivial bundle the Bianchi identity becomes $dF_\nabla = F_\nabla \wedge A - A \wedge F_\nabla$.

The curvature of induced connections on associated bundles can usually be expressed in terms of the curvature of the given connections. For the most frequent associated bundles they are given by the following proposition.

Proposition 4.3.7 *Let E_1, E_2 be vector bundles endowed with connections ∇_1 and ∇_2, respectively.*

i) *The curvature of the induced connection on the direct sum $E_1 \oplus E_2$ is given by*

$$F = F_{\nabla_1} \oplus F_{\nabla_2}.$$

ii) *On the tensor product $E_1 \otimes E_2$ the curvature is given by*

$$F_{\nabla_1} \otimes 1 + 1 \otimes F_{\nabla_2}.$$

iii) *For the induced connection ∇^* on the dual bundle E^* one has*

$$F_{\nabla^*} = -F_\nabla^t.$$

iv) *The curvature of the pull-back connection $f^*\nabla$ of a connection ∇ under a differentiable map $f : M \to N$ is*

$$F_{f^*\nabla} = f^* F_\nabla.$$

Proof. Let us prove ii). This is the following straightforward calculation

$$
\begin{aligned}
&F_\nabla(s_1 \otimes s_2) \\
&= \nabla(\nabla(s_1 \otimes s_2)) = \nabla(\nabla_1(s_1) \otimes s_2 + s_1 \otimes \nabla_2(s_2)) \\
&= \nabla_1^2(s_1) \otimes s_2 - \nabla_1(s_1) \otimes \nabla_2(s_2) + \nabla_1(s_1) \otimes \nabla_2(s_2) + s_1 \otimes \nabla_2^2(s_2) \\
&= F_{\nabla_1}(s_1) \otimes s_2 + s_1 \otimes F_{\nabla_2}(s_2).
\end{aligned}
$$

The sign appears, because ∇ is applied to the one-form $\nabla_1(s_1) \otimes s_2$. We leave i) and iii) to the reader (cf. Exercise 4.3.2).

iv) follows from the local situation, where ∇ is given as $d + A$. Then $F_{f^*\nabla} = F_{d+f^*A} = d(f^*A) + f^*(A) \wedge f^*(A) = f^*(d(A) + A \wedge A) = f^* F_\nabla$. $\quad\square$

Next we will study the curvature of the special connections introduced in Section 4.2.

Proposition 4.3.8 i) *The curvature of an hermitian connection ∇ on an hermitian vector bundle (E, h) satisfies $h(F_\nabla(s_1), s_2) + h(s_1, F_\nabla(s_2)) = 0$, i.e.*

$$F_\nabla \in \mathcal{A}^2(M, \mathrm{End}(E, h)).$$

ii) *The curvature F_∇ of a connection ∇ on a holomorphic vector bundle E over a complex manifold X with $\nabla^{0,1} = \bar{\partial}$ has no $(0,2)$-part, i.e.*

$$F_\nabla \in (\mathcal{A}^{2,0} \oplus \mathcal{A}^{1,1})(X, \mathrm{End}(E)).$$

iii) *Let E be a holomorphic bundle endowed with an hermitian structure h. The curvature of the Chern connection ∇ is of type $(1,1)$, real, and skew-hermitian, i.e. $F_\nabla \in \mathcal{A}_{\mathbb{R}}^{1,1}(X, \mathrm{End}(E, h))$. (Recall that $\mathrm{End}(E, h)$ is only a real vector bundle.)*

Proof. i) A local argument goes as follows: Choose a trivialization such that (E, h) is isomorphic to the trivial bundle $M \times \mathbb{C}^r$ endowed with the constant standard hermitian product. Then, $\nabla = d + A$ with $\bar{A}^t = -A$. For its curvature $F_\nabla = d(A) + A \wedge A$ one obtains

$$\bar{F}_\nabla^t = d(\bar{A}^t) + (\bar{A} \wedge \bar{A})^t = d(\bar{A}^t) - \bar{A}^t \wedge \bar{A}^t$$
$$= d(-A) - (-A) \wedge (-A) = -F_\nabla.$$

For a global argument one first observes that the assumption that ∇ is hermitian yields for form-valued sections $s_i \in \mathcal{A}^{k_i}(E)$ the equation

$$dh(s_1, s_2) = h(\nabla(s_1), s_2) + (-1)^{k_1} h(s_1, \nabla(s_2)).$$

Recall that $h(\alpha_1 \otimes t_1, \alpha_2 \otimes t_2) = (\alpha_1 \wedge \bar{\alpha}_2) h(t_1, t_2)$ for local forms α_1, α_2 and sections t_1, t_2.

Hence, for $s_1, s_2 \in \mathcal{A}^0(E)$ one has

$$dh(\nabla(s_1), s_2) = h(F_\nabla(s_1), s_2) - h(\nabla(s_1), \nabla(s_2))$$
$$dh(s_1, \nabla(s_2)) = h(\nabla(s_1), \nabla(s_2)) + h(s_1, F_\nabla(s_2))$$

and on the other hand

$$dh(\nabla(s_1), s_2)) + dh(s_1, \nabla(s_2)) = d(dh(s_1, s_2)) = 0.$$

This yields the assertion.

ii) Here, one first observes that the extension $\nabla : \mathcal{A}^k(E) \to \mathcal{A}^{k+1}(E)$ splits into a $(1, 0)$-part and a $(0, 1)$-part, the latter of which is $\bar{\partial}$. Then one computes $\nabla^2 = (\nabla^{1,0})^2 + \nabla^{1,0} \circ \bar{\partial} + \bar{\partial} \circ \nabla^{1,0}$, as $\bar{\partial}^2 = 0$. Hence, $F_\nabla(s) \in (\mathcal{A}^{2,0} \oplus \mathcal{A}^{1,1})(E)$ for all $s \in \mathcal{A}^0(E)$.

Locally one could argue as follows: Since $\nabla = d + A$ with A of type $(1, 0)$, the curvature $d(A) + A \wedge A = \bar{\partial}(A) + (\partial A + A \wedge A)$ is a sum of a $(1, 1)$-form and a $(2, 0)$-form.

iii) Combine i) and ii). By comparing the bidegree of $\bar{F}_\nabla^t$ and F_∇ in local coordinates we find that F_∇ is of type $(1, 1)$.

More globally, due to ii) one knows that $h(F_\nabla(s_1), s_2)$ and $h(s_1, F_\nabla(s_2))$ are of bidegree $(2, 0) + (1, 1)$ respectively $(1, 1) + (0, 2)$. Recall the convention $h(s_1, \alpha_2 \otimes s_2) = \bar{\alpha}_2 h(s_1, s_2)$. Using i), i.e. $h(F_\nabla(s_1), s_2) + h(s_1, F_\nabla(s_2)) = 0$, and comparing the bidegree shows that $h(F_\nabla(s_1), s_2)$ has trivial $(2, 0)$-part for all sections s_1, s_2. Hence, F_∇ is of type $(1, 1)$. $\qquad\square$

Examples 4.3.9 i) Suppose E is the trivial vector bundle $M \times \mathbb{C}^r$ with the constant standard hermitian structure. If $\nabla = d + A$ is an hermitian connection, then i) says $d(A + \bar{A}^t) + (A \wedge A + \overline{(A \wedge A)}^t) = 0$. For $r = 1$ this means that the real part of A is constant.

ii) If (L, h) is an hermitian holomorphic line bundle, then the curvature F_∇ of its Chern connection is a section of $\mathcal{A}_\mathbb{R}^{1,1}(X, \text{End}(L, h))$, which can be identified with the imaginary $(1, 1)$-forms on X. Indeed, $\text{End}(L, h)$ is the purely imaginary line bundle $i \cdot \mathbb{R} \times X$ (cf. page 176).

iii) If the hermitian structure on the holomorphic vector bundle E is locally given by the matrix H, then the Chern connection is of the form $d + \bar{H}^{-1}\partial(\bar{H})$. Hence, the curvature is

$$F = \bar{\partial}(\bar{H}^{-1}\partial(\bar{H})).$$

Indeed, a priori $F = \bar{\partial}(\bar{H}^{-1}\partial(\bar{H})) + \partial(\bar{H}^{-1}\partial(\bar{H}) + (\bar{H}^{-1}\partial(\bar{H})) \wedge (\bar{H}^{-1}\partial(\bar{H}))$, but the sum of the last two terms must vanish, as both are of type $(2,0)$.

If E is a line bundle, then the hermitian matrix is just a positive real function h. In this case one writes $F = \bar{\partial}\partial \log(h)$. Once again, we see that F can be considered as a purely imaginary two-form

iv) Let (X, g) be an hermitian manifold. Then the tangent bundle is naturally endowed with an hermitian structure. The curvature of the Chern connection on $\mathcal{T}_X$ is called the *curvature of the hermitian manifold* (X, g). In Section 4.A we shall see that the curvature of a Kähler manifold (X, g) is nothing but the usual curvature of the underlying Riemannian manifold.

For the Chern connection on a holomorphic hermitian bundle the Bianchi identity yields

$$0 = (\nabla(F_\nabla))^{1,2} = \bar{\partial}(F_\nabla),$$

i.e. F_∇ as an element of $\mathcal{A}^{1,1}(X, \mathrm{End}(E))$ is $\bar{\partial}$-closed. Thus, in this case the curvature yields a natural Dolbeault cohomology class $[F_\nabla] \in H^1(X, \Omega_X \otimes \mathrm{End}(E))$ for any holomorphic vector bundle E. A priori, this cohomology class might depend on the chosen hermitian structure or, equivalently, on the connection. That this is not the case is an immediate consequence of the following description of it as the Atiyah class of E.

Proposition 4.3.10 *For the curvature F_∇ of the Chern connection on an hermitian holomorphic vector bundle (E, h) one has*

$$[F_\nabla] = A(E) \in H^1(X, \Omega_X \otimes \mathrm{End}(E)).$$

Proof. The comparison of Dolbeault and Čech cohomology can be done by chasing through the following commutative diagram of sheaves:

$$
\begin{array}{ccccc}
\Omega_X \otimes \mathrm{End}(E) & \longrightarrow & \mathcal{C}^0(\{U_i\}, \Omega_X \otimes \mathrm{End}(E)) & \longrightarrow & \mathcal{C}^1(\{U_i\}, \Omega_X \otimes \mathrm{End}(E)) \\
\downarrow & & \downarrow & & \downarrow {\scriptstyle i} \\
\mathcal{A}^{1,0}(\mathrm{End}(E)) & \longrightarrow & \mathcal{C}^0(\{U_i\}, \mathcal{A}^{1,0}(\mathrm{End}(E))) & \xrightarrow{\ \delta_1\ } & \mathcal{C}^1(\{U_i\}, \mathcal{A}^{1,0}(\mathrm{End}(E))) \\
\downarrow & & \downarrow {\scriptstyle \bar{\partial}} & & \\
\mathcal{A}^{1,1}(\mathrm{End}(E)) & \xrightarrow{\ \delta_0\ } & \mathcal{C}^0(\{U_i\}, \mathcal{A}^{1,1}(\mathrm{End}(E))) & &
\end{array}
$$

Here, $X = \bigcup U_i$ is an open covering of X trivializing E via $\psi_i : E|_{U_i} \cong U_i \times \mathbb{C}^r$. With respect to ψ_i the hermitian structure h on U_i is given by an hermitian matrix H_i. Then the curvature F_∇ of the Chern connection on the holomorphic bundle E on U_i is given by $F_\nabla|_{U_i} = \psi^{-1}(\bar\partial(\bar H_i^{-1}\partial\bar H_i))\psi_i$. Thus,

$$\delta_0(F_\nabla) = \{U_i, \psi_i^{-1} \circ (\bar\partial(\bar H_i^{-1}\partial\bar H_i)) \circ \psi_i\}$$
$$\doteq \bar\partial\{U_i, \psi_i^{-1} \circ (\bar H_i^{-1}\partial\bar H_i) \circ \psi_i\},$$

as the maps ψ_i are holomorphic trivializations.

Hence, it suffices to show that

$$i\{U_{ij}, \psi_j^{-1} \circ (\psi_{ij}^{-1}d\psi_{ij}) \circ \psi_j\} = \delta_1\{U_i, \psi_i^{-1} \circ (\bar H_i^{-1}\partial\bar H_i) \circ \psi_i\},$$

because the cocycle $\{U_i, \psi_i^{-1} \circ (\psi_{ij}^{-1}d\psi_{ij}) \circ \psi_j\}$ represents by definition the Atiyah class of E. Using the definition of δ_1 one shows that the term on the right hand side equals

$$\{U_{ij}, \psi_j^{-1} \circ (\bar H_j^{-1}\partial\bar H_j) \circ \psi_j - \psi_i^{-1} \circ (\bar H_i^{-1}\partial\bar H_i) \circ \psi_i)\}.$$

Since

$$\psi_j^{-1} \circ (\bar H_j^{-1}\partial\bar H_j) \circ \psi_j - \psi_i^{-1} \circ (\bar H_i^{-1}\partial\bar H_i) \circ \psi_i$$
$$= \psi_j^{-1} \circ (\bar H_j^{-1}\partial\bar H_j - \psi_{ij}^{-1} \circ (\bar H_i^{-1}\partial\bar H_i) \circ \psi_{ij}) \circ \psi_j,$$

it suffices to prove

$$\bar H_j^{-1}\partial\bar H_j - \psi_{ij}^{-1} \circ (\bar H_i^{-1}\partial\bar H_i) \circ \psi_{ij} = \psi_{ij}^{-1}\partial\psi_{ij}.$$

The latter is a consequence of the compatibility $\psi_{ij}^t H_i \bar\psi_{ij} = H_j$ or, equivalently, $\bar\psi_{ij}^t \bar H_i \psi_{ij} = \bar H_j$ and the chain rule

$$\bar H_j^{-1}\partial\bar H_j = \psi_{ij}^{-1}\bar H_i^{-1}(\bar\psi_{ij}^t)^{-1} \left(\bar\psi_{ij}^t(\partial\bar H_i)\psi_{ij} + \bar\psi_{ij}^t \bar H_i \partial\psi_{ij}\right)$$
$$= \psi_{ij}^{-1}(\bar H_i^{-1}\partial\bar H_i)\psi_{ij} + \psi_{ij}^{-1}\partial\psi_{ij},$$

where we have used $\partial\bar\psi_{ij} = 0$. $\qquad\square$

Remark 4.3.11 Here is a more direct argument to see that $[F_\nabla]$ does not depend on the connection ∇. Indeed, any other connection is of the form $\nabla + a$ and $F_{\nabla+a} = F_\nabla + \nabla(a) + a \wedge a$. If both connections ∇ and $\nabla + a$ are Chern connections with respect to certain hermitian structures, then F_∇ and $F_{\nabla+a}$ are $(1,1)$-forms and $a \in \mathcal{A}^{1,0}(\mathrm{End}(E))$. Thus, $a \wedge a \in \mathcal{A}^{2,0}(\mathrm{End}(E))$ and, therefore, $\nabla(a) + a \wedge a = (\nabla(a) + a \wedge a)^{1,1} = \nabla(a)^{1,1} = \bar\partial(a)$, i.e. $F_{\nabla+a} = F_\nabla + \bar\partial(a)$ and hence

$$[F_{\nabla+a}] = [F_\nabla].$$

We have used here that the induced connection on the endomorphism bundle is again compatible with the holomorphic structure.

Example 4.3.12 Let us consider the standard homogeneous linear coordinates $z_0, \ldots, z_n$ on $\mathbb{P}^n$ as sections of $\mathcal{O}(1)$. According to Example 4.1.2 one associates with them a natural hermitian metric $h(t) = |\psi(t)|^2/(\sum |\psi(z_i)|^2)$ on $\mathcal{O}(1)$.

We claim that the curvature F of the Chern connection on the holomorphic line bundle $\mathcal{O}(1)$ endowed with this hermitian metric is

$$\frac{i}{2\pi} F = \omega_{\mathrm{FS}},$$

where ω_{FS} is the Fubini–Study Kähler form (see Example 3.1.9, i)).

This can be verified locally. On a standard open subset $U_i \subset \mathbb{P}^n$ one has $\omega_{\mathrm{FS}} = \frac{i}{2\pi} \partial\bar\partial \log(1 + \sum |w_i|^2)$.

The hermitian structure of $\mathcal{O}(1)|_{U_i}$ with respect to the natural trivialization is given by the scalar function $h = (1 + \sum |w_i|^2)^{-1}$. By Example iii) in 4.3.9 we have $F = \bar\partial\partial \log(h) = \partial\bar\partial \log(1 + \sum |w_i|^2)$.

Remark 4.3.13 In the general situation, i.e. of a holomorphic line bundle L on a complex manifold X, the hermitian structure $h_{\{s_i\}}$ induced by globally generating sections $s_0, \ldots, s_n$ is the pull-back of the hermitian structure on $\mathcal{O}(1)$ of the previous example under the induced morphism $\varphi : X \to \mathbb{P}^n$. Thus, for the curvature F_∇ of the Chern connection ∇ on L one has: $(i/2\pi)F_\nabla = \varphi^*\omega_{\mathrm{FS}}$, where ω_{FS} is the Fubini–Study form on $\mathbb{P}^n$.

Suppose now that L is not only holomorphic, but also endowed with an hermitian structure h. What is the relation between h and $h_{\{s_i\}}$? In general, they are not related at all, but if the sections $s_0, \ldots, s_n$ are chosen such that they form an orthonormal base of $H^0(X, L)$ then one might hope to approximate h by $h_{\{s_i\}}$. (Here, $H^0(X, L)$ is equipped with the hermitian product defined in 4.1.11.) This circle of questions is intensively studied at the moment and there are many open questions.

Definition 4.3.14 A real $(1,1)$-form α is called *(semi-)positive* if for all holomorphic tangent vectors $0 \neq v \in T^{1,0}X$ one has

$$-i\alpha(v, \bar v) > 0 \text{ (resp. } \geq).$$

At a point $x \in X$ any semi-positive $(1,1)$-form is a positive linear combination of forms of the type $i\beta \wedge \bar\beta$, where β is a $(1,0)$-form. The standard example of a positive form is provided by a Kähler form $\omega = \frac{i}{2} \sum h_{ij} dz_i \wedge d\bar z_j$. In this case (h_{ij}) is a positive hermitian matrix and thus $v^t(h_{ij})\bar v > 0$ for all non-zero holomorphic tangent vectors. Together with Example 4.3.12 this shows that iF_∇ of the Chern connection ∇ on $\mathcal{O}(1)$ (with the standard hermitian structure) is positive.

Clearly, the pull-back of a semi-positive form is again semi-positive. For the curvature of a globally generated line bundle L this implies

$$F_\nabla(v, \bar v) \geq 0.$$

We next wish to generalize this observation to the higher rank case. Let (E, h) be an hermitian vector bundle over a complex manifold X and let ∇ be an hermitian connection that satisfies $F_\nabla \in \mathcal{A}^{1,1}(X, \operatorname{End}(E))$.

Definition 4.3.15 The curvature F_∇ is (Griffiths-)*positive* if for any $0 \neq s \in E$ one has

$$h(F_\nabla(s), s)(v, \bar{v}) > 0$$

for all non-trivial holomorphic tangent vectors v. Semi-positivity (negativity, semi-negativity) is defined analogously.

Remark 4.3.16 Let L be an hermitian holomorphic line bundle and let ∇ be the Chern connection. Its curvature F_∇ is an imaginary $(1, 1)$-form. Then the curvature F_∇ is positive in the sense of Definition 4.3.15 if and only if the real $(1, 1)$-form iF_∇ is positive in the sense of Definition 4.3.14. The extra factor i is likely to cause confusion at certain points, but these two concepts of positivity (and others) are met frequently in the literature.

Before proving the semi-positivity of the curvature of any globally generated vector bundle, we need to relate the curvature and the second fundamental forms of a split vector bundle $E = E_1 \oplus E_2$. Let ∇ be a connection on E. We denote the induced connections by ∇_i and the second fundamental forms by b_i, $i = 1, 2$. Thus,

$$\nabla = \begin{pmatrix} \nabla_1 & b_2 \\ b_1 & \nabla_2 \end{pmatrix}$$

which immediately shows

Lemma 4.3.17 *The curvature of the induced connection ∇_1 on E_1 is given by $F_{\nabla_1} = \operatorname{pr}_1 \circ F_\nabla - b_2 \circ b_1$.* $\square$

Now let E_1 be a holomorphic subbundle of the trivial holomorphic vector bundle $E = \mathcal{O}^{\oplus r}$ endowed with the trivial constant hermitian structure. The curvature of the Chern connection ∇ on E is trivial, as ∇ is just the exterior differential. Hence, $F_{\nabla_1} = -b_2 \circ b_1$. This leads to

Proposition 4.3.18 *The curvature F_{∇_1} of the Chern connection ∇_1 of a subbundle $E_1 \subset E = \mathcal{O}^{\oplus r}$ (with the induced hermitian structure) is semi-negative.*

Proof. By the previous lemma we have $F_{\nabla_1} = -b_2 \circ b_1$. Thus, if h_2 is the induced hermitian structure on the quotient E/E_1 then

$$h_1(F_{\nabla_1}(s), s) = -h_1(b_2(b_1(s)), s) = -h_2(b_1(s), b_1(s)).$$

Here we use properties of the second fundamental form proved in Section 4.2.

More precisely, one has

$$
\begin{aligned}
h_1(b_2(\beta \otimes t), s) = h_1(-\beta \otimes b_2(t), s) &= -\beta h_1(b_2(t), s) \\
&= \beta h_2(t, b_1(s)) = h_2(\beta \otimes t, b_1(s))
\end{aligned}
$$

for any one-form β.

Now, it suffices to show that $h(\alpha, \alpha) \geq 0$ for any $\alpha \in \mathcal{A}^{1,0}(E)$. Fix an orthonormal basis s_i of E and write $\alpha = \sum \alpha_i s_i$. Then $h(\alpha, \alpha) = \sum \alpha_i \wedge \bar{\alpha}_i$. But $\alpha_i \wedge \bar{\alpha}_i$ is semi-positive due to

$$
(\alpha_i \wedge \bar{\alpha}_i)(v, \bar{v}) = \alpha_i(v) \cdot \bar{\alpha}_i(\bar{v}) = \alpha_i(v) \cdot \overline{\alpha_i(v)} \geq 0.
$$

(Note that we did use that b_1 is of type $(1, 0)$.) $\qquad\qquad \square$

A holomorphic vector bundle E is *globally generated* if there exist global holomorphic sections $s_1, \ldots, s_r \in H^0(X, E)$ such that for any $x \in X$ the values $s_1(x), \ldots, s_r(x)$ generate the fibre $E(x)$. In other words, the sections $s_1, \ldots, s_r$ induce a surjection $\mathcal{O}^{\oplus r} \twoheadrightarrow E$. The standard constant hermitian structure on $\mathcal{O}^{\oplus r}$ induces via this surjection an hermitian structure on E and dualizing this surjection yields an inclusion of vector bundles $E^* \subset \mathcal{O}^{\oplus r}$.

Using Exercise 4.3.3 the proposition yields

Corollary 4.3.19 *The curvature of a globally generated vector bundle is semi-positive.* $\qquad\qquad \square$

Here, the curvature is the curvature of the Chern connection with respect to a hermitian structure on E defined by the choice of the globally generating holomorphic sections.

Example 4.3.20 The Euler sequence on $\mathbb{P}^n$ twisted by $\mathcal{O}(-1)$ is of the form

$$
0 \longrightarrow \mathcal{O}(-1) \longrightarrow \mathcal{O}^{\oplus n+1} \longrightarrow \mathcal{T}_{\mathbb{P}^n}(-1) \longrightarrow 0 .
$$

Hence, $\mathcal{T}_{\mathbb{P}^n}(-1)$ admits an hermitian structure such that the curvature of the Chern connection is semi-positive. Twisting by $\mathcal{O}(1)$ yields a connection on $\mathcal{T}_{\mathbb{P}^n}$ with positive curvature. In fact, this is the curvature of the Fubini–Study metric on $\mathbb{P}^n$. See Example 4.1.5.

Remark 4.3.21 As for line bundles, one could have first studied the universal case. Recall that on the Grassmannian $\mathrm{Gr}_r(V)$ there exists a universal short exact sequence

$$
0 \longrightarrow S \longrightarrow \mathcal{O} \otimes V \longrightarrow Q \longrightarrow 0.
$$

By definition, the fibre of S over a point of $\mathrm{Gr}(V)$ that corresponds to $W \subset V$ is naturally isomorphic to W. Fixing an hermitian structure on V induces hermitian structures on S and Q (and therefore on the tangent bundle $\mathcal{T}$ of $\mathrm{Gr}_r(V)$, which is $\mathrm{Hom}(S, Q)$).

If E is any holomorphic vector bundle of rank r on a complex manifold X generated by global sections spanning $V \subset H^0(X, E)$, then there exists a morphism $\varphi : X \to \mathrm{Gr}_r(V)$ with $\varphi^* Q = E$ (see Exercise 2.4.11). Again, choosing an hermitian product on V induces an hermitian structure on E which coincides with the pull-back of the hermitian structure on Q under φ. This shows that the curvature of E and Q can be compared, namely $\varphi^* F_Q = F_E$.

Thus, showing the positivity of F_Q proves the semi-positivity for any globally generated vector bundle E. Moreover, if $\varphi : X \to \mathrm{Gr}_r(V)$ is an embedding, then the curvature F_E is positive.

Exercises

4.3.1 Show that $\nabla^2 : \mathcal{A}^k(E) \to \mathcal{A}^{k+2}(E)$ is given by taking the exterior product with the form part of the curvature $F_\nabla \in \mathcal{A}^2(M, \mathrm{End}(E))$ and applying its endomorphism part to E.

4.3.2 Prove i) and iii) of Proposition 4.3.7. Compute the connection and the curvature of the determinant bundle.

4.3.3 Let (E_1, h_1) and (E_2, h_2) be two hermitian holomorphic vector bundles endowed with hermitian connections ∇_1, ∇_2 such that the curvature of both is (semi-)positive. Prove the following assertions.
    i) The curvature of the induced connection ∇^* on the dual vector bundle E_1^* is (semi-)negative.
    ii) The curvature on $E_1 \otimes E_2$ is (semi-)positive and it is positive if at least one of the two connections has positive curvature.
    iii) The curvature on $E_1 \oplus E_2$ is (semi-)positive.

4.3.4 Find an example of two connections ∇_1 and ∇_2 on a vector bundle E, such that F_{∇_1} is positive and F_{∇_2} is negative.

4.3.5 Let L be a holomorphic line bundle on a complex manifold. Suppose L admits an hermitian structure whose Chern connection has positive curvature. Show that X is Kähler. If X is in addition compact prove $\int_X A(L)^n > 0$.

4.3.6 Show that the canonical bundle of $\mathbb{P}^n$ comes along with a natural hermitian structure such that the curvature of the Chern connection is negative.

4.3.7 Show that the curvature of a complex torus $\mathbb{C}^n/\Gamma$ endowed with a constant hermitian structure is trivial. Is this true for any hermitian structure on $\mathbb{C}^n/\Gamma$?

4.3.8 Show that the curvature of a the natural metric on the ball quotient introduced in Example 3.1.9, iv) is negative. The one-dimensional case is a rather easy calculation.

4.3.9   Let X be a compact Kähler manifold with $b_1(X) = 0$. Show that there exists a unique flat connection ∇ on the trivial holomorphic line bundle $\mathcal{O}$ with $\nabla^{0,1} = \bar{\partial}$. Moreover, up to isomorphism $\mathcal{O}$ is the only line bundle with trivial Chern class $c_1 \in H^2(X, \mathbb{Z})$.

4.3.10   Let ∇ be a connection on a (complex) line bundle L on a manifold M. Show that L locally admits trivializing parallel sections if and only if $F_\nabla = 0$.

This is the easiest case of the general fact that a connection on a vector bundle is flat if and only if parallel frames can be found locally. (Frobenius integrability.)

Comments: - Various notions for the positivity of vector bundles can be found in the literature. Usually they all coincide for line bundles, but the exact relations between them is not clear for higher rank vector bundles. Positivity is usually exploited to control higher cohomology groups. This will be illustrated in Section 5.2. We shall also see that, at least for line bundles on projective manifolds, an algebraic description of bundles admitting an hermitian structure whose Chern connection has positive curvature can be given. For a more in depth presentation of the material we refer to [35, 100] and the forthcoming book [83].

- The problem alluded to in Remark 4.3.13 is subtle. See [37, 107].

- The curvature of Kähler manifold will be dealt with in subsequent sections, in particular its comparison with the well-known Riemannian curvature shall be explained in detail.

- In [75, Prop. 1.2.2] one finds a cohomological version of the Bianchi identity in terms of the Atiyah class.

4.4 Chern Classes

Connection and curvature are not only objects that are naturally associated with any vector bundle, they also provide an effective tool to define cohomological and numerical invariants, called characteristic classes and numbers. These invariants are in fact topological, but the topological aspects will not be treated here. The goal of this sections is first to discuss the multilinear algebra needed for the definition of Chern forms and classes and then to present the precise definition of Chern classes and Chern and Todd characters. We conclude the section by comparing the various definitions of the first Chern class encountered throughout the text.

Let ∇ be a connection on a complex vector bundle E and let F_∇ denote its curvature. Recall that $F_\nabla \in \mathcal{A}^2(M, \mathrm{End}(E))$. We would like to apply certain multilinear operations to the linear part of F_∇, i.e. to the components in $\mathrm{End}(E)$, in order to obtain forms of higher degree. Let us start out with a discussion of the linear algebra behind this approach.

Let V be a complex vector space. A k-multilinear symmetric map

$$P : V \times \ldots \times V \longrightarrow \mathbb{C}$$

corresponds to an element in $\mathrm{S}^k(V)^*$. To each such P one associates its polarized form $\tilde{P} : V \to \mathbb{C}$, which is a homogeneous polynomial of degree k given by $\tilde{P}(B) := P(B, \ldots, B)$. Conversely, any homogeneous polynomial is uniquely obtained in this way. In our situation, we will consider $V = \mathfrak{gl}(r, \mathbb{C})$, the space of complex (r, r)-matrices.

Definition 4.4.1 A symmetric map

$$P : \mathfrak{gl}(r, \mathbb{C}) \times \ldots \times \mathfrak{gl}(r, \mathbb{C}) \longrightarrow \mathbb{C}$$

is called *invariant* if for all $C \in \mathrm{Gl}(r, \mathbb{C})$ and all $B_1, \ldots, B_k \in \mathfrak{gl}(r, \mathbb{C})$ one has

$$P(CB_1C^{-1}, \ldots, CB_kC^{-1}) = P(B_1, \ldots, B_k). \tag{4.5}$$

This condition can also be expressed in terms of the associated homogeneous polynomial as $\tilde{P}(CBC^{-1}) = \tilde{P}(B)$ for all $C \in \mathrm{Gl}(r, \mathbb{C})$ and all $B \in \mathfrak{gl}(r, \mathbb{C})$.

Lemma 4.4.2 *The k-multilinear symmetric map P is invariant if and only if for all $B, B_1, \ldots, B_k \in \mathfrak{gl}(r, \mathbb{C})$ one has*

$$\sum_{j=1}^{k} P(B_1, \ldots, B_{j-1}, [B, B_j], B_{j+1}, \ldots, B_k) = 0.$$

Proof. Use the invertible matrix $C = e^{tB}$ and differentiate the invariance equation (4.5) at $t = 0$. The converse is left as an exercise (cf. Exercise 4.4.8). $\square$

Proposition 4.4.3 *Let P be an invariant k-multilinear symmetric form on $\mathfrak{gl}(r, \mathbb{C})$. Then for any vector bundle E of rank r and any partition $m = i_1 + \ldots + i_k$ there exists a naturally induced k-linear map:*

$$P : \left(\bigwedge^{i_1} M \otimes \operatorname{End}(E) \right) \times \ldots \times \left(\bigwedge^{i_k} M \otimes \operatorname{End}(E) \right) \longrightarrow \bigwedge_\mathbb{C}^m M$$

defined by $P(\alpha_1 \otimes t_1, \ldots, \alpha_k \otimes t_k) = (\alpha_1 \wedge \ldots \wedge \alpha_k) P(t_1, \ldots, t_k)$.

Proof. Once a trivialization $E(x) \cong \mathbb{C}^r$ is fixed, the above definition makes sense. Since P is invariant, it is independent of the chosen trivialization. $\square$

Clearly, the k-linear map defined in this way induces also a k-multilinear map on the level of global sections

$$P : \mathcal{A}^{i_1}(M, \operatorname{End}(E)) \times \ldots \times \mathcal{A}^{i_k}(M, \operatorname{End}(E)) \longrightarrow \mathcal{A}_\mathbb{C}^m(M).$$

Note that P applied to form-valued endomorphisms is only graded symmetric, but restricted to even forms it is still a k-multilinear symmetric map. In particular, P restricted to $\mathcal{A}^2(M, \operatorname{End}(E)) \times \ldots \times \mathcal{A}^2(M, \operatorname{End}(E))$ can be recovered from its polarized form $\tilde{P}(\alpha \otimes t) = P(\alpha \otimes t, \ldots, \alpha \otimes t)$. In the following, this polarized form shall be applied to the curvature form F_∇ of a connection ∇ on E. We will need the following

Lemma 4.4.4 *For any forms $\gamma_j \in \mathcal{A}^{i_j}(M, \operatorname{End}(E))$ one has*

$$dP(\gamma_1, \ldots, \gamma_k) = \sum_{j=1}^k (-1)^{\sum_{\ell=1}^{j-1} i_\ell} P(\gamma_1, \ldots, \nabla(\gamma_j), \ldots, \gamma_k),$$

where ∇ also denotes the induced connection on $\operatorname{End}(E)$.

Proof. This can be seen by a local calculation. We write $\nabla = d + A$, where A is the local connection matrix of ∇. The induced connection on $\operatorname{End}(E)$ is of the form $\nabla = d + A$ with A acting as $\gamma \mapsto [A, \gamma]$. Using the usual Leibniz formula for the exterior differential one finds

$$dP(\gamma_1, \ldots, \gamma_k) = \sum_{j=1}^k (-1)^{\sum_{\ell=1}^{j-1} i_\ell} P(\gamma_1, \ldots, d\gamma_j, \ldots, \gamma_k)$$

$$= \sum_{j=1}^k (-1)^{\sum_{\ell=1}^{j-1} i_\ell} P(\gamma_1, \ldots, (\nabla - A)(\gamma_j), \ldots, \gamma_k).$$

By Lemma 4.4.2 the invariance of P proves the assertion. $\square$

Corollary 4.4.5 *Let F_∇ be the curvature of an arbitrary connection ∇ on a vector bundle E of rank r. Then for any invariant k-multilinear symmetric polynomial P on $\mathfrak{gl}(r, \mathbb{C})$ the induced $2k$-form $\tilde{P}(F_\nabla) \in \mathcal{A}_\mathbb{C}^{2k}(M)$ is closed.*

Proof. This is an immediate consequence of the Bianchi identity (Lemma 4.3.5), which says $\nabla(F_\nabla) = 0$, and the previous lemma. ☐

Thus, to any invariant k-multilinear symmetric map P on $\mathfrak{gl}(r, \mathbb{C})$ and any vector bundle E of rank r one can associate a cohomology class $[\tilde{P}(F_\nabla)] \in H^{2k}(M, \mathbb{C})$. In fact, due to the following lemma, this class is independent of the chosen connection.

Lemma 4.4.6 *If ∇ and ∇' are two connections on the same bundle E, then $[\tilde{P}(F_\nabla)] = [\tilde{P}(F_{\nabla'})]$.*

Proof. The space of all connections is an affine space over $\mathcal{A}^1(M, \mathrm{End}(E))$, i.e. if ∇ is given, then any other connection is of the form $\nabla' = \nabla + A$ for some $A \in \mathcal{A}^1(\mathrm{End}(E))$ (see Corollary 4.2.4). Thus, it suffices to show that the induced map

$$\mathcal{A}^1(M, \mathrm{End}(E)) \longrightarrow H^{2k}(M, \mathbb{C})$$

is constant. We use that $F_{\nabla+A} = F_\nabla + A \wedge A + \nabla(A)$.

The assertion can be proven by an infinitesimal calculation, i.e. in the following calculation we only consider terms of order at most one in t:

$$\tilde{P}(F_{\nabla+tA}) = \tilde{P}(F_\nabla) + ktP(F_\nabla, \ldots, F_\nabla, \nabla(A)).$$

Now the assertion follows from Lemma 4.4.4 and the Bianchi identity:

$$P(F_\nabla, \ldots, F_\nabla, \nabla(A)) = dP(F_\nabla, \ldots, F_\nabla, A) - P(\nabla(F_\nabla), F_\nabla, \ldots, F_\nabla, A) - \ldots$$
$$- P(F_\nabla, \ldots, F_\nabla, \nabla(F_\nabla), A)$$
$$= dP(F_\nabla, \ldots, F_\nabla, A).$$

☐

Remark 4.4.7 If we denote by $(S^k \mathfrak{gl}(r, \mathbb{C}))^{\mathrm{Gl}(r)}$ the invariant k-multilinear polynomials, then the above construction induces a canonical homomorphism

$$(S^k \mathfrak{gl}(r, \mathbb{C}))^{\mathrm{Gl}(r)} \longrightarrow H^{2k}(M, \mathbb{C})$$

for any vector bundle E of rank r. In fact, we actually obtain an algebra homomorphism

$$(S^* \mathfrak{gl}(r, \mathbb{C}))^{\mathrm{Gl}(r)} \longrightarrow H^{2*}(M, \mathbb{C})$$

which is called the *Chern–Weil homomorphism*.

So far, everything was explained for arbitrary invariant polynomials, but some polynomials are more interesting than others, at least regarding their applications to geometry. In the following we discuss the most frequent ones.

Examples 4.4.8 i) **Chern classes.** Let $\{\tilde{P}_k\}$ be the homogeneous polynomials with $\deg(\tilde{P}_k) = k$ defined by

$$\det(\mathrm{Id} + B) = 1 + \tilde{P}_1(B) + \ldots + \tilde{P}_r(B).$$

Clearly, these $\tilde{P}_k$ are invariant. The *Chern forms* of a vector bundle of rank r endowed with a connection are

$$c_k(E, \nabla) := \tilde{P}_k\left(\frac{i}{2\pi} F_\nabla\right) \in \mathcal{A}_{\mathbb{C}}^{2k}(M).$$

The *k-th Chern class* of the vector bundle E is the induced cohomology class

$$c_k(E) := [c_k(E, \nabla)] \in H^{2k}(M, \mathbb{C}).$$

Note that $c_0(E) = 1$ and $c_k(E) = 0$ for $k > \mathrm{rk}(E)$. The *total Chern class* is $c(E) := c_0(E) + c_1(E) + \ldots + c_r(E) \in H^{2*}(M, \mathbb{C})$.

ii) **Chern characters.** In order to define the Chern character of E one uses the invariant homogeneous polynomials $\tilde{P}_k$ of degree k defined by

$$\mathrm{tr}(e^B) = \tilde{P}_0(B) + \tilde{P}_1(B) + \ldots.$$

Then the *k-th Chern character* $\mathrm{ch}_k(E) \in H^{2k}(M, \mathbb{C})$ of E is defined as the cohomology class of

$$\mathrm{ch}_k(E, \nabla) := \tilde{P}_k\left(\frac{i}{2\pi} F_\nabla\right) \in \mathcal{A}_{\mathbb{C}}^{2k}(M).$$

Note that $\mathrm{ch}_0(E) = \mathrm{rk}(E)$. The *total Chern character* is $\mathrm{ch}(E) := \mathrm{ch}_0(E) + \ldots + \mathrm{ch}_r(E) + \mathrm{ch}_{r+1}(E) + \ldots$.

iii) **Todd classes.** The homogeneous polynomials used to define the Todd classes are given by

$$\frac{\det(tB)}{\det(\mathrm{Id} - e^{-tB})} = \sum_k \tilde{P}_k(B) t^k.$$

(The additional variable t is purely formal and is supposed to indicate that the left hand side can be developed as a power series in t with coefficients P_k which are of degree k. This could also have been done for the other characteristic classes introduced earlier.)

Then $\mathrm{td}_k(E, \nabla) := \tilde{P}_k((i/2\pi) F_\nabla)$ and

$$\mathrm{td}_k(E) := [\mathrm{td}_k(E, \nabla)] \in H^{2k}(M, \mathbb{C}).$$

The *total Todd class* is $\mathrm{td}(E) := \mathrm{td}_0(E) + \mathrm{td}_1(E) + \ldots$.

Note that the Todd classes are intimately related to the Bernoulli numbers B_k. In fact, by definition

$$\frac{t}{1 - e^{-t}} = 1 + \frac{t}{2} + \sum_{k=1}^{\infty} (-1)^{k+1} \frac{B_k}{(2k)!} t^{2k}.$$

Let us study some of the natural operations for vector bundles and see how the characteristic classes behave in these situations.

• Let $E = E_1 \oplus E_2$ be endowed with the direct sum ∇ of connections ∇_1 and ∇_2 on E_1 and E_2, respectively. The curvature F_∇ is again the direct sum $F_{\nabla_1} \oplus F_{\nabla_2}$ (Proposition 4.3.7) and since $\det\left((\mathrm{Id}_{E_1} + \frac{i}{2\pi}F_{\nabla_1}) \oplus (\mathrm{Id}_{E_2} + \frac{i}{2\pi}F_{\nabla_2})\right) = \det\left(\mathrm{Id}_{E_1} + \frac{i}{2\pi}F_{\nabla_1}\right) \cdot \det\left(\mathrm{Id}_{E_2} + \frac{i}{2\pi}F_{\nabla_2}\right)$, we obtain the *Whitney product formula*:

$$c(E, \nabla) = c(E_1, \nabla_1) \cdot c(E_2, \nabla_2).$$

Of course, this relation then also holds true for the total Chern class. In particular, one has $c_1(E_1 \oplus E_2) = c_1(E_1) + c_1(E_2)$ and $c_2(E) = c_2(E_1) + c_2(E_2) + c_1(E_1) \cdot c_1(E_2)$. A similar calculation shows

$$\mathrm{ch}(E_1 \oplus E_2) = \mathrm{ch}(E_1) + \mathrm{ch}(E_2).$$

• Consider two vector bundles E_1 and E_2 endowed with connections ∇_1 and ∇_2, respectively. Let ∇ be the induced connection $\nabla_1 \otimes 1 + 1 \otimes \nabla_2$ on $E = E_1 \otimes E_2$. Then $F_\nabla = F_{\nabla_1} \otimes 1 + 1 \otimes F_{\nabla_2}$ (see Proposition 4.3.7) and hence $\mathrm{tr}(e^{(i/2\pi)F_\nabla}) = \mathrm{tr}(e^{(i/2\pi)F_{\nabla_1}} \otimes e^{(i/2\pi)F_{\nabla_2}}) = \mathrm{tr}(e^{(i/2\pi)F_{\nabla_1}}) \cdot \mathrm{tr}(e^{(i/2\pi)F_{\nabla_2}})$. Therefore,

$$\mathrm{ch}(E_1 \otimes E_2) = \mathrm{ch}(E_1) \cdot \mathrm{ch}(E_2).$$

If $E_2 = L$ is a line bundle one finds $c_1(E_1 \otimes L) = c_1(E_1) + \mathrm{rk}(E_1) \cdot c_1(L)$ and $c_2(E_1 \otimes L) = c_2(E_1) + (\mathrm{rk}(E_1) - 1) \cdot c_1(E_1) \cdot c_1(L) + \binom{\mathrm{rk}(E_1)}{2}c_1^2(L)$. See Exercise 4.4.5 for the first few terms of the Chern character and Exercise 4.4.6 for the general formula for the Chern classes of $E_1 \otimes L$.

• The curvature F_{∇^*} of the naturally associated connection ∇^* on the dual bundle E^* is $F_{\nabla^*} = -F_\nabla^t$ (see Proposition 4.3.7). Thus, $c(E^*, \nabla^*) = \det(\mathrm{Id} + \frac{i}{2\pi}F_{\nabla^*}) = \det(\mathrm{Id} - \frac{i}{2\pi}F_\nabla^t) = \det(\mathrm{Id} - \frac{i}{2\pi}F_\nabla)$. Hence,

$$c_k(E^*, \nabla^*) = (-1)^k c_k(E, \nabla).$$

• Let $f : M \to N$ be a differentiable map and let E be a vector bundle on N endowed with a connection ∇. By Proposition 4.3.7 we know that $F_{f^*\nabla} = f^*F_\nabla$. This readily yields

$$c_k(f^*E, f^*\nabla) = f^*c_k(E, \nabla).$$

• The first Chern class of the line bundle $\mathcal{O}(1)$ on $\mathbb{P}^1$ satisfies the normalization

$$\int_{\mathbb{P}^1} c_1(\mathcal{O}(1)) = 1.$$

Indeed, in Example 4.3.12 we have shown that the Chern connection on $\mathcal{O}(1)$ on $\mathbb{P}^n$ with respect to the natural hermitian structure has curvature $F = (2\pi/i)\omega_{\mathrm{FS}}$ and by Example 3.1.9, i) we know $\int_{\mathbb{P}^1} \omega_{\mathrm{FS}} = 1$.

Remarks 4.4.9 i) In fact, all the characteristic classes introduced above are real. This can be seen as follows. Pick an hermitian metric on the vector bundle E and consider an hermitian connection ∇, which always exists. Then locally and with respect to an hermitian trivialization of E the curvature satisfies the equation $F_\nabla^* = \bar{F}_\nabla^t = -F_\nabla$. Hence, $\overline{(i/2\pi)F_\nabla} = (i/2\pi)F_\nabla^t$ and, therefore, $c(E, \nabla) = \det(\mathrm{Id} + (i/2\pi)F_\nabla) = \det(\mathrm{Id} + (i/2\pi)F_\nabla^t) = \overline{\det(\mathrm{Id} + (i/2\pi)F_\nabla)} = \overline{c(E, \nabla)}$, i.e. $c(E, \nabla)$ is a real form. Thus,

$$c(E) \in H^*(M, \mathbb{R}).$$

The same argument works for $\mathrm{ch}(E)$ and $\mathrm{td}(E)$.

ii) If E is a holomorphic vector bundle over a complex manifold X, then we may use an hermitian connection ∇ that is in addition compatible with the holomorphic structure of E (cf. Proposition 4.2.14). In this case the curvature F_∇ is a $(1,1)$-form, i.e. $F_\nabla \in \mathcal{A}^{1,1}(X, \mathrm{End}(E))$. But then also the Chern forms $c_k(E, \nabla)$ are of type (k, k) for all k.

If X is compact and Kähler, we have the decomposition $H^{2k}(X, \mathbb{C}) = \bigoplus_{p+q=2k} H^{p,q}(X)$ (Corollary 3.2.12) and the Chern classes of any holomorphic bundle are contained in the (k, k)-component, i.e.

$$c_k(E) \in H^{2k}(X, \mathbb{R}) \cap H^{k,k}(X).$$

iii) We are going to explain an ad hoc version of the *splitting principle*. The geometric splitting principle works on the level of cohomology and tries to construct for a given vector bundle E a ring extension $H^*(M, \mathbb{R}) \subset A^*$ and elements $\gamma_i \in A^2$, $i = 1, \ldots, \mathrm{rk}(E)$, such that $c(E) = \prod(1 + \gamma_i)$. Moreover, A^* is constructed geometrically as the cohomology ring of a manifold N such that the inclusion $H^*(M, \mathbb{R}) \subset A^* = H^*(N, \mathbb{R})$ is induced by a submersion $\pi : N \to M$ (e.g. N can be taken as the full flag manifold associated to E). The map π is constructed such that $\pi^* E$ is a direct sum $\bigoplus L_i$ of line bundles L_i with $\gamma_i = c_1(L_i)$.

We propose to study a similar construction on the level of forms. This approach is less geometrical but sufficient for many purposes.

Consider $\mathbb{C}^r$ with the standard hermitian structure and let $B \in \mathfrak{gl}(r, \mathbb{C})$ be a self-adjoint (or, hermitian matrix), i.e. $B^t = \bar{B}$. Then, there exists an orthonormal basis with respect to which B takes diagonal form with eigenvalues $\lambda_1, \ldots, \lambda_r$. Clearly, every expression of the form $\tilde{P}(B)$, with P an invariant symmetric map, can be expressed in terms of $\lambda_1, \ldots, \lambda_r$. E.g. $\mathrm{tr}(B) = \lambda_1 + \ldots + \lambda_r$.

Let us now consider the curvature matrix F_∇ of an hermitian connection ∇ on an hermitian vector bundle (E, h) of rank r on a manifold M. At a fixed point $x \in M$ we may trivialize (E, h) such that it becomes isomorphic to $\mathbb{C}^r$ with the standard hermitian structure. Then $i \cdot F_\nabla$ in x corresponds to an hermitian matrix B, but with coefficients not in $\mathbb{C}$ but in $R := \mathbb{C}[\bigwedge_x^2 M]$.

Diagonalizing $B = i \cdot F_\nabla(x)$ can still be achieved, but in general only over a certain ring extension $R \subset R'$. (One has to adjoint certain eigenvalues, to assure that a vector of length one can be completed to an orthonormal basis, etc.) Let us suppose this has been done, i.e. in the new basis one has $B = \text{diag}(\gamma_1, \ldots, \gamma_r)$ with $\gamma_i \in R'$. Then for any invariant symmetric map P one finds $\tilde{P}(iF_\nabla) = \tilde{P}(\text{diag}(\gamma_1, \ldots, \gamma_r))$, where P is extended R'-linearly from $\mathbb{C}^r$ to $\mathbb{C}^r \otimes_{\mathbb{C}} R'$. The result $\tilde{P}(iF_\nabla)$ is of course contained in $\mathbb{C}[\bigwedge_x^2 M]$ and can thus be projected to $\bigwedge_x^{2*} M$.

In general, this procedure will not work globally, but it often suffices to have at one's disposal the splitting principle in this form. The elements $\gamma_1, \ldots, \gamma_r$ (up to the scalar factor $(1/2\pi)$, which we suppress) are called the *formal Chern roots* of ∇ on E.

The primary use of this construction is to verify various formal identities. As an example, let us show how the rather elementary identity $\text{ch}_2(E) = (1/2)c_1^2(E) - c_2(E)$ can be proved. This can be done pointwise and so we may assume that the ring extension $R = \mathbb{C}[\bigwedge_x^2 M] \subset R'$ and the formal Chern roots of E have been found. Hence,

$$\text{ch}_2(E, \nabla)(x) = \frac{1}{8\pi^2}\text{tr}\,(iF_\nabla(x)) = \frac{1}{8\pi^2}\text{tr}\,(\text{diag}(\gamma_1^2, \ldots, \gamma_r^2)) = \frac{1}{8\pi^2}\sum_i \gamma_i^2$$

and

$$c_1^2(E, \nabla)(x) - 2c_2(E, \nabla)(x) = \frac{1}{4\pi^2}\left(\left(\sum_i \gamma_i\right)^2 - 2\sum_{i<j}\gamma_i\gamma_j\right) = \frac{1}{4\pi^2}\sum_i \gamma_i^2.$$

Another example for this type of argument can be found in Section 5.1. See also Exercises 4.4.5 and 4.4.9.

iv) There is also an axiomatic approach to Chern classes which shows that the Whitney product formula, the compatibility under pull-back, and the normalization $\int c_1(\mathcal{O}(1)) = 1$ determine the Chern classes uniquely.

Definition 4.4.10 The *Chern classes of a complex manifold X* are

$$c_k(X) := c_k(\mathcal{T}_X) \in H^{2k}(X, \mathbb{R}),$$

where $\mathcal{T}_X$ is the holomorphic tangent bundle. Similarly, one defines $\text{ch}_k(X)$ and $\text{td}_k(X)$ by means of $\mathcal{T}_X$.

Note that we actually only need an almost complex structure in order to define the Chern classes of the manifold. Also note that it might very well happen that two different complex structures on the same differentiable manifold yield different Chern classes, but in general counter-examples are not easy to construct. A nice series of examples of complex structures on the product of a K3 surface with S^2 with different Chern classes can be found in the recent paper [84].

Example 4.4.11 Let us compute the characteristic classes of a hypersurface $Y \subset X$. The normal bundle sequence in this case takes the form

$$0 \longrightarrow \mathcal{T}_Y \longrightarrow \mathcal{T}_X|_Y \longrightarrow \mathcal{O}_Y(Y) \longrightarrow 0.$$

Since any short exact sequence of holomorphic vector bundles splits as a sequence of complex vector bundles, the Whitney product formula yields $i^*c(X) = c(Y) \cdot i^*c(\mathcal{O}(Y))$. Therefore,

$$c(Y) = i^* \left(c(X) \cdot (1 - c_1(\mathcal{O}(Y)) + c_1(\mathcal{O}(Y))^2 \pm \ldots) \right).$$

In particular, $c_1(Y) = i^*(c_1(X) - c_1(\mathcal{O}(Y))$ which reflects the adjunction formula 2.2.17.

For a quartic hypersurface $Y \subset \mathbb{P}^3$ this yields $c_1(Y) = 0$ and $c_2(Y) = i^*c_2(\mathbb{P}^3)$. Hence, $\int_Y c_2(Y) = \int_{\mathbb{P}^3} c_2(\mathbb{P}^3)(4c_1(\mathcal{O}(1))) = 24$. Here we use $c_2(\mathbb{P}^3) = 6c_1^2(\mathcal{O}(1))$ which follows from the Euler sequence and the Whitney formula (cf. Exercise 4.4.4).

So far, we have encountered various different ways to define the first Chern class of a complex or holomorphic line bundle. We will now try to summarize and compare these.

For a holomorphic line bundle L on a complex manifold X we have used the following three definitions:

i) Via the curvature as $c_1(L) = [c_1(L, \nabla)] \in H^2(X, \mathbb{R}) \subset H^2(X, \mathbb{C})$, where ∇ is a connection on L.

ii) Via the Atiyah class $A(L) \in H^1(X, \Omega_X)$. See Remarks 4.2.20 ii).

iii) Via the exponential sequence and the induced boundary operator $\delta : H^1(X, \mathcal{O}_X^*) \to H^2(X, \mathbb{Z})$. See Definition 2.2.13.

By Proposition 4.3.10 the first two definitions are compatible in the sense that $A(L) = [F_\nabla]$ if ∇ is the Chern connection on L endowed with an hermitian structure. In case that X is a compact Kähler manifold we can naturally embed $H^1(X, \Omega_X) = H^{1,1}(X) \subset H^2(X, \mathbb{C})$ and thus obtain

$$\frac{i}{2\pi} A(L) = c_1(L).$$

The comparison of i) and iii) will be done more generally for complex line bundles L on a differentiable manifold M.

A complex line bundle L on a differentiable manifold M is described by its cocycle $\{U_{ij}, \psi_{ij}\} \in H^1(M, \mathcal{C}_\mathbb{C}^*)$ (see Appendix B). The invertible complex valued differentiable functions $\psi_{ij} \in \mathcal{C}_\mathbb{C}^*(U_{ij})$ are given as $\psi_{ij} = \psi_i \circ \psi_j^{-1}$, where $\psi_i : L|_{U_i} \cong U_i \times \mathbb{C}$ are trivializations.

In the present context we work with the smooth exponential sequence

$$0 \longrightarrow \mathbb{Z} \longrightarrow \mathcal{C}_{\mathbb{C}} \longrightarrow \mathcal{C}_{\mathbb{C}}^* \longrightarrow 0$$

which induces a boundary isomorphism $\delta : H^1(M, \mathcal{C}_{\mathbb{C}}^*) \cong H^2(M, \mathbb{Z})$, for $\mathcal{C}_{\mathbb{C}}$ is a soft sheaf (Appendix B). In other words, complex line bundles on a manifold M are parametrized by the (discrete) group $H^2(M, \mathbb{Z})$. Since $H^2(M, \mathbb{Z})$ maps naturally to $H^2(M, \mathbb{R}) \subset H^2(M, \mathbb{C})$, one may compare $\delta(L)$ and $c_1(L)$. In Proposition 4.4.12 we will see that they only differ by a sign.

Clearly, the exponential sequence on a complex manifold X and the exponential sequence on the underlying real manifold M are compatible. Thus, we show at the same time that i) and iii) above are compatible. We therefore obtain the following commutative diagram

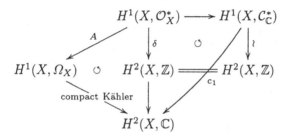

Proposition 4.4.12 *Let L be a complex line bundle over a differentiable manifold M. Then the image of $\delta(L) \in H^2(M, \mathbb{Z})$ under the natural map $H^2(M, \mathbb{Z}) \to H^2(M, \mathbb{C})$ equals $-c_1(L)$. Here, δ is the boundary map of the exponential sequence.*

The annoying sign is due to various conventions, e.g. in the definition of the boundary operator. Often, it is dropped altogether, as it is universal and of no importance.

Proof. In order to prove this, we have to consider the two resolutions of the constant sheaf $\mathbb{C}$ on M given by the de Rham complex and the Čech complex, respectively. They are compared as follows:

$$
\begin{array}{ccccccc}
\mathbb{C} & \longrightarrow & \mathcal{C}^0(\{U_i\}, \mathbb{C}) & \longrightarrow & \mathcal{C}^1(\{U_i\}, \mathbb{C}) & \longrightarrow & \mathcal{C}^2(\{U_i\}, \mathbb{C}) \\
\downarrow & & & & & & \downarrow i \\
\mathcal{A}^0 & & & & \mathcal{C}^1(\{U_i\}, \mathcal{A}^0) & \xrightarrow{\ \delta_2\ } & \mathcal{C}^2(\{U_i\}, \mathcal{A}^0) \\
\downarrow & & & & \downarrow d & & \\
\mathcal{A}^1 & & \mathcal{C}^0(\{U_i\}, \mathcal{A}^1) & \xrightarrow{\ \delta_1\ } & \mathcal{C}^1(\{U_i\}, \mathcal{A}^1) & & \\
\downarrow & & \downarrow d & & & & \\
\mathcal{A}^2 & \xrightarrow{\ \delta_0\ } & \mathcal{C}^0(\{U_i\}, \mathcal{A}^2) & & & &
\end{array}
$$

Let $M = \bigcup U_i$ be an open covering trivializing L and such that $U_{ij} = U_i \cap U_j$ are simply connected. Choose trivialization $\psi_i : L|_{U_i} \cong U_i \times \mathbb{C}$. Then, $\psi_{ij} = \psi_i \circ \psi_j^{-1}$ are sections of $\mathcal{C}_{\mathbb{C}}^*(U_{ij})$. Furthermore, by choosing a branch of the logarithm for any U_{ij} we find $\varphi_{ij} \in \mathcal{C}_{\mathbb{C}}(U_{ij})$ with $e^{2\pi i \varphi_{ij}} = \psi_{ij}$. The boundary $\delta(L) = \delta\{\psi_{ij}\}$ is given by $\{U_{ijk}, \varphi_{jk} - \varphi_{ik} + \varphi_{ij}\}$ which takes values in the locally constant sheaf $\mathbb{Z}$.

Now choose an arbitrary connection ∇ on L. Locally with respect to the trivialization ψ_i it can be written as $\nabla = d + A_i$, where the connection matrices A_i are one-forms on U_i. The compatibility condition ensures $\psi_{ij}^{-1} d(\psi_{ij}) + \psi_{ij}^{-1} A_i \psi_{ij} = A_j$ (see Remark 4.2.5), i.e.

$$A_j - A_i = \psi_{ij}^{-1} d(\psi_{ij}) = (2\pi i) d(\varphi_{ij}),$$

since in the rank one case one has $\psi_{ij}^{-1} A_i \psi_{ij} = A_i$.

The curvature F_∇ of the line bundle L in terms of the connection forms A_i is given as $F_\nabla = d(A_i)$. With these information we can now easily go through the above diagram:

$$\delta_0(\tfrac{i}{2\pi} F_\nabla) = \{U_i, \tfrac{i}{2\pi} d(A_i)\} = d\{U_i, \tfrac{i}{2\pi} A_i\}$$

$$\delta_1\{U_i, \tfrac{i}{2\pi} A_i\} = \{U_{ij}, \tfrac{i}{2\pi}(A_j - A_i)\} = -d\{U_{ij}, \varphi_{ij}\}$$

$$-\delta_2\{U_{ij}, \varphi_{ij}\} = -\{U_{ijk}, \varphi_{jk} - \varphi_{ik} + \varphi_{ij}\}.$$

This proves the claim. $\square$

There is yet another way to associate a cohomology class to a line bundle in case the line bundle is given in terms of a divisor. Let X be a compact complex manifold and let $D \subset X$ be an irreducible hypersurface. Its fundamental class $[D] \in H^2(X, \mathbb{R})$ is in fact contained in the image of $H^2(X, \mathbb{Z}) \to H^2(X, \mathbb{R})$ (cf. Remark 2.3.11).

Proposition 4.4.13 *Under the above assumptions one has* $c_1(\mathcal{O}(D)) = [D]$.

Proof. Let $L = \mathcal{O}(D)$ and let ∇ be the Chern connection on L with respect to a chosen hermitian structure h. In order to prove that $[\tfrac{i}{2\pi} F_\nabla] = c_1(L)$ equals $[D]$ one needs to show that for any closed real form α one has

$$\frac{i}{2\pi} \int_X F_\nabla \wedge \alpha = \int_D \alpha.$$

Let us fix an open covering $X = \bigcup U_i$ and holomorphic trivializations $\psi_i : L|_{U_i} \cong U_i \times \mathbb{C}$. On U_i the hermitian structure h shall be given by the function $h_i : U_i \to \mathbb{R}_{>0}$, i.e. $h(s(x)) = h(s(x), s(x)) = h_i(x) \cdot |\psi_i(s(x))|^2$ for any local section s.

If s is holomorphic on U_i vanishing along D one has (Examples 4.3.9, iii)):

$$\bar{\partial}\partial \log(h \circ s) = \bar{\partial}\partial \log(h_i) = F_\nabla|_{U_i} \quad \text{on } U_i \setminus D.$$

Here we have used that $\bar{\partial}\partial \log(\psi_i \circ s) = \bar{\partial}\partial \log(\bar{\psi}_i \circ s) = 0$, since ψ_i is holomorphic.

Let now $s \in H^0(X, L)$ be the global holomorphic section defining D and denote by D_ε the tubular neighbourhood $D_\varepsilon := \{x \in X \mid |h(s(x))| < \varepsilon\}$. Then

$$\frac{i}{2\pi} \int_X F_\nabla \wedge \alpha = \lim_{\varepsilon \to 0} \frac{i}{2\pi} \int_{X \setminus D_\varepsilon} F_\nabla \wedge \alpha$$

$$= \lim_{\varepsilon \to 0} \frac{i}{2\pi} \int_{X \setminus D_\varepsilon} \bar{\partial}\partial \log(h \circ s) \wedge \alpha$$

$$= \lim_{\varepsilon \to 0} \frac{i}{4\pi} \int_{X \setminus D_\varepsilon} d(\partial - \bar{\partial}) \log(h \circ s) \wedge \alpha$$

$$= \lim_{\varepsilon \to 0} \frac{i}{4\pi} \int_{\partial D_\varepsilon} (\partial - \bar{\partial}) \log(h \circ s) \wedge \alpha \quad \text{by Stokes and using } d\alpha = 0.$$

On the open subset U_i we may write

$$(\partial - \bar{\partial}) \log(h \circ s)$$
$$= (\partial - \bar{\partial}) \log(\psi_i \circ s) + (\partial - \bar{\partial}) \log(\bar{\psi}_i \circ s) + (\partial - \bar{\partial}) \log(h_i)$$
$$= \partial \log(\psi_i \circ s) - \overline{\partial \log(\psi_i \circ s)} + (\partial - \bar{\partial}) \log(h_i)$$
$$= (2i) \cdot \mathrm{Im}(\partial \log(\psi_i \circ s)) + (\partial - \bar{\partial}) \log(h_i).$$

The second summand does not contribute to the integral for $\varepsilon \to 0$ as h_i is bounded from below by some $\delta > 0$. Thus, it suffices to show

$$\lim_{\varepsilon \to 0} \frac{1}{2\pi} \int_{\partial D_\varepsilon \cap U_i} \mathrm{Im}(\partial \log(\psi_i \circ s)) \wedge \alpha = -\int_{D \cap U_i} \alpha.$$

This is a purely local statement. In order to prove it, we may assume that D_ε is given by $z_1 = 0$ in a polydisc B. Moreover, $|h(z_1)| = h_i \cdot |z_1|$ if h on U_i is given by h_i. Hence, $\partial D_\varepsilon = \{z \mid |z_1| = \varepsilon/h_i\}$. Furthermore, $\partial \log(\psi_i \circ s) = \partial \log(z_1) = z_1^{-1} dz_1$ and $\alpha = f(dz_2 \wedge \ldots \wedge dz_n) \wedge (d\bar{z}_2 \wedge \ldots \wedge d\bar{z}_n) + dz_1 \wedge \alpha' + d\bar{z}_1 \wedge \bar{\alpha}'$. Notice that $\partial \log(\psi_i \circ s) \wedge (dz_1 \wedge \alpha') = 0$ and that $\partial \log(\psi_i \circ s) \wedge (d\bar{z}_1 \wedge \bar{\alpha}') = (dz_1 \wedge d\bar{z}_1) \wedge ((1/z_1)\bar{\alpha}')$ does not contribute to the integral over ∂D_ε.

Thus,

$$\int_{z_1 = 0} \alpha = \int f(0, z_2, \ldots, z_n)(dz_2 \wedge \ldots \wedge dz_n) \wedge (d\bar{z}_2 \wedge \ldots \wedge d\bar{z}_n)$$

and

$$\int_{\partial D_\varepsilon} \partial \log(\psi_i \circ s) \wedge \alpha = -\int_{|h(z_1)|=\varepsilon} f z_1^{-1} dz_1 \wedge (dz_2 \wedge \ldots \wedge dz_n) \wedge (d\bar{z}_2 \wedge \ldots \wedge d\bar{z}_n).$$

The minus sign appears as we initially integrated over the exterior domain.

Eventually, one applies the Cauchy integral formula (1.4):

$$\lim_{\varepsilon \to 0} \int_{\partial D_\varepsilon} \partial \log(\psi_i \circ s) \wedge \alpha$$

$$= -\lim_{\varepsilon \to 0} \int_{|z_1|=\varepsilon/h_i} \left(\int_{z_i > 1} f \cdot (dz_2 \wedge \ldots \wedge dz_n \wedge d\bar{z}_2 \wedge \ldots \wedge d\bar{z}_n) \right) \frac{dz_1}{z_1}$$

$$= (-2\pi i) \cdot \int_{z_1=0} f(0, z_2, \ldots, z_n)(dz_2 \wedge \ldots \wedge dz_n \wedge d\bar{z}_2 \wedge \ldots \wedge d\bar{z}_n)$$

$$= -2\pi i \int_{z_1=0} \alpha$$

and hence

$$\lim_{\varepsilon \to 0} \frac{1}{2\pi} \int_{\partial D_\varepsilon} \mathrm{Im} \left(\partial \log(\psi_i \circ s) \wedge \alpha \right) = -\mathrm{Im} \left(\int_{z_1=0} i \cdot \alpha \right) = -\int_{z_1=0} \alpha.$$

$\square$

Remark 4.4.14 Since taking the first Chern class c_1 and taking the fundamental class are both linear operations, the assertion of Proposition 4.4.13 holds true for arbitrary divisors, i.e. $c_1(\mathcal{O}(\sum n_i D_i)) = \sum n_i [D_i]$.

The reader may have noticed that in the proof above we have, for simplicity, assumed that D is smooth. The argument might easily be adjusted to the general case.

Exercises

4.4.1 Let C be a connected compact curve. Then there is a natural isomorphism $H^2(C, \mathbb{Z}) \cong \mathbb{Z}$. Show that with respect to this isomorphism (or, rather, its $\mathbb{R}$-linear extension) one has $c_1(L) = \deg(L)$ for any line bundle L on C.

4.4.2 Show that for a base-point free line bundle L on a compact complex manifold X the integral $\int_X c_1(L)^n$ is non-negative.

4.4.3 Show that $\mathrm{td}(E_1 \oplus E_2) = \mathrm{td}(E_1) \cdot \mathrm{td}(E_2)$.

4.4.4 Compute the Chern classes of (the tangent bundle of) $\mathbb{P}^n$ and $\mathbb{P}^n \times \mathbb{P}^m$. Try to interpret $\int_{\mathbb{P}^n} c_n(\mathbb{P}^n)$ and $\int_{\mathbb{P}^n \times \mathbb{P}^m} c_n(\mathbb{P}^n \times \mathbb{P}^m)$.

4.4.5 Prove the following explicit formulae for the first three terms of $\mathrm{ch}(E)$ and $\mathrm{td}(E)$ in terms of $c_i(E)$:

$$\mathrm{ch}(E) = \mathrm{rk}(E) + c_1(E) + \frac{c_1^2(E) - 2c_2(E)}{2} + \frac{c_1^3(E) - 3c_1(E)c_2(E) + 3c_3(E)}{6} + \ldots$$

$$\mathrm{td}(E) = 1 + \frac{c_1(E)}{2} + \frac{c_1^2(E) + c_2(E)}{12} + \frac{c_1(E)c_2(E)}{24} + \ldots$$

4.4.6 Let E be a vector bundle and L a line bundle. Show

$$c_i(E \otimes L) = \sum_{j=0}^{i} \binom{\text{rk}(E) - j}{i - j} c_j(E) c_1(L)^{i-j}.$$

This generalizes the computation for the first two Chern classes of $E \otimes L$ on page 197.

4.4.7 Show that on $\mathbb{P}^n$ one has $c_1(\mathcal{O}(1)) = [\omega_{\text{FS}}] \in H^2(\mathbb{P}^n, \mathbb{R})$. Consider first the case of $\mathbb{P}^1$ and then the restriction of $\mathcal{O}(1)$ and the Fubini–Study metric to $\mathbb{P}^1$ under a linear embedding $\mathbb{P}^1 \subset \mathbb{P}^n$.

4.4.8 Prove that a polynomial P of degree k on the space of $r \times r$ matrices is invariant if and only if $\sum P(A_1, \ldots, A_{i-1}, [A, A_i], A_{i+1}, \ldots, A_k) = 0$ for all matrices $A_1, \ldots, A_k, A$ (cf. Lemma 4.4.2).

4.4.9 Show that $c_1(\text{End}(E)) = 0$ on the form level and compute $c_2(\text{End}(E))$ in terms of $c_i(E)$, $i = 1, 2$. Compute $(4c_2 - c_1^2)(L \oplus L)$ for a line bundle L. Show that $c_{2k+1}(E) = 0$, if $E \cong E^*$.

4.4.10 Let L be a holomorphic line bundle on a compact Kähler manifold X. Show that for any closed real $(1,1)$-form α with $[\alpha] = c_1(L)$ there exists an hermitian structure on L such that the curvature of the Chern connection ∇ on L satisfies $(i/2\pi)F_\nabla = \alpha$. (Hint: Fix an hermitian structure on h_0 on L. Then any other is of the form $e^f \cdot h_0$. Compute the change of the curvature. We will give the complete argument in Remark 4.B.5.

4.4.11 Let X be a compact Kähler manifold. Show that via the natural inclusion $H^k(X, \Omega_X^k) \subset H^{2k}(X, \mathbb{C})$ one has

$$\text{ch}_k(E) = \frac{1}{k!} \left(\frac{i}{2\pi} \right)^k \text{tr} \left(A(E)^{\otimes k} \right).$$

Here, $A(E)^{\otimes k}$ is obtained as the image of $A(E) \otimes \ldots \otimes A(E)$ under the natural map $H^1(X, \Omega_X \otimes \text{End}(E)) \times \ldots \times H^1(X, \Omega_X \otimes \text{End}(E)) \to H^k(X, \Omega_X^k \otimes \text{End}(E))$ which is induced by composition in $\text{End}(E)$ and exterior product in Ω_X^*.

4.4.12 Let X be a compact Kähler manifold and let E be a holomorphic vector bundle admitting a holomorphic connection $D : E \to \Omega_X \otimes E$. Show that $c_k(E) = 0$ for all $k > 0$.

Comments: Chern classes were first defined by Chern in [26]. A more topological approach to characteristic classes, also in the real situation, can be found in [91]. Since Chern classes are so universal, adapted versions appear in many different areas, e.g. algebraic geometry, Arakelov theory, etc.

Appendix to Chapter 4

4.A Levi-Civita Connection and Holonomy on Complex Manifolds

In this section we will compare our notion of a connection with the notion used in Riemannian geometry. In particular, we will clarify the relation between the Chern connection on the holomorphic tangent bundle of an hermitian manifold and the Levi-Civita connection on the underlying Riemannian manifold. Very roughly, we will see that these connections coincide if and only if the hermitian manifold is Kähler.

We will end this appendix with a discussion of the holonomy group of a Riemannian manifold and the interpretation of a Kähler structure in terms of the holonomy group of its underlying Riemannian structure.

Let us first review some basic concepts from Riemannian geometry. For this purpose we consider a Riemannian manifold (M, g). A connection on M by definition is a connection on the real tangent bundle TM, i.e. an $\mathbb{R}$-linear map $D : \mathcal{A}^0(TM) \to \mathcal{A}^1(TM)$ satisfying the Leibniz rule 4.2. For any two vector fields u and v we denote by $D_u v$ the one-form Dv with values in TM applied to the vector field u. Note that $D_u v$ is $\mathbb{R}$-linear in v and $\mathcal{A}^0$-linear in u. With this notation, the Leibniz rule reads $D_u(f \cdot v) = f \cdot D_u(v) + (df)(u) \cdot v$.

A connection is *metric* if $dg(u, v) = g(Du, v) + g(u, Dv)$. In other words, D is metric if and only if g is parallel, i.e. $D(g) = 0$, where D is the induced connection on $T^*M \otimes T^*M$ (cf. Exercise 4.2.8).

Before defining the torsion of a connection recall that the Lie bracket is an $\mathbb{R}$-linear skew-symmetric map $[\, , \,] : \mathcal{A}^0(TM) \times \mathcal{A}^0(TM) \to \mathcal{A}^0(TM)$ which locally for $u = \sum_i a_i \frac{\partial}{\partial x_i}$ and $v = \sum_i b_i \frac{\partial}{\partial x_i}$ is defined by

$$[u, v] = \sum_j \sum_i \left(a_i \frac{\partial b_j}{\partial x_i} - b_i \frac{\partial a_j}{\partial x_i} \right) \frac{\partial}{\partial x_j}$$
$$= \sum_j (db_j(u) - da_j(v)) \frac{\partial}{\partial x_j}.$$

In particular, one has $[f \cdot u, v] = f \cdot [u, v] - df(v) \cdot u$.

Definition 4.A.1 The *torsion* of a connection D is given by

$$T_D(u, v) := D_u v - D_v u - [u, v]$$

for any two vector fields u and v.

The first thing one observes is that T_D is skew-symmetric, i.e. $T_D : \bigwedge^2 TM \to TM$. Moreover, T_D is $\mathcal{A}^0$-linear and can therefore be considered as an element of $\mathcal{A}^2(TM)$. Indeed, $T_D(f \cdot u, v) = f \cdot D_u v - (f \cdot D_v u + df(v) \cdot u) - (f \cdot [u, v] - df(v) \cdot u) = f \cdot T_D(u, v)$. A connection ∇ is called *torsion free* if $T_D = 0$.

Let us try to describe the torsion in local coordinates. So we may assume that our connection is of the form $D = d + A$. Here, A is a one-form with values in $\text{End}(TM)$. In the following we will write $A(u) \in \mathcal{A}^0(\text{End}(TM))$ for the endomorphism that is obtained by applying the one-form part of A to the vector field u. If u is constant, then $A(u) = D_u$. On the other hand, $A \cdot u \in \mathcal{A}^1(TM)$ is obtained by applying the endomorphism part of A to u. Confusion may arise whenever we use the canonical isomorphism $\mathcal{A}^1(TM) \cong \mathcal{A}^0(\text{End}(TM))$.

Lemma 4.A.2 *If $D = d + A$ then $T_D(u, v) = A(u) \cdot v - A(v) \cdot u$.*

Proof. If $u = \sum_i a_i \frac{\partial}{\partial x_i}$ and $v = \sum_i b_i \frac{\partial}{\partial x_i}$ then

$$T_D(u, v) = \left(\sum db_i(u) \frac{\partial}{\partial x_i} + A(u) \cdot v \right) - \left(\sum da_i(v) \frac{\partial}{\partial x_i} + A(v) \cdot u \right) - [u, v]$$

$$= A(u) \cdot v - A(v) \cdot u$$

$\square$

Classically, one expresses the connection matrix A in terms of the *Christoffel symbols* Γ_{ij}^k as

$$A\left(\frac{\partial}{\partial x_i} \right) \cdot \frac{\partial}{\partial x_j} = \sum_k \Gamma_{ij}^k \frac{\partial}{\partial x_k}.$$

Then

$$T_D\left(\frac{\partial}{\partial x_i}, \frac{\partial}{\partial x_j} \right) = \sum_k \left(\Gamma_{ij}^k - \Gamma_{ji}^k \right) \frac{\partial}{\partial x_k}.$$

In particular, D is torsion free if and only if $\Gamma_{ij}^k = \Gamma_{ji}^k$ for all i, j, k. The following result is one of the fundamental statements in Riemannian geometry and can be found in most textbooks on the subject, see e.g. [79].

Theorem 4.A.3 *Let (M, g) be a Riemannian manifold. Then there exists a unique torsion free metric connection on M; the* Levi-Civita *connection.* $\square$

Why the notion of a torsion free connection is geometrically important is not evident from the definition. But in any case torsion free connections turn out to behave nicely in many ways. E.g. the exterior differential can be expressed in terms of torsion free connections.

Proposition 4.A.4 *If D is a torsion free connection on M then the induced connection on the space of forms satisfies*

$$(d\alpha)(v_1, \ldots v_{k+1}) = \sum_{i=0}^{k} (-1)^i (D_{v_i} \alpha)(v_1, \ldots, \hat{v}_i, \ldots, v_{k+1})$$

for any k-form α and vector fields $v_1, \ldots v_{k+1}$.

Proof. We leave the complete proof to the reader (cf. Exercise 4.A.1). One has to use the following definition of the exterior differential (see Appendix A):

$$d\alpha(v_1,\ldots,v_{k+1}) = \sum_{i=1}^{k+1}(-1)^{i+1}v_i(\alpha(v_1,\ldots,\hat{v}_i,\ldots,v_{k+1}))$$
$$+ \sum_{1\le i<j\le k+1}(-1)^{i+j}\alpha([v_i,v_j],v_1,\ldots,\hat{v}_i,\ldots,\hat{v}_j,\ldots v_{k+1})$$

$\square$

A form $\alpha \in \mathcal{A}^k(M)$ is parallel if $D(\alpha) = 0$. Thus, Proposition 4.A.4 implies

Corollary 4.A.5 *Let D be a torsion free connection on M. Any D-parallel form is closed.* $\square$

Let us now turn to hermitian manifolds. By definition, an hermitian structure on a complex manifold X is just a Riemannian metric g on the underlying real manifold compatible with the complex structure I defining X (see Definition 3.1.1). Recall that the complexified tangent bundle $T_{\mathbb{C}}X$ decomposes as $T_{\mathbb{C}}X = T^{1,0}X \oplus T^{0,1}X$ and that the bundle $T^{1,0}X$ is the complex bundle underlying the holomorphic tangent bundle $\mathcal{T}_X$ (Proposition 2.6.4). Moreover, the hermitian extension $g_{\mathbb{C}}$ of g to $T_{\mathbb{C}}X$ restricted to $T^{1,0}X$ is $\frac{1}{2}(g - i\omega)$, where the complex vector bundles $T^{1,0}X$ and (TX, I) are identified via the isomorphism

$$\xi : TX \longrightarrow T^{1,0}X, \quad u \longmapsto \frac{1}{2}(u - iI(u))$$

and ω is the fundamental form $g(I(\ ),(\ ))$ (cf. Section 1.2).

We will compare hermitian connections ∇ on $(T^{1,0}X, g_{\mathbb{C}})$ with the Levi-Civita connection D on TX via the isomorphism ξ. One first observes the following easy

Lemma 4.A.6 *Under the natural isomorphism ξ any hermitian connection ∇ on $T^{1,0}X$ induces a metric connection D on the Riemannian manifold (X,g).*

Proof. By assumption $dg_{\mathbb{C}}(u,v) = g_{\mathbb{C}}(\nabla u,v) + g_{\mathbb{C}}(u,\nabla v)$. Taking real parts of both sides yields $dg(u,v) = g(Du,v) + g(u,Dv)$, i.e. the induced connection D is metric. $\square$

In general, an hermitian connection ∇ on $(T^{1,0}X, g_{\mathbb{C}})$ will not necessarily induce the Levi-Civita connection on the Riemannian manifold (X,g). In fact, this could hardly be true, as the Levi-Civita connection is unique, but there are many hermitian connections $(T^{1,0}X, g_{\mathbb{C}})$. But even for the Chern connection on the holomorphic tangent bundle $(\mathcal{T}_X, g_{\mathbb{C}})$, which is unique, the induced connection is not the Levi-Civita connection in general.

In order to state the relevant result comparing these two notions, we need to introduce the torsion $T_\nabla \in \mathcal{A}^2(X)$ of an hermitian connection ∇. By definition, T_∇ of ∇ is the torsion of the induced connection D on TX, i.e. for $u, v \in TX$ one has $T_\nabla(u, v) = \xi^{-1}(\nabla_u \xi(v) - \nabla_v \xi(u)) - [u, v] = D_u v - D_v u - [u, v] = T_D(u, v)$. As before we call the hermitian connection torsion free if its torsion is trivial.

Proposition 4.A.7 *Let ∇ be a torsion free hermitian connection on the hermitian bundle $(T^{1,0}X, g_{\mathbb{C}})$.*
   i) Then ∇ is the Chern connection on the holomorphic bundle T_X endowed with the hermitian structure $g_{\mathbb{C}}$.
   ii) The induced connection D on the underlying Riemannian manifold is the Levi-Civita connection.
   iii) The hermitian manifold (X, g) is Kähler.

Proof. If we write the connection ∇ with respect to a local holomorphic base $\frac{\partial}{\partial z_i}$ as $\nabla = d + A$ then we have to show that the assumption $T_\nabla = 0$ implies $A \in \mathcal{A}^{1,0}(\mathrm{End}(T^{1,0}X))$. By definition the latter condition is equivalent to the vanishing of $A(u + iI(u))$ for any $u \in TX$ or, equivalently, to $A(u + iI(u)) \cdot \xi(v) = 0$ for all v. Using the analogue of Lemma 4.A.2 for the torsion of the connection ∇, one computes

$$A(u + iI(u)) \cdot \xi(v)$$
$$= A(u + iI(u)) \cdot \xi(v) - A(v) \cdot (\xi(u) + iI(\xi(u))), \quad \text{since } \xi = -iI\xi$$
$$= (A(u) \cdot \xi(v) - A(v) \cdot \xi(u)) + i\,(A(I(u)) \cdot \xi(v) - A(v) \cdot \xi(I(u)))$$
$$= \xi(T(u, v)) + i\xi(T(I(u), v)) = 0,$$

as ξ and I are $\mathbb{C}$-linear. This proves i).
   ii) is a consequence of what has been said before. In order to show iii), i.e. that the fundamental form ω is closed, one may use Corollary 4.A.5. Thus, it suffices to show that ω is parallel. This is the following straightforward calculation:

$$(D\omega)(u, v) = d(\omega(u, v)) - \omega(Du, v) - \omega(u, Dv)$$
$$= dg(Iu, v) - g(DI(u), v) - g(I(u), D(v)) = 0,$$

as the connection is metric.   □

Note that in the proof we have tacitly assumed that D commutes with the complex structure I, which is obvious as the hermitian connection ∇ on the complex vector bundle $(T^{1,0}X, g_{\mathbb{C}})$ is in particular $\mathbb{C}$-linear and $T^{1,0}X \cong (TX, I)$ is an isomorphism of complex vector bundles. However, if we try to associate to a connection on the underlying real manifold a connection on the holomorphic tangent bundle, then the compatibility with the complex structure is the crucial point. If D is a connection on the tangent bundle TX. Then $DI = ID$ if and only if I is a parallel section of $\mathrm{End}(TX)$ with respect to the induced connection on the endomorphism bundle.

Proposition 4.A.8 *Let D be the Levi-Civita connection on the Riemannian manifold (X, g) and assume that the complex structure I is parallel.*

i) *Under the isomorphism $\xi : TX \cong T^{1,0}X$ the connection D induces the Chern connection ∇ on the holomorphic tangent bundle $T^{1,0}X$.*

ii) *The manifold is Kähler and, moreover, the Kähler form ω is parallel.*

Proof. Since I is parallel, we do obtain a connection ∇ on the complex vector bundle $T^{1,0}X$. This connection is hermitian if and only if $dg_{\mathbb{C}}(u, v) = g_{\mathbb{C}}(\nabla u, v) + g_{\mathbb{C}}(u, \nabla v)$. Since the Levi-Civita connection is metric, the real parts of both sides are equal. The imaginary parts are (up to a factor) $d\omega(u, v) = dg(I(u), v)$ respectively $g(I(Du), v) + g(I(u), Dv)$. Using $DI = ID$ and again that ∇ is metric, one sees that they also coincide.

The Levi-Civita connection is by definition torsion free and using Proposition 4.A.7 this proves i) and the first assertion of ii). That $D(\omega) = 0$ follows from $D(I) = 0$ and $D(g) = 0$ as in the proof of Proposition 4.A.7.   □

As a partial converse of Proposition 4.A.7, one proves.

Proposition 4.A.9 *Let (X, g) be a Kähler manifold. Then under the isomorphism $\xi : TX \cong T^{1,0}X$ the Chern connection ∇ on the holomorphic tangent bundle $T_X = T^{1,0}X$ corresponds to the Levi-Civita connection D.*

Proof. We have to show that under the assumption that (X, g) is Kähler the Chern connection is torsion free. This is done in local coordinates. Locally we write $\omega = \frac{1}{2} \sum h_{ij} dz_i \wedge dz_j$ and $A = (h_{ji})^{-1}(\partial h_{ji})$ for the connection form. The fundamental form is closed if and only if $\frac{\partial h_{ij}}{\partial z_k} = \frac{\partial h_{kj}}{\partial z_i}$. The latter can then be used to prove the required symmetry of the torsion form. We leave the details to the reader (cf. Exercise 4.A.2)   □

From this slightly lengthy discussion the reader should only keep in mind that the following four conditions are equivalent:

i) The complex structure is parallel with respect to the Levi-Civita connection.

ii) (X, g) is Kähler.

iii) Levi-Civita connection D and Chern connection ∇ are identified by ξ.

iv) The Chern connection is torsion free.

For this reason we will in the following not distinguish anymore between the Levi-Civita connection D and the Chern connection ∇ provided the manifold is Kähler.

Let us now turn to the curvature tensor of a Riemannian manifold (M, g). Classically it is defined as

$$R(u, v) := D_u D_v - D_v D_u - D_{[u,v]},$$

where D is the Levi-Civita connection. How does this definition compare with the one given in Section 4.3? Not surprisingly they coincide. This is shown by the following direct calculation. We assume for simplicity that $Ds = \alpha \otimes t$:

$$
\begin{aligned}
R(u, v)(s) &= D_u D_v s - D_v D_u s - D_{[u,v]} s \\
&= D_u(\alpha(v)t) - D_v(\alpha(u)t) - \alpha([u, v])t \\
&= u(\alpha(v))t + \alpha(v)D_u t - v(\alpha(u))t - \alpha(u)D_v t - \alpha([u, v])t \\
&= (d\alpha)(u, v) \cdot t + \alpha(v)D_u t - \alpha(u)D_v t = ((d\alpha) \cdot t + \alpha \cdot D(t)) (u, v) \\
&= D(\alpha \otimes t)(u, v) = F(u, v)(s),
\end{aligned}
$$

where we have used a special case of the formula that describes the exterior differential in terms of the Lie bracket (cf. Exercise 4.A.1).

In Riemannian geometry one also considers the *Ricci tensor*

$$
r(u, v) := \operatorname{tr}(w \mapsto R(w, u)v) = \operatorname{tr}(w \longmapsto R(w, v, u))
$$

(see [79]). Combined with the complex structure one has

Definition 4.A.10 The *Ricci curvature* $\operatorname{Ric}(X, g)$ of a Kähler manifold (X, g) is the real two-form
$$
\operatorname{Ric}(u, v) := r(I(u), v).
$$
The Kähler metric is called *Ricci-flat* if $\operatorname{Ric}(X, g) = 0$.

The Ricci curvature can be computed by means of the curvature F_∇ of the Levi-Civita (or, equivalently, the Chern) connection and the Kähler form ω. This goes as follows.

The contraction of the curvature $F_\nabla \in \mathcal{A}^{1,1}(\operatorname{End}(T^{1,0}X))$ with the Kähler form ω yields an element $\Lambda_\omega F \in \mathcal{A}^0(X, \operatorname{End}(T^{1,0}X))$ or, equivalently, an endomorphism $T^{1,0}X \to T^{1,0}X$. Its composition with the isomorphism $T^{1,0}X \to \bigwedge^{0,1} X$ induced by the Kähler form will be denoted

$$
\tilde{\omega}(\Lambda_\omega F) : T^{1,0}X \xrightarrow{\Lambda_\omega F_\nabla} T^{1,0}X \xrightarrow{\omega} \bigwedge^{0,1} X.
$$

One easily verifies that $\tilde{\omega}(\Lambda_\omega F) \in \mathcal{A}^{1,1}(X)$.

Proposition 4.A.11 *Let (X, g) be a Kähler manifold and ∇ the Levi-Civita or, equivalently, the Chern connection. Then the following two identities hold true:*

i) $\operatorname{Ric}(X, g) = i \cdot \tilde{\omega} (\Lambda_\omega F_\nabla)$.

ii) $\operatorname{Ric}(X, g) = i \cdot \operatorname{tr}_{\mathbb{C}}(F_\nabla)$, *where the trace is taken in the endomorphism part of the curvature.*

Proof. We shall use the following well-known identities from Riemannian geometry (see [12, page 42]):

$$g(R(u,v)x, y) = g(R(x,y)u, v) \tag{4.6}$$

$$R(u,v)w + R(v,w)u + R(w,u)v = 0 \tag{4.7}$$

$$g(R(u,v)w, x) + g(w, R(u,v)x) = 0 \tag{4.8}$$

The third one is clearly due to the fact that the Levi-Civita connection is metric. Equation (4.7), which is the algebraic Bianchi identity, and equation (4.6) are more mysterious.

The proof consists of computing all three expressions explicitly in terms of an orthonormal basis of the form $x_1, \ldots, x_n, y_1 = I(x_1), \ldots, y_n = I(x_n)$:

$$
\begin{aligned}
\mathrm{Ric}(u,v) &= \mathrm{tr}\,(w \mapsto R(w,v)I(u)) \\
&= \sum g(R(x_i, v)I(u), x_i) + g(R(y_i, v)I(u), y_i) \\
&\overset{(4.8)}{=} -\sum g(I(u), R(x_i, v)x_i) + g(I(u), R(y_i, v)y_i) \\
&= \sum g(u, R(x_i, v)y_i) - g(u, R(y_i, v)x_i) \\
&\overset{(4.7)}{=} -\sum g(u, R(y_i, x_i)v) \\
&\overset{(4.8)}{=} -\sum g(R(x_i, y_i)u, v),
\end{aligned}
$$

where we use twice the compatibility of g and I and the fact that $R(u,v)$ is skew-symmetric.

Furthermore, using the notation and convention of Section 1.2 one calculates:

$$
\begin{aligned}
\mathrm{tr}_{\mathbb{C}}(F_\nabla)(u,v) &= \mathrm{tr}_{\mathbb{C}}\,(w \mapsto F_\nabla(u,v)w) \quad \text{with } w \in T^{1,0} \\
&= \mathrm{tr}_{\mathbb{C}}\,(w \mapsto R(u,v)w) \quad \text{with } w \in T \\
&= \sum (R(u,v)x_i, x_i) \quad \text{since } x_1, \ldots, x_n \text{ is an orthonormal basis} \\
&\qquad\qquad\qquad \text{of the hermitian vector space } (T, (\ ,\ )) \\
&= \sum g(R(u,v)x_i, x_i) - i \cdot \omega(R(u,v)x_i, x_i) \\
&\qquad\qquad\qquad \text{as } (\ ,\ ) = g - i \cdot \omega \\
&= \sum g(R(u,v)x_i, x_i) + i \cdot g(R(u,v)x_i, y_i) \\
&\overset{(4.6)}{=} \sum g(R(x_i, x_i)u, v) + i \cdot g\,(R(x_i, y_i)u, v) \\
&= 0 + i \cdot \sum g(R(x_i, y_i)u, v)
\end{aligned}
$$

Both computations together show ii).

In order to see i), let us write $\omega = \sum x^i \wedge y^i$. Then a straightforward computation shows $\Lambda_\omega \alpha = \sum \alpha(x_i, y_i)$ and

$$2i \cdot \tilde{\omega}(\Lambda_\omega F)(u, v) = i \cdot \omega(\Lambda_\omega F(\xi(u)), \overline{\xi(v)}) = \omega(\Lambda_\omega(F(\xi(I(u)), \overline{\xi(v)}))$$

$$= \omega\left(\sum F(x_i, y_i)\xi(I(u)), \overline{\xi(v)}\right)$$

$$= \omega\left(\xi\left(\sum R(x_i, y_i)I(u)\right), \overline{\xi(v)}\right)$$

$$= 2g\left(I\left(\sum R(x_i, y_i)I(u)\right), v\right) = -2\sum g(R(x_i, y_i)u, v).$$

$\square$

In the following, we will sketch the relation between the holonomy of a Riemannian manifold and its complex geometry.

To any Riemannian metric g on a manifold M of dimension m there is associated, in a unique way, the Levi-Civita connection D. By means of D one can define the parallel transport of tangent vectors along a path in M. This goes as follows.

Let $\gamma : [0,1] \to M$ be a path connecting two points $x := \gamma(0)$ and $y := \gamma(1)$. The pull-back connection $\gamma^* D$ on $\gamma^*(TM)$ is necessarily flat over the one-dimensional base $[0,1]$ and $\gamma^*(TM)$ can therefore be trivialized by flat sections. In this way, one obtains an isomorphism, the *parallel transport along the path* γ:

$$P_\gamma : T_x M \longrightarrow T_x M .$$

In other words, for any $v \in T_x M$ there exists a unique vector field $v(t)$ with $v(t) \in T_{\gamma(t)}M$, $v(0) = v$, and such that $v(t)$ is a flat section of $\gamma^*(TM)$. Then $P_\gamma(v) = v(1)$.

The first observation concerns the compatibility of P_γ with the scalar products on $T_x M$ and $T_y M$ given by the chosen Riemannian metric g.

Lemma 4.A.12 P_γ *is an isometry.* $\square$

In particular, if γ is a closed path, i.e. $\gamma(0) = \gamma(1) = x$, then $P_\gamma \in O(T_x M, g_x) \cong O(m)$.

Definition 4.A.13 For any point $x \in M$ of a Riemannian manifold (M, g) the *holonomy group* $\mathrm{Hol}_x(M, g) \subset O(T_x M)$ is the group of all parallel transports P_γ along closed paths $\gamma : [0,1] \to M$ with $\gamma(0) = \gamma(1) = x$.

If two points $x, y \in M$ can be connected at all, e.g. if M is connected, then the holonomy groups $\mathrm{Hol}_x(M, g)$ and $\mathrm{Hol}_y(M, g)$ are conjugate and thus isomorphic. More precisely, if $\gamma : [0,1] \to M$ is a path connecting x with y then $\mathrm{Hol}_y(M, g) = P_\gamma \circ \mathrm{Hol}_x(M, g) \circ P_\gamma^{-1}$.

Hence, if M is connected then one can define the group $\mathrm{Hol}(M, g)$ as a subgroup of $O(m)$ up to conjugation.

There is a further technical issue if M is not simply connected. Then there is a difference between $\mathrm{Hol}(M,g)$ and the *restricted holonomy group* $\mathrm{Hol}^\circ(M,g) \subset \mathrm{Hol}(M,g)$ of all parallel transports P_γ along contractible paths γ (i.e. $1 = [\gamma] \in \pi_1(M)$). For simplicity, we will assume throughout that M is simply connected.

One fundamental problem in Riemannian geometry is the classification of holonomy groups. What groups $\mathrm{Hol}(M,g) \subset O(m)$ can arise?

Firstly, the holonomy of a product $(M,g) = (M_1,g_1) \times (M_2,g_2)$ is the product $\mathrm{Hol}(M,g) = \mathrm{Hol}(M_1,g_1) \times \mathrm{Hol}(M_2,g_2) \subset O(m_1) \times O(m_2) \subset O(m_1 + m_2)$.

Thus, in order to be able to classify all possible holonomy groups we shall assume that (M,g) is *irreducible*, i.e. cannot be written (locally) as a product. This is indeed reflected by an algebraic property of the holonomy group:

Proposition 4.A.14 *If (M,g) is irreducible Riemannian manifold, then the inclusion $\mathrm{Hol}(M,g) \subset O(m)$ defines an irreducible representation on $\mathbb{R}^m$.* $\square$

This proposition is completed by the following theorem ensuring the existence of a decomposition into irreducible factors.

Theorem 4.A.15 (de Rham) *If (M,g) is a simply connected complete (e.g. compact) Riemannian manifold then there exists a decomposition $(M,g) = (M_1,g_1) \times \ldots \times (M_k,g_k)$ with irreducible factors (M_i,g_i).* $\square$

Secondly, many groups can occur as holonomy groups of symmetric spaces, a special type of homogeneous spaces. For the precise definition and their classification see [12]. If symmetric spaces are excluded then, surprisingly, a finite list of remaining holonomy groups can be given.

Theorem 4.A.16 (Berger) *Let (M,g) be a simply connected, irreducible Riemannian manifold of dimension m and let us assume that (M,g) is irreducible and not locally symmetric. Then the holonomy group $\mathrm{Hol}(M,g)$ is isomorphic to one of the following list:*
  i) $SO(m)$.
  ii) $U(n)$ *with* $m = 2n$.
  iii) $SU(n)$ *with* $m = 2n$, $n \geq 3$
  iv) $Sp(n)$ *with* $m = 4n$.
  v) $Sp(n)Sp(1)$ *with* $m = 4n$, $n \geq 2$.
  vi) G_2, *with* $m = 7$.
  vii) $\mathrm{Spin}(7)$, *with* $m = 8$. $\square$

We don't go into any detail here, in particular we don't define G_2 or explain the representation of $\mathrm{Spin}(7)$. Very roughly, $SO(m)$ is the case of a general Riemannian metric and vi) and vii) are very special. In fact compact examples for vi) have been found only recently.

The irreducible holonomy groups that are relevant in complex geometry are ii), iii), iv) and, a little less, v).

In the following we shall discuss the case $\mathrm{Hol}(M, g) = \mathrm{U}(n)$ and its subgroup $\mathrm{Hol}(M, g) = \mathrm{SU}(n)$. We will see that this leads to Kähler respectively Ricci-flat Kähler manifolds. The case $\mathrm{Hol}(M, g) = \mathrm{Sp}(n)$, not discussed here, is related to so-called hyperkähler or, equivalently, holomorphic symplectic manifolds.

How is the holonomy $\mathrm{Hol}(M, g)$ of a Riemannian manifold related to the geometry of M at all?

This is explained by the *holonomy principle* :

Choose a point $x \in M$ and identify $\mathrm{Hol}(M, g) = \mathrm{Hol}_x(M, g) \subset \mathrm{O}(T_x M)$. This representation of $\mathrm{Hol}(M, g)$ induces representations on all tensors associated with the vector space $T_x M$, e.g. on $\mathrm{End}(T_x M)$. Suppose α_x is a tensor invariant under $\mathrm{Hol}(M, g)$. We will in particular be interested in the case of an almost complex structure $\alpha_x = I_x \in \mathrm{End}(T_x M)$.

One way to obtain such an invariant tensor α_x is by starting out with a parallel tensor field α on M. (Recall that the Levi-Civita connection induces connections on all tensor bundles, e.g. on $\mathrm{End}(TM)$, so that we can speak about parallel tensor fields.) Clearly, any parallel tensor field α yields a tensor α_x which is invariant under the holonomy group. As an example take a Kähler manifold (M, g, I). Then I is a parallel section of $\mathrm{End}(TM)$ and the induced $I_x \in \mathrm{End}(T_x M)$ is thus invariant under the holonomy group. Hence, $\mathrm{Hol}(M, g) \subset \mathrm{O}(2n) \cap \mathrm{Gl}(n, \mathbb{C}) = \mathrm{U}(n)$.

We are more interested in the other direction which works equally well:

Holonomy principle. If α_x is an $\mathrm{Hol}(M, g)$-invariant tensor on $T_x M$ then α_x can be extended to a parallel tensor field α over M.

Let us consider the case of an almost complex structure $I_x \in \mathrm{End}(T_x M)$ invariant under $\mathrm{Hol}(M, g)$. Then there exists a parallel section I of $\mathrm{End}(TM)$. Since $\mathrm{id} \in \mathrm{End}(TM)$ is parallel and $I_x^2 = -\mathrm{id}$, we have $I^2 = -\mathrm{id}$ everywhere, i.e. I is an almost complex structure on M.

Moreover, since g and I are both parallel, also the Kähler form $g(I(\ ),\ )$ is parallel and, in particular, closed. There is an additional argument that shows that I is in fact integrable. (One uses the fact that the Nijenhuis tensor, which we have not defined but which determines whether an almost complex structure is integrable, is a component of $\nabla(I)$ in case the connection is torsion free. But in our case, I is parallel. See [61].) Thus, we obtain an honest Kähler manifold (M, g, I).

This yields the following proposition. The uniqueness statement is left to the reader.

Proposition 4.A.17 *If $\mathrm{Hol}(M, g) \subset \mathrm{U}(n)$ with $m = 2n$, then there exists a complex structure I on M with respect to which g is Kähler. If $\mathrm{Hol}(M, g) = \mathrm{U}(n)$, then I is unique.*

In the next proposition we relate $\mathrm{SU}(n)$-holonomy to Ricci-flat Kähler metrics, which will be explained in the next section in more detail. So the reader might prefer to skip the following proposition at first reading.

Proposition 4.A.18 *If* $\mathrm{Hol}(M, g) \subset \mathrm{SU}(n)$ *with* $m = 2n$, *then there exists a complex structure* I *on* M *with respect to which* g *is a Ricci-flat Kähler metric. If* $\mathrm{Hol}(M, g) = \mathrm{SU}(n)$ *with* $n \geq 3$ *then* I *is unique.*

Proof. The existence of the Kähler metric and its uniqueness are proved as in the previous proposition. The assumption that the holonomy group is contained in $\mathrm{SU}(n)$ says that any chosen trivialization of $\bigwedge_x M = \det(T_x M^*)$ is left invariant by parallel transport. Hence, there exists a parallel section Ω of $\det_{\mathbb{C}}(\bigwedge M) = \det(\Omega_X) = K_X$ (where $X = (M, I)$). A parallel section of K_X is holomorphic and, if not trivial, without zeros, as the zero section itself is parallel. Thus, K_X is trivialized by a holomorphic volume form Ω.

Moreover, as Ω and ω^n are both parallel, they differ by a constant. Using Corollary 4.B.23 this shows that the Kähler metric g is Ricci-flat.   □

Exercises

4.A.1   Complete the proof of Proposition 4.A.4, i.e. prove that for a torsion free connection on a differentiable manifold M one has for any k-form α

$$(d\alpha)(v_1, \ldots, v_{k+1}) = \sum_{i=0}^{k} (-1)^i (D_{v_i} \alpha)(v_0, \ldots, \hat{v}_i, \ldots, v_{k+1}).$$

4.A.2   Complete the proof of Proposition 4.A.9.

4.A.3   Let (X, g) be a compact Kähler manifold. Show that $i \cdot \mathrm{Ric}(X, g)$ is the curvature of the Chern connection on K_X with respect to the induced hermitian metric.

Comments: - For a thorough discussion of most of this section we refer to [12]. A short introduction to holonomy with special emphasize on the relations to algebraic geometry can be found in [9].
- We have not explained the relation between the curvature and the holonomy group. Roughly, the curvature tensor determines the Lie algebra of Hol. See [12] for more details.
- A detailed account of the more recent results on the holonomy of (compact) manifolds can be found in [61] or [72]. Joyce was also the first one to construct compact G_2 manifolds.

4.B Hermite–Einstein and Kähler–Einstein Metrics

Interesting metrics on compact manifolds are not easy to construct. This appendix discusses two types of metrics which are of importance in Kähler geometry.

If (X, g) is an hermitian manifold, $\omega := g(I(\ ), (\ ))$ is its fundamental form. By definition (X, g) is Kähler if and only if ω is closed, i.e. $d\omega = 0$. The hermitian structure on X can be viewed as an hermitian structure on its holomorphic tangent bundle. So we might look more generally for interesting metrics on an arbitrary holomorphic vector bundle E on X. We will discuss special hermitian metrics on E, so called Hermite–Einstein metrics, by comparing the curvature F_∇ of the Chern connection ∇ on E with the fundamental form ω. In the special case that E is the holomorphic tangent bundle T_X and the hermitian structure h is induced by g this will lead to the concept of Kähler–Einstein metrics on complex manifolds.

In some of our examples, e.g. the Fubini–Study metric on $\mathbb{P}^n$, we have already encountered this special type of Kähler metrics. However, on other interesting manifolds, like K3 surfaces, concrete examples of Kähler metrics have not been discussed. Of course, if a manifold is projective one can always consider the restriction of the Fubini–Study metric, but this usually does not lead to geometrically interesting structures (at least not directly).

For the time being, we let E be an arbitrary holomorphic vector bundle with an arbitrary hermitian metric h. Recall that the curvature F_∇ of the Chern connection on (E, h) is of type $(1, 1)$, i.e. $F_\nabla \in \mathcal{A}^{1,1}(X, \mathrm{End}(E))$. The fundamental form ω induces an element of the same type $\omega \cdot \mathrm{id}_E \in \mathcal{A}^{1,1}(X, \mathrm{End}(E))$. These two are related to each other by the Hermite–Einstein condition:

Definition 4.B.1 An hermitian structure h on a holomorphic vector bundle E is called *Hermite–Einstein* if

$$i \cdot \Lambda_\omega F_\nabla = \lambda \cdot \mathrm{id}_E$$

for some constant scalar $\lambda \in \mathbb{R}$. Here, Λ_ω is the contraction by ω.

In this case, we will also say that the connection ∇ is Hermite–Einstein or even that the holomorphic bundle E is Hermite–Einstein. Note that the Hermite–Einstein condition strongly depends on the hermitian structure on the manifold X. It may happen that a vector bundle E admits an Hermite–Einstein structure with respect to one hermitian structure g on X, but not with respect to another g'.

Example 4.B.2 The easiest example of an Hermite–Einstein bundle is provided by flat bundles. In this case the curvature F_∇ is trivial and the Hermite–Einstein condition is, therefore, automatically satisfied with $\lambda = 0$.

Let us discuss a few equivalent formulations of the Hermite–Einstein condition. Firstly, one can always write the curvature of the Chern connection ∇ on any bundle E as

$$F_\nabla = \frac{\mathrm{tr}(F_\nabla)}{\mathrm{rk}(E)} \cdot \mathrm{id} + F_\nabla^\circ$$

where F_∇° is the trace free part of the curvature. Let us now assume that g is a Kähler metric, i.e. that ω is closed (and thus, automatically, harmonic). Then the connection is Hermite–Einstein if and only if $\mathrm{tr}(F_\nabla)$ is an harmonic $(1,1)$-form and F_∇° is (locally) a matrix of primitive $(1,1)$-forms. Indeed, if ∇ is Hermite–Einstein, then $i \cdot \mathrm{tr}(F_\nabla) = (\mathrm{rk}(E) \cdot \lambda) \cdot \omega + \alpha$ with α a primitive $(1,1)$-form. Since the trace is closed (Bianchi identity), the form α is closed and hence harmonic (see Exercise 3.1.12). The other assertion and the converse are proved analogously.

Secondly, the Hermite–Einstein condition for the curvature of a Chern connection is equivalent to writing

$$i \cdot F_\nabla = (\lambda/n) \cdot \omega \cdot \mathrm{id}_E + F_\nabla',$$

where F_∇' is locally a matrix of ω-primitive $(1,1)$-forms. Here, $n = \dim_{\mathbb{C}} X$. The factor $(1/n)$ is explained by the commutator relation $[L, \Lambda] = H$, which yields $\Lambda L(1) = n$ (cf. Proposition 1.2.26).

Using standard results from Section 1.2 one easily finds that h is Hermite–Einstein if and only if

$$i \cdot F_\nabla \wedge \omega^{n-1} = (\lambda/n) \cdot \omega^n \cdot \mathrm{id}_E.$$

In particular,

$$i \cdot \mathrm{tr}(F_\nabla) \wedge \omega^{n-1} = \frac{\mathrm{rk}(E) \cdot \lambda}{n} \cdot \omega^n. \tag{4.9}$$

If X is compact and Kähler, (4.9) can be used to show that λ depends only on the first Chern class of E and its rank. Indeed, integrating (4.9) yields

$$\lambda \cdot \int_X [\omega]^n = n \cdot \frac{\int_X i \cdot \mathrm{tr}(F_\nabla) \wedge \omega^{n-1}}{\mathrm{rk}(E)}.$$

Hence, $\lambda = (2\pi) \cdot n \cdot \left(\int_X [\omega]^n \right)^{-1} \mu(E)$, where the slope $\mu(E)$ of E is defined as follows:

Definition 4.B.3 The *slope* of a vector bundle E with respect to the Kähler form ω is defined by

$$\mu(E) := \frac{\int_X c_1(E) \wedge [\omega]^{n-1}}{\mathrm{rk}(E)}.$$

In general, Hermite–Einstein metrics are not easy to describe, but they exist quite frequently and those holomorphic bundles that admit Hermite–Einstein metrics can be described algebraically (see Theorem 4.B.9).

Lemma 4.B.4 *Any holomorphic line bundle L on a compact Kähler manifold X admits an Hermite–Einstein structure.*

Proof. The curvature $i \cdot F_\nabla$ of the Chern connection on the holomorphic line bundle L endowed with an hermitian structure h is a real $(1,1)$-form. Hence, $i \cdot \Lambda_\omega F_\nabla$ is a real function φ, which can be written as $\lambda - \partial^* \bar\partial f$ for some function f and some constant λ. Since $\partial^* \bar\partial f = (1/2)d^* df$, we can assume that the function f is real.

Then define a new hermitian structure h' on L by $h' = e^f \cdot h$. The curvature of the induced connection ∇' is $F_{\nabla'} = F_\nabla + \bar\partial\partial f$ (see iii), Examples 4.3.9). Using the Kähler identity $[\Lambda, \bar\partial] = -i\partial^*$ on $\mathcal{A}^0(X)$ one computes $\Lambda_\omega \bar\partial\partial f = -i\partial^*\partial f$. Hence, $i \cdot \Lambda_\omega(F_{\nabla'}) = (\varphi + \partial^*\partial f) = \lambda$. $\qquad\square$

Remark 4.B.5 Clearly, the first Chern class $c_1(L) \in H^2(X, \mathbb{R})$ can uniquely be represented by an harmonic form. The lemma shows that one actually finds an hermitian structure on L such that the first Chern form $c_1(L, \nabla)$ of the associated Chern connection ∇ is this harmonic representative. The same argument can be used to solve Exercise 4.4.10.

From here one can go on and construct many more vector bundles admitting Hermite–Einstein structures. E.g. the tensor product $E_1 \otimes E_2$ of two Hermite–Einstein bundles is again Hermite–Einstein, as well as the dual bundle E_1^*. However, the direct sum $E_1 \oplus E_2$ admits an Hermite–Einstein structure if and only if $\mu(E_1) = \mu(E_2)$ (cf. Exercise 4.B.2). Indeed, it is not hard to see that the direct sum of the two Hermite–Einstein connections is Hermite–Einstein under this condition. The other implication is slightly more complicated.

Vector bundles admitting Hermite–Einstein metrics satisfy surprising topological restrictions.

Proposition 4.B.6 *Let E be a holomorphic vector bundle of rank r on a compact hermitian manifold (X, g). If E admits an Hermite–Einstein structure then*

$$\int_X \left(2rc_2(E) - (r-1)c_1^2(E)\right) \wedge \omega^{n-2} \geq 0.$$

Proof. The bundle $G := \mathrm{End}(E)$ with the naturally induced connection ∇_G is Hermite–Einstein (cf. Exercise 4.B.2) and has vanishing first Chern form $c_1(G, \nabla_G)$ (cf. Exercise 4.4.9). In particular, $\Lambda_\omega F_{\nabla_G} = 0$.

For such a bundle G we will show that $\mathrm{ch}_2(G, \nabla_G)\omega^{n-2} \leq 0$ (pointwise!). Since $\mathrm{ch}_2(\mathrm{End}(E), \nabla) = -(2rc_2(E, \nabla) - (r-1)c_1^2(E, \nabla))$ (cf. Exercise 4.4.9), this then proves the assertion.

The rest of the proof is pure linear algebra applied to the tangent space at an arbitrary point. By definition

$$\mathrm{ch}_2(G, \nabla_G) = \frac{1}{2}\mathrm{tr}\left(\left(\frac{i}{2\pi}F_{\nabla_G}\right)^2\right) = \mathrm{tr}\left(\frac{i}{2\pi}F_{\nabla_G} \wedge \overline{\left(\frac{i}{2\pi}F_{\nabla_G}\right)}^{\,t}\right),$$

where the expression on the right hand side is meant with respect to a local orthonormal basis.

But by the Hodge–Riemann bilinear relation 1.2.36 any matrix of primitive $(1,1)$-forms $A = (a_{ij})$ satisfies $\mathrm{tr}(A\bar{A}^{\mathrm{t}}) \wedge \omega^{n-2} = \sum(a_{ij}\bar{a}_{ij}) \wedge \omega^{n-2} \leq 0.$ □

Remarks 4.B.7 i) In this context the inequality is due to Lübke [87]. It is often called the Bogomolov–Lübke inequality, as its algebraic version was first observed by Bogomolov [14].

ii) Also note that the above Chern class combination is the only natural one (up to scaling) among those that involve only the first two Chern classes, as it is the only one that remains unchanged when passing from E to a line bundle twist $E \otimes L$, which also carries a Hermite–Einstein structure (cf. Exercise 4.B.2).

iii) In the proof we have actually shown the pointwise inequality. Thus, in this sense the assertion holds true also for non-compact manifolds X. Moreover, from the proof one immediately deduces that equality (global or pointwise) implies that the endomorphism bundle has vanishing curvature.

iv) Furthermore, we only used the weak Hermite–Einstein condition where the scalar λ is replaced by a function. Due to the Exercise 4.B.3 this does not really generalize the statement as formulated above.

When does a holomorphic bundle that satisfies the above inequality really admit a Hermite–Einstein metric? This is a difficult question, but a complete answer is known due to the spectacular results of Donaldson, Uhlenbeck, and Yau. It turns out that the question whether E admits an Hermite–Einstein metric can be answered by studying the algebraic geometry of E. In particular, one has to introduce the concept of stability.

Definition 4.B.8 A holomorphic vector bundle E on a compact Kähler manifold X is *stable* if and only if

$$\mu(F) < \mu(E)$$

for any proper non-trivial $\mathcal{O}_X$-subsheaf $F \subset E$.

A few comments are needed here. First of all, the notion depends on the chosen Kähler structure of X or, more precisely, on the Kähler class $[\omega]$. Secondly, the slope was defined only for vector bundles F and not for arbitrary $\mathcal{O}_X$-sheaves $F \subset E$, but it is not difficult to find the correct definition in this more general context. E.g. one could first define the determinant of F and then define the first Chern class of F as the first Chern class of its determinant. Another possibility would be to use the Atiyah class to define Chern classes of coherent sheaves in general (cf. Remark 5.1.5).

Unfortunately, already from dimension three on one really needs to check arbitrary subsheaves of E and not just locally free subsheaves (let alone subvector bundles, i.e. locally free subsheaves with locally free quotient). There is however one condition that is slightly stronger than stability and which uses only vector bundles. A holomorphic bundle E is stable if for any $0 < s < \mathrm{rk}(E)$ and any line bundle $L \subset \bigwedge^s E$ one has $\mu(L) < s \cdot \mu(E)$.

Also note that other stability concepts for holomorphic vector bundles exist. The one we use is usually called *slope-stability* or *Mumford–Takemoto stability* .

One can also define polystability for holomorphic vector bundles (not to be confused with semi-stability, which we shall not define). A holomorphic vector bundle E is *polystable* if $E = \bigoplus E_i$ with E_i stable vector bundles all of the same slope $\mu(E) = \mu(E_i)$. The following beautiful result shows that algebraic geometry of a vector bundle determines whether an Hermite–Einstein metric exists. The proof is a pure existence result and it allows to deduce the existence of Hermite–Einstein metrics without ever actually constructing any Hermite–Einstein metric explicitly.

Theorem 4.B.9 (Donaldson, Uhlenbeck, Yau) *A holomorphic vector bundle E on a compact Kähler manifold X admits an Hermite–Einstein metric if and only if E is polystable.*   □

Remark 4.B.10 One direction of the theorem is not very hard, any holomorphic bundle endowed with an Hermite–Einstein metric is polystable. Donaldson in [36] proved the converse for algebraic surfaces. This was a generalization of an old result of Narasimhan and Seshadri [94] for vector bundles on curves. On curves, Hermite–Einstein metrics are intimately related to unitary representation of the fundamental group of the curve. Uhlenbeck and Yau [112] generalized Donaldson's result to arbitrary compact Kähler manifolds and Buchdahl managed to adjust the proof to the case of compact hermitian manifolds. This kind of result is nowadays known as Kobayashi–Hitchin correspondence (cf. [88]).

Next, we shall study the case that E is the holomorphic tangent bundle $\mathcal{T}_X$. This leads to a much more restrictive notion. Any hermitian structure g on the complex manifold X induces an hermitian structure on the holomorphic tangent bundle $\mathcal{T}_X$. So, it would not be very natural to look for another unrelated hermitian structure on $\mathcal{T}_X$. Recall that the Hermite–Einstein condition intertwines the hermitian structure on X with the hermitian structure on the vector bundle in question.

Definition 4.B.11 An hermitian manifold (X, g) is called *Kähler–Einstein* if (X, g) is Kähler and the naturally induced hermitian structure on the holomorphic tangent bundle is Hermite–Einstein. In this case the metric g is called

Kähler–Einstein. If a Kähler–Einstein metric g on X exists, the complex manifold X is also called a Kähler–Einstein manifold.

Explicitly, this means that the curvature of the Levi-Civita connection satisfies

$$i \cdot \Lambda_\omega F_\nabla = \lambda \cdot \mathrm{id}_{T_X} \tag{4.10}$$

for some constant scalar factor λ.

One may try to avoid the condition that the manifold is Kähler, but as we explained in Section 4.A the Chern connection on the holomorphic tangent bundle coincides with the Levi-Civita connection if and only if the manifold (X, g) is Kähler. If not, the Hermite–Einstein condition would not seem very natural for the Riemannian metric g.

Note that the holomorphic tangent bundle on an hermitian (or Kähler) manifold (X, g) might very well admit a Hermite–Einstein metric without X being Kähler–Einstein.

Usually the Kähler–Einstein condition is introduced via the Ricci curvature (cf. Definition 4.A.10). Let us begin with the Riemannian version.

Definition 4.B.12 A Riemannian metric g on a differentiable manifold M is *Einstein* if its Ricci tensor $r(M, g)$ satisfies

$$r(M, g) = \lambda \cdot g$$

for some constant scalar factor λ.

If g is a Kähler metric on the complex manifold $X = (M, I)$, then the Ricci curvature $\mathrm{Ric}(X, g)$ is defined by $\mathrm{Ric}(u, v) = r(I(u), v)$ (see Definition 4.A.10). This is in complete analogy to the definition of the Kähler form ω as $\omega(u, v) = g(I(u), v)$. Thus, a metric g on a complex manifold $X = (M, g)$ is an Einstein metric on M if and only if $\mathrm{Ric}(X, g) = \lambda \cdot \omega$ for some constant scalar λ.

Recall that Proposition 4.A.11 shows $i \cdot \tilde{\omega}(\Lambda_\omega F_\nabla) = \mathrm{Ric}(X, g)$. This leads to:

Corollary 4.B.13 *Let g be a Kähler metric on the complex manifold $X = (M, I)$. Then g is an Einstein metric on M if and only if g is a Kähler–Einstein metric on X.*

Proof. Indeed, if g is a Kähler–Einstein metric, then for the curvature F of the Levi-Civita connection one has $i \cdot \Lambda_\omega F = \lambda \cdot \mathrm{id}$ and hence $\mathrm{Ric}(X, g) = i \cdot \tilde{\omega}(\Lambda_\omega F) = \lambda \cdot \omega$. Thus, $r(M, g) = \lambda \cdot g$.

For the converse, go the argument backwards. $\qquad\square$

Applying the argument explained before for Hermite–Einstein metrics (see page 218) and using $c_1(X) = (i/2\pi)[\mathrm{tr}(F)]$, one finds that the scalar factor λ in the Kähler–Einstein condition (4.10) can be computed as

$$\lambda = \frac{i \cdot \int \mathrm{tr}(F) \wedge \omega^{n-1}}{\int \omega^n} = \frac{(2\pi) \int c_1(X) \wedge \omega^{n-1}}{\int \omega^n}.$$

In other words, $c_1(X) \in H^2(X, \mathbb{R})$ and $[\omega] \in H^2(X, \mathbb{R})$ satisfy the linear equation $(c_1(X) - (\lambda/2\pi)[\omega])[\omega]^{n-1} = 0$. In fact, one can prove more. Namely, if g is a Kähler–Einstein metric on X, then for its associated Kähler form ω one has

$$c_1(X) = \frac{\lambda}{2\pi} \cdot [\omega]$$

with λ the scalar factor occurring in the Kähler–Einstein condition (4.10). Indeed, $c_1(X) = c_1(T_X) = [(i/2\pi)\mathrm{tr}(F_\nabla)]$ and by Proposition 4.A.11 and the Kähler–Einstein condition one has $(i/2\pi)\mathrm{tr}(F_\nabla) = (1/2\pi)\mathrm{Ric} = (\lambda/2\pi)\omega$.

Also note that in the decomposition $F_\nabla = (\lambda/n) \cdot \omega \cdot \mathrm{id} + F'$ the primitive part F' is traceless.

Corollary 4.B.14 *If (X, g) is a Kähler–Einstein manifold, then one of the three conditions holds true:*
  i) $c_1(X) = 0$,
  ii) $c_1(X)$ *is a Kähler class*,
  iii) $-c_1(X)$ *is a Kähler class*.   □

In other words, the first Chern class of the canonical bundle K_X of a compact Kähler–Einstein manifold is either trivial, negative, or positive.

If $c_1(X) = 0$, e.g. if the canonical bundle K_X is trivial, and g is a Kähler–Einstein metric then $\mathrm{Ric}(X, g) = 0$. Indeed, in this case the scalar factor λ is necessarily trivial and hence $\mathrm{Ric}(X, g) = \lambda \cdot \omega = 0$, i.e. the Kähler metric g is Ricci-flat.

Remark 4.B.15 Let us emphasize that there are two types of symmetries satisfied by the curvature of the Levi-Civita connection of a Kähler manifold, both stated in Proposition 4.A.11.

The first one allows to show that the primitive part of the curvature is traceless and hence $c_1(X, g) = (i/2\pi)\mathrm{tr}(F) = (\lambda/2\pi)\omega$ if g is a Kähler–Einstein metric. (Recall that $c_1(X, g) = (1/2\pi)\mathrm{Ric}(X, g)$ holds for any Kähler metric g.)

The second relation, which is ii) in Proposition 4.A.11, was used to prove the equivalence of the Einstein condition for the Kähler metric g and the Hermite–Einstein condition for the induced hermitian structure on T_X (Corollary 4.B.13).

Examples 4.B.16 In the following we will give one example of a Kähler–Einstein manifold in each of the three classes in Corollary 4.B.14.

i) **Projective space.** It turns out that the Fubini–Study metric, the only Kähler metric on $\mathbb{P}^n$ that has been introduced, is indeed Kähler–Einstein. By Exercise 4.4.7 we know that $c_1(\mathcal{O}(1)) = [\omega_{FS}] \in H^2(\mathbb{P}^n, \mathbb{R})$. Thus, if the Fubini–Study metric g_{FS} is indeed Kähler–Einstein then the scalar factor can be computed by $c_1(\mathbb{P}^n) = (n+1) \cdot c_1(\mathcal{O}(1)) = \lambda \cdot [\omega_{FS}]$, i.e. $\lambda = n + 1$.

In order to see that g_{FS} is Kähler–Einstein, recall that the induced hermitian structure on $\det(T_{\mathbb{P}^n}) \cong \mathcal{O}(n+1)$ is $h^{\otimes n+1}$, where h is the standard hermitian structure on $\mathcal{O}(1)$ determined by the choice of the basis $z_0, \ldots, z_n \in H^0(\mathbb{P}^n, \mathcal{O}(1))$ (see Example 4.1.5). For the latter we have computed in Example 4.3.12 that the Chern connection ∇ satisfies $c_1(\mathcal{O}(1), \nabla) = \omega_{FS}$. Hence, $c_1(\mathbb{P}^n, g_{FS}) = (i/2\pi)\mathrm{tr}(F_{\nabla_{FS}}) = (i/2\pi)(n+1)\omega_{FS}$.

Clearly, any other Fubini–Study metric obtained by applying a linear coordinate change is Kähler–Einstein as well.

ii) **Complex tori.** The holomorphic tangent bundle of a complex torus $X = \mathbb{C}^n/\Gamma$ is trivial. The Chern connection for any constant Kähler structure on X is flat. Thus, the Kähler–Einstein condition is satisfied with the choice of the scalar $\lambda = 0$.

Complex tori are trivial examples of Ricci-flat manifolds. Any other example is much harder to come by.

iii) **Ball quotients.** The standard Kähler structure $\omega = (i/2)\partial\bar{\partial}(1 - \|z\|^2)$ on the unit disc $D^n \subset \mathbb{C}^n$ is Kähler–Einstein with $\lambda < 0$. For simplicity we consider only the one-dimensional case. Then $\omega = (i/2)(1 - |z|^2)^{-2} dz \wedge d\bar{z}$ and, hence, the hermitian metric is $h = (1 + |z|^2)^{-2}$. The curvature of its Chern connection is thus given by $F = -\partial(h^{-1}\bar{\partial}h) = \frac{-2}{(1-|z|^2)^2} dz \wedge d\bar{z}$. Therefore, $i \cdot F = -4 \cdot \omega$.

This way, one obtains negative Kähler–Einstein structures on all ball quotients.

As the holomorphic tangent bundle of a Kähler–Einstein manifold is in particular Hermite–Einstein, any Kähler–Einstein manifold satisfies the Bogomolov–Lübke inequality 4.B.6. In fact, a stronger inequality can be proved by using the additional symmetries of the curvature of a Kähler manifold. As for the Bogomolov–Lübke inequality, there are algebraic and analytic proofs of this inequality. The first proof, using the Kähler–Einstein condition was given by Chen and Oguie [24]. An algebraic version of it was proved by Miyaoka. The inequality is usually called the *Miyaoka–Yau inequality*.

Proposition 4.B.17 *Let X be a Kähler–Einstein manifold of dimension n and let ω be a Kähler–Einstein form. Then*

$$\int_X \left(2(n+1)c_2(X) - nc_1^2(X)\right) \wedge \omega^{n-2} \geq 0. \qquad (4.11)$$

Remark 4.B.18 It might be instructive to consider the Miyaoka–Yau inequality in the case of a compact surface. Here it says $3c_2(X) \geq c_1^2(X)$. Since $c_1^2(X) \geq 0$

for a Kähler–Einstein surface, this inequality is stronger than the Bogomolov–Lübke inequality. It is noteworthy that for the projective plane and a complex torus the inequality becomes an equality.

In fact, one can prove that equality in (4.11) for a Kähler–Enstein manifold X implies that the universal cover of X is isomorphic to $\mathbb{P}^n$, $\mathbb{C}^n$, or a ball. For a proof of this result see e.g. Tian's lecture notes [106].

In many examples it can easily be checked whether the canonical bundle is negative, trivial, or positive and whether the Miyaoka–Yau inequality is satisfied. In fact, often this can be done without even constructing any Kähler metric on X just by using an embedding of X in a projective space and pulling-back the Fubini–Study metric on $\mathcal{O}(1)$. But even when this necessary condition holds, we still don't know whether X admits a Kähler–Einstein structure and if how many.

The key result that is behind many others in this area is the following fundamental theorem of Calabi and Yau. Perhaps, it is worth emphasizing that this result works for arbitrary compact Kähler manifolds without any condition on the canonical bundle. It will also lead to the fundamental result that any form representing $c_1(X)$ is the Ricci curvature of a unique Kähler metric with given Kähler class (cf. Proposition 4.B.21).

Theorem 4.B.19 (Calabi–Yau) *Let (X, g_0) be a compact Kähler manifold of dimension n and let ω_0 be its Kähler form. For any real differentiable function f on X with*

$$\int_X e^f \cdot \omega_0^n = \int_X \omega_0^n$$

there exists a unique Kähler metric g with associated Kähler form ω such that

$$[\omega] = [\omega_0] \quad \text{and} \quad \omega^n = e^f \cdot \omega_0^n.$$

Proof. The proof of the existence is beyond the scope of these notes (see [61]), but for the uniqueness an easy argument, due to Calabi, goes as follows:

Suppose ω_1 and ω_2 are two Kähler forms with $\omega_1^n = \omega_2^n$. If they are cohomologous, there exists a real function f on M with $\omega_2 = \omega_1 + i\partial\bar{\partial}f$ (see Exercise 3.2.16). Hence, $0 = \omega_2^n - \omega_1^n = \gamma \wedge (\omega_2 - \omega_1) = \gamma \wedge (i\partial\bar{\partial}f)$ with $\gamma = \omega_2^{n-1} + \omega_2^{n-2} \wedge \omega_1 + \ldots + \omega_2 \wedge \omega_1^{n-2} + \omega_1^{n-1}$.

The form γ is a positive linear combination of positive forms $\omega_2^k \wedge \omega_1^{n-1-k}$ and, hence, itself positive. The equation $0 = \gamma \wedge \partial\bar{\partial}f$ together with the maximum principle imply that f is constant and hence $\omega_1 = \omega_2$.

For the convenience of the reader we spell out how the maximum principle is applied here. Since M is compact, there exists a point $x \in M$ where f attains its maximum. For simplicity we will assume that the Hessian of f in x is negative definite. (If not one has to perturb by a quadratic function as in the proof of the maximum principle for harmonic functions.)

Now, let us choose local coordinates $(z_1 = x_1 + iy_1, \ldots, z_n = x_n + iy_n)$ around $x \in M$ such that ω_1 and ω_2 are simultaneously diagonalized in $x \in M$. We may assume $\omega_1(x) = (i/2) \sum dz_i \wedge d\bar{z}_i = \sum dx_i \wedge y_i$ and $\omega_2(x) = (i/2) \sum \lambda_i dz_i \wedge d\bar{z}_i = \sum \lambda_i dx_i \wedge dy_i$. Since ω_2 is positive, $\lambda_i > 0$. Thus, $0 = \gamma \wedge \partial\bar{\partial} f$ in $x \in M$ yields an equation of the form $0 = \sum \mu_i \left(\frac{\partial^2 f}{\partial x_i^2}(x) + \frac{\partial^2 f}{\partial y_i^2}(x) \right)$, where the coefficients μ_i are positive linear combinations of terms of the form $\lambda_{i_1} \ldots \lambda_{i_k}$. But this contradicts the fact that the Hessian of f is negative definite. $\square$

The theorem can be rephrased as follows: If X is a compact Kähler manifold with a given volume form vol which is compatible with the natural orientation, then there exists a unique Kähler metric g on X with $\omega^n = $ vol and prescribed $[\omega] \in H^2(X, \mathbb{R})$.

Yet another way to say this uses the Kähler cone $\mathcal{K}_X \subset H^{1,1}(X, \mathbb{R})$ of all Kähler classes (see Definition 3.2.14) and the set $\widetilde{\mathcal{K}}_X$ of all Kähler forms ω with $\omega^n = \lambda \cdot$ vol for some $\lambda \in \mathbb{R}_{>0}$. Then the natural map that projects a closed form to its cohomology class induces the following diagram:

$$
\begin{array}{ccc}
\mathcal{A}^{1,1}(X)_{\mathrm{cl}} & \longrightarrow & H^{1,1}(X) \\
\cup & & \cup \\
\widetilde{\mathcal{K}}_X & \xrightarrow{\ \sim\ } & \mathcal{K}_X
\end{array}
$$

Lemma 4.B.20 *Let ω and ω' be two Kähler forms on a compact Kähler manifold. If $\omega^n = e^f \cdot \omega'^n$ for some real function f, then $\mathrm{Ric}(X, \omega) = \mathrm{Ric}(X, \omega') + i\bar{\partial}\partial f$.*

Proof. The two Kähler forms correspond to Kähler metrics g and g', respectively, which are locally given by matrices (g_{ij}) and (g'_{ij}). The induced volume forms are thus given by the functions $\det(g_{ij})$ respectively $\det(g'_{ij})$. Hence, $\det(g_{ij}) = e^f \cdot \det(g'_{ij})$.

On the other hand, the two metrics induce hermitian structures h respectively h' on T_X and thus on $\det(T_X)$. The curvature forms of the latter are $\bar{\partial}\partial \log(\det(h))$ and $\bar{\partial}\partial \log(\det(h'))$, respectively. Since $\det(h)$ and $\det(h')$ differ again by the scalar factor e^f, this yields $\mathrm{Ric}(X, \omega) = \mathrm{Ric}(X, \omega') + i\bar{\partial}\partial \log(e^f)$, as the Ricci curvature is the curvature of the induced connection on $\det(T_X)$ (see Proposition 4.A.11). $\square$

This lemma together with the Calabi–Yau theorem 4.B.19 yields:

Proposition 4.B.21 *Let X be a compact Kähler manifold and let $\alpha \in \mathcal{K}_X$ be a Kähler class. Assume β is a closed real $(1,1)$-form with $[\beta] = c_1(X)$. Then there exists a unique Kähler structure g on X such that*
   i) $\mathrm{Ric}(X, g) = (2\pi) \cdot \beta$ *and*
   ii) $[\omega] = \alpha$ *for the Kähler form ω of the Kähler metric g.*

Proof. Let ω_0 be an arbitrary Kähler form on X. Then $\mathrm{Ric}(X, \omega_0)$ represents $(2\pi) \cdot c_1(X)$ and hence is cohomologous to $(2\pi) \cdot \beta$. Thus, since X is Kähler, one finds a real function f with $(2\pi) \cdot \beta = \mathrm{Ric}(X, \omega_0) + i\bar{\partial}\partial f$.

By the Calabi–Yau theorem 4.B.19 there exists a unique Kähler metric with associated Kähler form ω such that $[\omega] = \alpha$ and $\omega^n = e^{f+c} \cdot \omega_0^n$, where the constant c is chosen such that $\int_X \alpha^n = e^c \int_X e^f \cdot \omega_0^n$.

Using Lemma 4.B.20, we find that the Ricci curvature of g is given by

$$\mathrm{Ric}(X, \omega) = \mathrm{Ric}(X, \omega_0) + i\bar{\partial}\partial f = (2\pi) \cdot \beta.$$

Using again the lemma and the uniqueness part of the Calabi–Yau theorem, we find that ω is unique. □

Corollary 4.B.22 *If X is a compact Kähler manifold with $c_1(X) = 0$ then there exists a unique Ricci-flat Kähler structure g on X with given Kähler class $[\omega]$. The volume form up to a scalar does not depend on the chosen Ricci-flat metric or the Kähler class $[\omega]$.*

Proof. Choosing $\beta = 0$ in Proposition 4.B.21 yields a unique Kähler structure in each Kähler class with vanishing Ricci curvature. The uniqueness of the volume is easily deduced from Lemma 4.B.20. □

Thus, any compact Kähler manifold X with $c_1(X) = 0$ is Ricci-flat. Clearly, any compact Kähler manifold with trivial canonical bundle $K_X \cong \mathcal{O}_X$ has $c_1(X) = 0$. For this type of manifold, the Ricci-flatness of a Kähler form can be determined by the following criterion

Corollary 4.B.23 *Let X be a compact Kähler manifold of dimension n with trivial canonical bundle K_X. Fix a holomorphic volume form, i.e. a trivializing section $\Omega \in H^0(X, K_X)$. Then, a Kähler form ω is Ricci-flat if and only if*

$$\omega^n = \lambda \cdot (\Omega \wedge \overline{\Omega})$$

for some constant $\lambda \in \mathbb{C}^*$.

Proof. Suppose $\omega^n = \lambda \cdot (\Omega \wedge \overline{\Omega})$. Since ω is parallel, i.e. $\nabla(\omega) = 0$ for the Levi-Civita connection ∇ (see Proposition 4.A.8), also $\nabla(\omega^n) = 0$ and hence $\nabla(\Omega \wedge \overline{\Omega}) = 0$.

On the other hand, the Levi-Civita connection on a Kähler manifold is compatible with the complex structure. Since $\bar{\partial}\Omega = 0$, this shows that $\nabla(\Omega) = \alpha \otimes \Omega$ with $\alpha \in \mathcal{A}^{1,0}(X)$. Therefore, using the bidegree decomposition the equality $0 = \nabla(\Omega \wedge \overline{\Omega}) = \nabla(\Omega) \wedge \overline{\Omega} + \Omega \wedge \nabla(\overline{\Omega}) = (\alpha + \bar{\alpha})(\Omega \wedge \overline{\Omega})$ implies $\alpha = 0$. Thus, Ω is a parallel section of K_X and, in particular, the curvature of the Levi-Civita connection on K_X, which is the Ricci curvature, vanishes. Thus, ω is Ricci-flat.

Conversely, if a Ricci-flat Kähler form ω is given there exists a unique Kähler form ω' in the same cohomology class with $\omega'^n = \lambda \cdot (\Omega \wedge \overline{\Omega})$ for some

$\lambda \in \mathbb{C}^*$. By what has been said before, this yields that also ω' is Ricci-flat and the uniqueness of the Ricci-flat representative of a Kähler class proves $\omega = \omega'$.

Clearly, the constant λ is actually real and positive.   □

The other two cases for which Kähler–Einstein metrics could *a priori* exist are much harder. If $c_1(X)$ is negative, i.e. $-c_1(X)$ can be represented by a Kähler form, the question is completely settled by the following theorem, due to Aubin and Yau.

Theorem 4.B.24 (Aubin, Yau) *Let X be a compact Kähler manifold such that $c_1(X)$ is negative. Then X admits a unique Kähler–Einstein metric up to scalar factors.*

Proof. The uniqueness is again rather elementary. See [5, 12] for more comments.   □

Thus, Theorem 4.B.24 and Corollary 4.B.22 can be seen as the non-linear analogue of the Donaldson–Uhlenbeck–Yau description of Hermite–Einstein metrics, but clearly the situation here is more subtle. E.g. for $c_1(X)$ positive the situation is, for the time being, not fully understood. One knows that in this case a Kähler–Einstein metric need not exist. E.g. the Fubini–Study metric on $\mathbb{P}^2$ is Kähler–Einstein, but the blow-up of $\mathbb{P}^2$ in two points for which K_X^* is still ample does not admit any Kähler–Einstein metric. In order to ensure the existence of a Kähler–Einstein metric, a certain stability condition on X has to be added . There has been done a lot of work on this problem recently. See the survey articles [19] or [105].

Exercises

4.B.1   Verify that the only stable vector bundles on $\mathbb{P}^1$ are line bundles. Find a semi-stable vector bundle of rank two on an elliptic curve. (A semi-stable bundle satisfies only the weaker stability condition $\mu(F) \leq \mu(E)$ for all sub-bundles $F \subset E$.)

4.B.2   Let E_1, E_2 be holomorphic vector bundles endowed with Hermite–Einstein metrics h_1 and h_2, respectively. Show that the naturally induced metrics on $E_1 \otimes E_2$, $\mathrm{Hom}(E_1, E_2)$, and E_i^* are all Hermite–Einstein. If $\mu(E_1) = \mu(E_2)$, then also $h_1 \oplus h_2$ is Hermite–Einstein on $E_1 \oplus E_2$.

4.B.3   Let (E, h) be an hermitian holomorphic vector bundle on a compact Kähler manifold such that $i \cdot \Lambda_\omega F_\nabla = \lambda \cdot \mathrm{id}_E$ for the Chern connection ∇ and a function λ. Show that by changing h to $e^f \cdot h$ for some real function f, one finds an hermitian metric on E the Chern connection of which satisfies the Hermite–Einstein condition with constant factor λ.

4.B.4   Let E be a holomorphic vector bundle on a compact Kähler manifold X with a chosen Kähler structure ω. Without using Theorem 4.B.9, show that if E admits an Hermite–Einstein metric with respect to ω then E admits an Hermite–Einstein metric with respect to any other Kähler form ω' with $[\omega] = \lambda[\omega']$ for any $\lambda \in \mathbb{R}_{>0}$.

(This corresponds to the easy observation that stability only depends on the Kähler class (and not on the particular Kähler form) and that scaling by a constant does not affect the stability condition.)

4.B.5   Give an algebraic argument for the stability of the tangent bundle of $\mathbb{P}^n$.

Comments: - The Hermite–Einstein condition for holomorphic vector bundles is discussed in detail in [78].

- For the algebraic theory of stable vector bundles and their moduli see [70] and the references therein.

5

Applications of Cohomology

This chapter illustrates how cohomological methods can be applied to study the geometry of compact complex manifolds. The first section states and discusses the Hirzebruch–Riemann–Roch formula. This formula allows to compute the dimension of the space of global sections of a given vector bundle in terms of its Chern classes. In fact, the higher cohomology groups enter this formula as correction terms. If the vanishing of the higher cohomology groups can be ensured, e.g. by Kodaira's vanishing result which shall be explained in Section 5.2, then the formula gives the precise answer. Section 5.3 proves a cohomological criterion for the projectivity of a compact complex manifold. It turns out that the projectivity of a Kähler manifold is encoded by the position of its Kähler cone within the natural weight-two Hodge structure.

5.1 Hirzebruch–Riemann–Roch Theorem

One of the most useful formulae in global complex and algebraic geometry is the Hirzebruch–Riemann–Roch formula. It is needed for any sort of practical computation. In this section we state the theorem without saying anything about its proof. A few applications and special cases are discussed in detail. The reader is advised to work through the examples and exercises in order to get some feeling for the power of this technique.

Historically, the Hirzebruch–Riemann–Roch formula generalizes the Riemann–Roch formula

$$\chi(C, E) = \deg(E) + \mathrm{rk}(E) \cdot (1 - g(C)) \tag{5.1}$$

for a holomorphic vector bundle E on a compact curve C and an analogous formula for line bundles on surfaces. It yields an expression for the *Euler–Poincaré characteristic*

$$\chi(X, E) := \sum_{i=0}^{\dim(X)} (-1)^i h^i(X, E)$$

of a holomorphic vector bundle E on a compact complex manifold X in terms of the Chern classes of E and X. Combined with the various vanishing results (5.2.2 and 5.2.7) it can often effectively be used to determine the dimension of the space of global sections $H^0(X, E)$. This in turn is important when one wants to study the geometry of X. E.g. $h^0(X, L)$ of an ample line bundle L determines the dimension of the projective spaces in which X can be embedded via the associated morphism φ_L.

Almost every algebraic geometer uses the Hirzebruch–Riemann–Roch formula in one form or the other in daily life. A proof of this important result, however, shall not be given here, it would lead us astray from our main objectives. For algebraic proofs in the case of curves and surfaces one might consult [66]. The general situation is much more complicated. In fact, there are various far reaching generalizations of the Hirzebruch–Riemann–Roch formula, most notably the Grothendieck–Riemann–Roch formula and the Atiyah–Singer index theorem, and it might be more reasonable to prove one of these instead of just the particular case. But both theorems are far beyond any basic course in complex geometry.

Nevertheless, we encourage the reader to apply the Hirzebruch–Riemann–Roch formula. He will soon get used to it by observing how amazingly well and effectively this formula works in so many different situations.

Theorem 5.1.1 (Hirzebruch–Riemann–Roch) *Let E be a holomorphic vector bundle on a compact complex manifold X. Then its Euler-Poincaré characteristic is given by*

$$\chi(X, E) = \int_X \mathrm{ch}(E)\mathrm{td}(X). \tag{5.2}$$

□

A few comments are in order. Firstly, the class $\mathrm{ch}(E)\mathrm{td}(X)$ has in general non-trivial components in various degrees. What is meant by the integral on the right hand side, of course, is the evaluation of the top degree component $[\mathrm{ch}(E)\mathrm{td}(X)]_{2n} = \sum \mathrm{ch}_i(E)\mathrm{td}_{n-i}(X)$, where $n = \dim_{\mathbb{C}}(X)$.

Secondly, for any short exact sequence $0 \to E_1 \to E \to E_2 \to 0$ of holomorphic vector bundles one has $\chi(X, E) = \chi(X, E_1) + \chi(X, E_2)$ (see Corollary B.0.37). On the right hand side of the Hirzebruch–Riemann–Roch formula (5.2), this corresponds to the additivity of the Chern character (see page 197).

Thirdly, the probably most striking feature of the above formula is that the holomorphic Euler–Poincaré characteristic $\chi(X, E)$ turns out to be independent of the holomorphic structure of E. Indeed, the Chern character on the right hand side depends only on the complex vector bundle underlying E. So, it frequently happens that for two holomorphic line bundles L_1 and L_2 with $c_1(L_1) = c_1(L_2)$ one has $h^0(X, L_1) \neq h^0(X, L_2)$, but one always has $\chi(X, L_1) = \chi(X, L_2)$, for the right hand side of the Hirzebruch–Riemann–Roch formula for a line bundle L only depends on X and $c_1(L)$.

It takes some time before one gets really comfortable with the beautiful formulae provided by the Hirzebruch–Riemann–Roch theorem. Let us look at a few special cases:

Examples 5.1.2 i) **Line bundles on a curve.** Let C be a connected compact curve and let $L \in \mathrm{Pic}(C)$. Then the Hirzebruch–Riemann–Roch formula (5.2) reads

$$\chi(C, L) = \int_C c_1(L) + \frac{c_1(C)}{2} = \deg(L) + \frac{\deg(K_C^*)}{2}.$$

For the comparison of the degree and the first Chern class see Exercise 4.4.1.

The special case $L = \mathcal{O}_C$ yields $\chi(C, \mathcal{O}_C) = \frac{\deg(K_C^*)}{2}$ and, therefore, $\deg(K_C) = 2(h^1(C, \mathcal{O}_C) - 1)$. Hence, for the genus $g(C)$ of the curve as introduced in Exercise 4.1.2 one finds $g(C) = (1/2)\deg(K_C) + 1 = h^1(C, \mathcal{O}_C) = h^0(C, K_C)$, where we used Serre duality for the last equality. This yields the Riemann–Roch formula in the form of (5.1).

ii) **Line bundles on a surface.** Let us first consider the case of a trivial line bundle. Then the Hirzebruch–Riemann–Roch formula specializes to *Noether's formula*:

$$\chi(X, \mathcal{O}_X) = h^0(\mathcal{O}_X) - h^1(\mathcal{O}_X) + h^2(\mathcal{O}_X)$$
$$= \int_X \frac{c_1^2(X) + c_2(X)}{12}.$$

If L is any line bundle on a compact complex surface X. Then

$$\chi(X, L) = h^0(X, L) - h^1(X, L) + h^2(X, L)$$
$$= \int_X \frac{c_1(L)(c_1(L) + c_1(X))}{2} + \chi(X, \mathcal{O}_X).$$

Sometimes, this is also written as

$$\chi(X, L) = \frac{L.(L - K_X)}{2} + \chi(X, \mathcal{O}_X).$$

iii) **Line bundles on a torus.** If $X = \mathbb{C}^n/\Gamma$ is a complex torus, then all characteristic classes of X itself are trivial. Thus,

$$\chi(\mathbb{C}^n/\Gamma, L) = \int_{\mathbb{C}^n/\Gamma} \frac{c_1(L)^n}{n!}.$$

for an arbitrary line bundle L on X. Note that for an elliptic curve $\mathbb{C}/\Gamma$ the formula coincides with the Riemann–Roch formula, as $g(\mathbb{C}/\Gamma) = 1$ and hence $\deg(K_{\mathbb{C}/\Gamma}) = 0$.

Let us consider a particular relevant application of the Hirzebruch–Riemann–Roch formula, for which no extra vector bundle is chosen.

The formula

$$p_a(X) := (-1)^n \left(\chi(\mathcal{O}_X) - 1 \right)$$

defines the *arithmetic genus* of a compact complex manifold X of dimension n. If X is a connected curve C, then the arithmetic genus is $1 - (h^0(X, \mathcal{O}_X) - h^1(C, \mathcal{O}_C)) = g(C)$, the (geometric) genus.

We have also already considered the alternating sum $\sum \chi(X, \Omega_X^p) = \sum (-1)^q h^{p,q}(X)$ which computes the signature of the intersection form on the middle cohomology of an even dimensional complex manifold (Corollary 3.3.18). Both expressions, $\chi(X, \mathcal{O}_X)$ and $\sum \chi(X, \Omega_X^p)$, are special values of the Hirzebruch χ_y-genus:

Definition 5.1.3 Let X be a compact complex manifold of dimension n. The *Hirzebruch χ_y-genus* is the polynomial

$$\chi_y := \sum_{p=0}^{n} \chi(X, \Omega_X^p) y^p = \sum_{p,q=0}^{n} (-1)^q h^{p,q}(X) y^p.$$

We are interested in the following special values of the Hirzebruch χ_y-genus:

i) $y = 0$: Then $\chi_{y=0} = \chi(X, \mathcal{O}_X)$ is essentially, up to sign and the extra term ± 1, the arithmetic genus of X.

ii) $y = 1$: Suppose X is a Kähler manifold of even complex dimension. Then $\chi_{y=1} = \mathrm{sgn}(X)$ is the signature Use Corollary 3.3.18 and $h^{p,q} = h^{q,p}$.

iii) $y = -1$: Suppose is X compact and Kähler. Then

$$\chi_{y=-1} = \sum_{p,q=0}^{n} (-1)^{p+q} h^{p,q}(X) = \sum_{k=0}^{2n} (-1)^k b_k(X) = e(X)$$

is the *Euler number* of X.

Using the yoga of Chern roots explained in Section 4.4, the Hirzebruch–Riemann–Roch formula (5.2) allows us to calculate the χ_y-genus as follows:

Corollary 5.1.4 *Let γ_i denote the formal Chern roots of T_X. Then*

$$\chi_y = \int_X \prod_{i=1}^{n} \left(1 + y e^{-\gamma_i} \right) \frac{\gamma_i}{1 - e^{-\gamma_i}}.$$

Proof. This follows immediately from the definition of the Todd classes, Theorem 5.1.1 and the equality

$$\mathrm{ch} \left(\bigoplus_{p=0}^{n} \Omega_X^p y^p \right) = \prod_{i=1}^{n} \left(1 + y e^{-\gamma_i} \right),$$

the proof of which is left to the reader. □

Let us apply this to the three special values of χ_y discussed above.

i) For $y = 0$ this just gives back the ordinary Hirzebruch–Riemann–Roch formula $\chi(X, \mathcal{O}_X) = \int_X \operatorname{td}(X)$.

ii) For $y = 1$ it yields the so called *Hirzebruch signature theorem* for compact Kähler manifolds of even complex dimension $n = 2m$:

$$\operatorname{sgn}(X) = \chi\left(\bigoplus \Omega_X^p\right) = \int_X \operatorname{ch}\left(\bigoplus \Omega_X^p\right) \operatorname{td}(X)$$

$$= \int_X L(X),$$

where $L(X)$ is the *L-genus* which in terms of the Chern roots is just

$$L(X) = \prod_{i=1}^{n} \gamma_i \cdot \coth\left(\frac{\gamma_i}{2}\right).$$

Note that this result, with a different proof though, holds for any compact complex manifold of even dimension.

iii) For $y = -1$ and X a compact Kähler manifold of dimension n we obtain the following result (the *Gauss–Bonnet formula*), which also holds more generally for any compact complex manifold:

$$e(X) = \int_X c_n(X).$$

Indeed, the corollary yields $e(x) = \int_X \prod_{i=1}^{n} \gamma_i$, but clearly, $\int_X c_n(X) = \int_X \prod_{i=1}^{n} \gamma_i$.

Other *Chern numbers*, i.e. integrals of the form $\int_X c_{i_1}(X) \ldots c_{i_k}(X)$ with $\sum i_j = \dim(X)$, of the compact complex manifold X are equally interesting.

As mentioned before, the Hirzebruch–Riemann–Roch formula (5.2) can be considered a special case of the Grothendieck–Riemann–Roch formula or, likewise, of the Atiyah–Singer index theorem. For sake of completeness, we state these two results without even explaining the meaning of some of the ingredients. I hope that the reader nevertheless gets a vague idea how they might be related to the Hirzebruch–Riemann–Roch formula.

Grothendieck–Riemann–Roch formula. Let $f : X \to Y$ be a smooth projective morphism of smooth projective varieties. Then for any coherent sheaf $\mathcal{F}$ (e.g. a vector bundle) on X one has

$$\operatorname{ch}\left(\sum (-1)^i R^i f_* \mathcal{F}\right) \operatorname{td}(Y) = f_* \left(\operatorname{ch}(\mathcal{F}) \operatorname{td}(X)\right)$$

in the rational Chow group $\operatorname{CH}(Y)_{\mathbb{Q}}$ or, likewise, in $H^*(Y, \mathbb{R})$.

To deduce the Hirzebruch–Riemann–Roch formula from this statement, one considers the projection to a point $f : X \rightarrow \{pt\}$. Then f_* is nothing else than the integral $\int_X$ and the higher direct image sheaves $R^i f_* \mathcal{F}$ become the cohomology groups $H^i(X, \mathcal{F})$. Since $\mathrm{td}\{pt\} = 1$ and the Chern character of a vector bundle on $\{pt\}$, i.e. a vector space, is just the dimension, this yields (5.2).

Atiyah–Singer Index Theorem. Let $D : \Gamma(E) \rightarrow \Gamma(F)$ be an elliptic differential operator between vector bundles E and F on a compact oriented differentiable manifold M. Then the analytic index $\mathrm{index}(D) := \dim \mathrm{Ker}(D) - \dim \mathrm{Coker}(D)$ and the topological index $\gamma(D)$ satisfy

$$\mathrm{index}(D) = \gamma(D).$$

The topological index is usually expressed in characteristic classes of E and F. In order to see that the Atiyah–Singer formula as well implies (5.2) or the Hirzebruch signature theorem, one has to consider the appropriate elliptic differential operator. E.g. for (5.2) one takes the Laplacian $\Delta_{\bar{\partial}_E}$.

Remark 5.1.5 It frequently happens that one wants to compute $\chi(X, \mathcal{F})$ of a coherent sheaf $\mathcal{F}$, which is not locally free, e.g. $\mathcal{F} = \mathcal{I}_Z$ the ideal sheaf of a submanifold $Z \subset X$. If X is projective, then there exists a locally free resolution $0 \rightarrow E_n \rightarrow \ldots \rightarrow E_1 \rightarrow \mathcal{F} \rightarrow 0$. Applying the above formula to the sheaves E_i and using the additivity of the Euler–Poincaré characteristic and the Chern character, one immediately obtains a Hirzebruch–Riemann–Roch formula for $\mathcal{F}$.

In fact, for the Grothendieck–Riemann–Roch formula it is not even very natural to restrict to locally free sheaves, as already the direct image sheaves $R^i f_* E$ of a locally free sheaf E are in general not locally free anymore.

As an example one might consider the structure sheaf $\mathcal{O}_Y$ of a smooth hypersurface $Y \subset X$. The structure sheaf sequence $0 \rightarrow \mathcal{O}(-Y) \rightarrow \mathcal{O} \rightarrow \mathcal{O}_Y \rightarrow 0$ provides a locally free resolution of $\mathcal{O}_Y$. The Hirzebruch–Riemann–Roch formula (5.2) yields

$$\chi(Y, \mathcal{O}_Y) = \chi(X, \mathcal{O}_X) - \chi(X, \mathcal{O}(-Y)) = \int_X \mathrm{td}(X) - \int_X \mathrm{ch}\left(\mathcal{O}(-Y)\right) \mathrm{td}(X)$$

$$= \int_X \left(1 - e^{-[Y]}\right) \mathrm{td}(X),$$

where $1 - e^{-[Y]}$ could also be considered as $\mathrm{ch}(\mathcal{O}_Y)$.

What happens if X is not projective? *A priori*, neither the algebraic approach of the Grothendieck–Riemann–Roch formula nor the analytic one of the Atiyah–Singer index theorem, where one works with differential operators on vector bundles, seem to work. A very different technique was invented by O'Brian, Toledo, and Tong [96, 109] in order to prove (5.2) for arbitrary coherent sheaves on arbitrary compact complex manifolds. In fact, even for vector bundles they can avoid to choose hermitian structures. Moreover, they also succeeded to prove a version of the Grothendieck–Riemann–Roch formula for complex manifolds. Very roughly, instead of computing cohomology in terms of harmonic forms, they use Čech-cohomology. Thus, any of the above mentioned results holds true for arbitrary complex manifolds and coherent sheaves. (Warning: We have actually not even defined Chern classes of arbitrary coherent sheaves. This can be done by using the Atiyah class approach sketched in Exercise 4.4.11 (see [3]).)

Exercises

5.1.1   Let X be a K3 surface (cf. Exercise 2.5.5). Show that $b_2(X) = 22$. Prove that the Picard number $\rho(X)$ is bounded by 20.

5.1.2   Let $X = \mathbb{C}^n/\Gamma$ be a complex torus and $L \in \mathrm{Pic}(X)$. Consider $c_1(L) \in H^2(X, \mathbb{R})$ as an alternating form on $H^1(X, \mathbb{R})^*$ and choose a basis such that it corresponds to the matrix

$$\begin{pmatrix} 0 & \begin{pmatrix} \lambda_1 & & \\ & \ddots & \\ & & \lambda_n \end{pmatrix} \\ \begin{pmatrix} -\lambda_1 & & \\ & \ddots & \\ & & -\lambda_n \end{pmatrix} & 0 \end{pmatrix}.$$

Show that $\chi(X, L) = \lambda_1 \cdot \ldots \cdot \lambda_n$.

(Together with Kodaira's vanishing result for an ample line bundle L it yields Frobenius' theorem asserting $h^0(X, L) = \lambda_1 \cdot \ldots \cdot \lambda_n$. Here, $h^0(X, L)$ can also be interpreted as the dimensions of the space of certain theta functions.)

5.1.3   The *Hilbert polynomial* of a polarized manifold (X, L), i.e. L is an ample line bundle on X, is defined as the function

$$\mathbb{Z} \longrightarrow \mathbb{Z}, \quad m \longmapsto P_{(X,L)}(m) := \chi(X, L^{\otimes m}).$$

Show that $P_{(X,L)}$ is indeed a polynomial in m. Determine its degree and its leading coefficient.

(Using Proposition 5.2.7 one can prove that $P_{(X,L)}(m) = h^0(X, L^{\otimes m})$ for $m \gg 0$. Notice that $m \mapsto h^0(X, L^{\otimes m})$ is not a polynomial function, as $h^0(X, L^{\otimes m}) = 0$ for $m < 0$. See Exercise 5.2.11.)

5.1.4   Compute the Hilbert polynomial of a hypersurface $Y \subset \mathbb{P}^n$ of degree k.

5.1.5   Let L be a line bundle on a compact connected curve C with $\deg(L) > g(C) - 1$. Show that L admits non-trivial global holomorphic sections.

5.1.6   Let L be a line bundle on a compact connected curve C with $\deg(L) > 2g(C) - 1$. Show that L is globally generated.

5.1.7   Let L be a line bundle on a compact surface X with $\int_X c_1(L)^2 > 0$. Show that for $m \gg 0$ either $L^{\otimes m}$ or $L^{\otimes(-m)}$ admits non-trivial global holomorphic sections.

5.1.8   Let X be a compact surface such that $c_1(X) \in 2H^2(X, \mathbb{Z})$. Use (5.2) to show that $\int_X c_1(L)^2$ is even for any line bundle L on X.

5.1.9   Let X and Y be compact complex manifolds and let $f : X \to Y$ be a smooth finite morphism of degree d. In other words, $f : X \to Y$ is smooth surjective with $\dim(X) = \dim(Y)$ and every fibre $f^{-1}(y)$ contains d points.

   Show that $\mathrm{td}(X) = f^* \mathrm{td}(Y)$ and deduce $\chi(X, \mathcal{O}_X) = d \cdot \chi(Y, \mathcal{O}_Y)$. In particular, if X and Y are K3 surfaces, then $d = 1$.

Comments: - For the proof of the Hirzebruch–Riemann–Roch formula and the Atiyah–Singer index theorem we recommend [15, 54, 68]. The algebraic approach is explained in [47, 49].

   - The Hirzebruch χ_y-genus is yet a special case of a more general function, the *elliptic genus*. The elliptic genus is a mathematical analogue of the physicist's partition function.

   - The Hirzebruch signature theorem can also directly be deduced from the Atiyah–Singer index theorem. For historical comments we refer to [69].

   - A deep and important result of Kollár and Matsusaka [81] says that the two highest coefficients of the Hilbert polynomial of a polarized manifold (X, L) determine the whole Hilbert polynomial up to finitely many possibilities.

   - Under certain assumptions on the positivity of the curvature of a holomorphic line bundle, one can prove approximative formulae for the truncated Hilbert polynomials $\sum_{i=0}^{k}(-1)^i h^i(X, L^{\otimes m})$. See [35].

5.2 Kodaira Vanishing Theorem and Applications

Let X be a compact Kähler manifold of dimension n and let L be a holomorphic line bundle on X. Assuming a certain positivity of L, the higher cohomology groups of L can be controlled due to the so called Kodaira(–Nakano–Akizuki) vanishing theorem. In conjunction with the Hirzebruch–Riemann–Roch formula (5.2) this often yields effective (and topological) bounds for the dimension of the space of global holomorphic sections $H^0(X, L)$ of a holomorphic line or vector bundle. This sections contains, besides the proof of the Kodaira vanishing theorem, a discussion of important applications like the Weak Lefschetz theorem and Serre's theorem.

Let us start out by introducing the appropriate positivity concept.

Definition 5.2.1 A line bundle L is called *positive* if its first Chern class $c_1(L) \in H^2(X, \mathbb{R})$ can be represented by a closed positive real (1,1)-form.

Note that a compact complex manifold X that admits a positive line bundle L is automatically Kähler. Indeed, the closed positive real $(1, 1)$-form representing $c_1(L)$ defines a Kähler structure on X.

The notion of positive forms and positive curvature has been discussed in Section 4.3. Since any closed real $(1, 1)$-form representing $c_1(L)$ is the curvature of a Chern connection (modulo the factor $i/2\pi$) (cf. Exercise 4.4.10 or Remark 4.B.5), a line bundle L is positive if and only if it admits an hermitian structure such that the curvature of the induced Chern connection is positive in the sense of Definition 4.3.15.

The algebraic inclined reader might replace 'positive' by 'ample'. The equivalence of both concepts will be proved in the next section. Note that in particular the manifolds considered in the present section will usually be projective.

Proposition 5.2.2 (Kodaira vanishing) *Let L be a positive line bundle on a compact Kähler manifold X. Then*

$$H^q(X, \Omega_X^p \otimes L) = 0 \quad \text{for} \ \ p + q > n.$$

Before proving this result, we will state and prove a few lemmas which are valid without any assumption on L and some of them work even for vector bundles of arbitrary rank.

Let E be an arbitrary holomorphic vector bundle on X with a fixed hermitian structure. In addition to the two operators $\bar{\partial}_E$ and $\bar{\partial}_E^*$ introduced in Section 4.1, we shall use the linear Lefschetz operators L and Λ on $\mathcal{A}^{p,q}(E)$ depending on a chosen Kähler structure on X, i.e. $L = L \otimes 1$ and $\Lambda := \Lambda \otimes 1$ on $\bigwedge^{p,q} X \otimes E$.

Recall the following two Kähler identities on $\mathcal{A}^{p,q}(X)$ (cf. Propositions 1.2.26 and 3.1.12):

$$\text{i) } [\Lambda, L] = (n - (p+q)) \cdot \mathrm{id} \quad \text{and} \quad \text{ii) } [\Lambda, \bar{\partial}] = -i\, \partial^*.$$

The first one, which is linear, holds true for the corresponding operators on $\mathcal{A}^{p,q}(E)$. The second one is generalized by the following

Lemma 5.2.3 (Nakano identity) *Let ∇ be the Chern connection on E. Then*

$$[\Lambda, \bar{\partial}_E] = -i\,(\nabla^{1,0})^* = i\,(\bar{*}_{E^*} \circ \nabla^{1,0}_{E^*} \circ \bar{*}_E).$$

Proof. The second equality is the definition of the adjoint operator of $\nabla^{1,0}$.

The first equality is local and we may, therefore, use an orthonormal trivialization $\psi : E|_U \cong U \times \mathbb{C}^r$. With respect to such a trivialization the Hodge operator $\bar{*}_E$ becomes the complex conjugate $\bar{*}$ of the usual Hodge operator $*$. Writing the connection on E with respect to ψ as $\nabla_E = d + A$ one has $\nabla_{E^*} = d + A^* = d - A$, $\bar{\partial}_E = \bar{\partial} + A^{0,1}$, and $(\nabla^{1,0}_E)^* = -\bar{*} \circ (\partial - A^{1,0}) \circ \bar{*} = - * \circ \bar{\partial} \circ * - \bar{*} \circ A^{1,0} \circ \bar{*} = \partial^* - (A^{1,0})^*$.

This yields

$$\begin{aligned}
[\Lambda, \bar{\partial}_E] + i\,(\nabla^{1,0}_E)^* &= [\Lambda, \bar{\partial}] + i\partial^* + [\Lambda, A^{0,1}] - i\,(A^{1,0})^* \\
&= [\Lambda, A^{0,1}] - i\,(A^{1,0})^*,
\end{aligned}$$

where we used the original Kähler identity (Proposition 3.1.12). Thus, the global operator $[\Lambda, \bar{\partial}_E] + i\,(\nabla^{1,0}_E)^*$ turns out to be linear. In order to show that it vanishes, it thus suffices to choose the orthonormal trivialization ψ in a neighbourhood of $x \in X$ such that $A(x) = 0$. This can always be achieved (cf. Remark 4.2.5). $\qquad\square$

It should be clear from the proof that the compactness of X has not been used yet. If X is compact then $(\nabla^{1,0}_E)^* = -\bar{*}_{E^*} \circ \nabla^{1,0}_{E^*} \circ \bar{*}_E$ is indeed the formal adjoint of $\nabla^{1,0}_E$ with respect to the natural hermitian product on $\mathcal{A}^{p,q}(X, E)$. Copy the proof of Lemma 4.1.12.

Lemma 5.2.4 *Let (E, h) be an arbitrary hermitian holomorphic vector bundle on a compact Kähler manifold (X, g). Then for the curvature F_∇ of the Chern connection ∇ and an arbitrary harmonic form $\alpha \in \mathcal{H}^{p,q}(X, E)$ one has:*

$$\text{i) } \frac{i}{2\pi}\,(F_\nabla \Lambda(\alpha), \alpha) \leq 0 \quad \text{and} \quad \text{ii) } \frac{i}{2\pi}\,(\Lambda F_\nabla(\alpha), \alpha) \geq 0.$$

Proof. As before, $(\ ,\ )$ is the natural hermitian product on $\mathcal{A}^{p,q}(X, E)$ depending on both hermitian structures, on E and on X. Since $\Lambda(\alpha) \in \mathcal{A}^{p-1,q-1}(X, E)$ (which is not necessarily harmonic), the form $F_\nabla \Lambda(\alpha)$ is indeed in $\mathcal{A}^{p,q}(X, E)$.

Since the Chern connection is hermitian with $\nabla^{0,1} = \bar{\partial}_E$, one has $F_\nabla = \nabla^{1,0} \circ \bar{\partial}_E + \bar{\partial}_E \circ \nabla^{1,0}$. Also recall that the form part of $F_\nabla(\alpha)$, which is of bidegree $(p+1, q+1)$, is just the exterior product of of the form part of the curvature form F_∇ and the form part of α. The assumption that α is harmonic means $\bar{\partial}_E \alpha = \bar{\partial}_E^* \alpha = 0$.

Using this information one computes

$$
\begin{aligned}
(iF_\nabla \Lambda(\alpha), \alpha) &= i\left(\nabla^{1,0}\bar{\partial}_E\Lambda(\alpha), \alpha\right) + i\left(\bar{\partial}_E\nabla^{1,0}\Lambda(\alpha), \alpha\right) \\
&= i\left(\bar{\partial}_E\Lambda(\alpha), (\nabla^{1,0})^*\alpha\right) + i\left(\nabla^{1,0}\Lambda(\alpha), \bar{\partial}_E^*(\alpha)\right) \\
&= \left(\bar{\partial}_E\Lambda(\alpha), -i(\nabla^{1,0})^*(\alpha)\right) + 0, \text{ as } \alpha \text{ is harmonic} \\
&= \left(\bar{\partial}_E\Lambda(\alpha), [\Lambda, \bar{\partial}_E](\alpha)\right) \text{ by Lemma 5.2.3} \\
&= -\left(\bar{\partial}_E\Lambda(\alpha), \bar{\partial}_E\Lambda\alpha\right), \text{ as } \alpha \text{ is harmonic} \\
&= -\|\bar{\partial}_E\Lambda(\alpha)\|^2 \leq 0.
\end{aligned}
$$

Similarly,

$$
\begin{aligned}
(i\Lambda F_\nabla(\alpha), \alpha) &= i\left(\Lambda\bar{\partial}_E\nabla^{1,0}(\alpha), \alpha\right) \text{ by harmonicity of } \alpha \\
&= i\left([\Lambda, \bar{\partial}_E]\nabla^{1,0}(\alpha), \alpha\right) + i\left(\bar{\partial}_E\Lambda\nabla^{1,0}(\alpha), \alpha\right) \\
&= i\left(-i(\nabla^{1,0})^*\nabla^{1,0}(\alpha), \alpha\right) + i\left(\Lambda\nabla^{1,0}(\alpha), \bar{\partial}_E^*\alpha\right) \\
&= \|\nabla^{1,0}(\alpha)\|^2 + 0 \geq 0
\end{aligned}
$$

$\square$

Let us now come to the proof of the Kodaira vanishing theorem.

Proof. Choose an hermitian structure on the positive line bundle L such that the curvature of the Chern connection is positive, i.e. $\frac{i}{2\pi}F_\nabla$ is a Kähler form on X. Thus, with respect to this Kähler structure on X, which we will fix once and for all, the Lefschetz operator L is nothing but the curvature operator $\frac{i}{2\pi}F_\nabla$. Using Lemma 5.2.4 and the commutator relation $[\Lambda, L] = -H$, we obtain

$$0 \leq \left(\frac{i}{2\pi}[\Lambda, F_\nabla](\alpha), \alpha\right) = ([\Lambda, L](\alpha), \alpha) = (n - (p+q)) \cdot \|\alpha\|^2$$

for any $\alpha \in \mathcal{H}^{p,q}(X, L) = H^q(X, \Omega_X^p \otimes L)$. This proves the assertion. $\square$

Example 5.2.5 Consider $\mathcal{O}(1)$ on $\mathbb{P}^n$, which is positive due to Example 4.3.12 (see also page 197). Thus, $H^q(\mathbb{P}^n, \Omega^p \otimes \mathcal{O}(m)) = 0$ for $p + q > n$ and $m > 0$. In particular, $H^q(\mathbb{P}^n, \mathcal{O}(m)) = 0$ for $q > 0$ and $m \geq -n$, as $K_{\mathbb{P}^n} \cong \mathcal{O}(-n-1)$ by Proposition 2.4.3. Using Serre duality this yields

$$H^q(\mathbb{P}^n, \mathcal{O}(m)) = \begin{cases} 0 & \text{if } 0 < q < n \\ 0 & \text{if } q = 0, m < 0 \\ 0 & \text{if } q = n, m > -n - 1. \end{cases}$$

Due to Proposition 2.4.1 one also knows

$$H^0(\mathbb{P}^n, \mathcal{O}(m)) = \mathbb{C}[z_0, \dots, z_n]_m$$

for $m \geq 0$ and

$$H^n(\mathbb{P}^n, \mathcal{O}(m)) = H^0(\mathbb{P}^n, \mathcal{O}(-n-1-m))^*$$

for $m \leq -n-1$.

As another application of the Kodaira vanishing theorem, we prove the so called weak Lefschetz theorem. In fact, the injectivity below was already studied in Exercise 3.3.5.

Proposition 5.2.6 (Weak Lefschetz theorem) *Let X be a compact Kähler manifold of dimension n and let $Y \subset X$ be a smooth hypersurface such that the induced line bundle $\mathcal{O}(Y)$ is positive. Then the canonical restriction map*

$$H^k(X, \mathbb{C}) \longrightarrow H^k(Y, \mathbb{C})$$

is bijective for $k \leq n - 2$ and injective for $k \leq n - 1$.

Proof. For both manifolds we have the bidegree decomposition $H^k = \bigoplus H^{p,q}$ and the restriction map $H^k(X, \mathbb{C}) \to H^k(Y, \mathbb{C})$ is compatible with it. Hence, it suffices to show that the map $H^q(X, \Omega_X^p) \to H^q(Y, \Omega_Y^p)$ is bijective for $p + q \leq n - 2$ and injective for $p + q \leq n - 1$. For this we will use the two short exact sequences

$$0 \longrightarrow \mathcal{O}_X(-Y) \longrightarrow \mathcal{O}_X \longrightarrow \mathcal{O}_Y \longrightarrow 0$$

and

$$0 \longrightarrow \mathcal{O}_Y(-Y) \longrightarrow \Omega_X|_Y \longrightarrow \Omega_Y \longrightarrow 0,$$

where the latter is the dual of the normal bundle sequence. (See Definition 2.2.16 and use Exercise 2.3.2, which shows that the normal bundle of $Y \subset X$ is $\mathcal{O}_X(Y)$.) Twisting the first one with Ω_X^p and taking the p-th exterior product of the second one yields short exact sequences of the form:

$$0 \longrightarrow \Omega_X^p(-Y) \longrightarrow \Omega_X^p \longrightarrow \Omega_X^p|_Y \longrightarrow 0 \qquad (5.3)$$

and

$$0 \longrightarrow \Omega_Y^{p-1}(-Y) \longrightarrow \Omega_X^p|_Y \longrightarrow \Omega_Y^p \longrightarrow 0. \qquad (5.4)$$

(For (5.4) we use Exercise 2.2.2.) Kodaira vanishing 5.2.2, Serre duality 4.1.16, and the fact that $\Omega_X^{p*} \otimes K_X \cong \Omega_X^{n-p}$ (see Exercise 2.2.3) prove

$$\begin{aligned}
H^q(X, \Omega_X^p(-Y)) &= H^{n-q}(X, \Omega_X^{p*} \otimes \mathcal{O}(Y) \otimes K_X)^* \\
&= H^{n-q}(X, \Omega_X^{n-p} \otimes \mathcal{O}(Y))^* \\
&= 0 \text{ for } p + q < n.
\end{aligned}$$

Thus, using the long exact cohomology sequence induced by (5.3), one finds that the natural restriction map $H^q(X, \Omega_X^p) \to H^q(Y, \Omega_X^p|_Y)$ is bijective for $p + q + 1 < n$ and at least injective for $p + q < n$.

This map will be composed with the natural map $H^q(Y, \Omega_X^p|_Y) \to H^q(Y, \Omega_Y^p)$, whose kernel and cokernel are contained in cohomology groups of the bundle $\Omega_Y^{p-1} \otimes \mathcal{O}_Y(-Y)$. (In order to see this, we use the long exact cohomology sequence associated to (5.4).) Since the restriction of $\mathcal{O}(Y)$ to Y is again positive, we can apply the Kodaira vanishing theorem as before. Hence, $H^q(Y, \Omega_X^p|_Y) \to H^q(Y, \Omega_Y^p)$ is bijective for $p + q < n - 1$ and injective for $q + p < n$.

Both statements together prove the assertion. □

A slight modification of the proof of the Kodaira vanishing yields also the following result, which is known as Serre's theorem.

Proposition 5.2.7 *Let L be a positive line bundle on a compact Kähler manifold X. For any holomorphic vector bundle E on X there exists a constant m_0 such that*

$$H^q(X, E \otimes L^m) = 0$$

for $m \geq m_0$ and $q > 0$.

Proof. Choose hermitian structures on E and L and denote the associated Chern connections by ∇_E and ∇_L, respectively. By assumption we may suppose that $(i/2\pi)F_{\nabla_L}$ is a Kähler form ω. We endow X with the corresponding Kähler structure. By Lemma 5.2.4 we have

$$\frac{i}{2\pi} ([\Lambda, F_\nabla](\alpha), \alpha) \geq 0$$

for any $\alpha \in \mathcal{H}^{p,q}(X, E \otimes L^m)$. Here, $\nabla = \nabla_E \otimes 1 + 1 \otimes \nabla_{L^m}$, where, the connection ∇_{L^m} on L^m is induced by ∇_L on L. In particular, $(i/2\pi)F_{\nabla_{L^m}} = m \cdot \omega$. Hence,

$$\frac{i}{2\pi}F_\nabla = \frac{i}{2\pi}F_{\nabla_E} \otimes 1 + m(1 \otimes \omega)$$

and, therefore,

$$\frac{i}{2\pi} ([\Lambda, F_\nabla](\alpha), \alpha) = \frac{i}{2\pi} ([\Lambda, F_{\nabla_E}](\alpha), \alpha) + m ([\Lambda, L_\omega](\alpha), \alpha)$$

$$= \frac{i}{2\pi} ([\Lambda, F_{\nabla_E}](\alpha), \alpha) + m(n - (p+q)) \cdot \|\alpha\|^2.$$

The fibrewise Cauchy–Schwarz inequality $|([\Lambda, F_{\nabla_E}](\alpha), \alpha)| \leq \|[\Lambda, F_{\nabla_E}]\| \cdot \|\alpha\|^2$ provides a corresponding global inequality, where the operator norm $C := \|[\Lambda, F_{\nabla_E}]\|$ does not depend on m. Hence, if $C + 2\pi \cdot m(n - (p+q)) < 0$ one necessarily has $\alpha = 0$.

Therefore, if $m_0 > C/2\pi$, then $H^q(X, E \otimes K_X \otimes L^m) = 0$ for all $m \geq m_0$ and $q > 0$. To conclude, we apply these arguments to the bundle $E \otimes K_X^*$

instead of E. (The constant m_0 might change in the process.) This proves the assertion. $\square$

As an application of Serre's vanishing theorem we will prove the following classification result for vector bundles on $\mathbb{P}^1$.

Corollary 5.2.8 (Grothendieck lemma) *Every holomorphic vector bundle E on $\mathbb{P}^1$ is isomorphic to a holomorphic vector bundle of the form $\bigoplus \mathcal{O}(a_i)$. The ordered sequence $a_1 \geq a_2 \geq \ldots \geq a_r$ is uniquely determined.*

Proof. If E is a rank one vector bundle, i.e. a line bundle, then the assertion is known already (Exercise 3.2.11). For arbitrary rank r the assertion is proved by induction.

Let a_1 be maximal among all a with $\operatorname{Hom}(\mathcal{O}(a), E) = H^0(\mathbb{P}^1, E(-a)) \neq 0$. First of all, there always exists a with $H^0(\mathbb{P}^1, E(-a)) \neq 0$, since by Serre vanishing $H^1(\mathbb{P}^1, E(-a)) = 0$ for $a \ll 0$ and $\chi(\mathbb{P}^1, E(-a)) = \deg(E) - (1 - a) \cdot \operatorname{rk}(E)$ by the Riemann–Roch formula for curves (5.1). Hence, the Euler–Poincaré characteristic is positive for $a \ll 0$. There also exists a maximal such a, because again by Serre vanishing 5.2.7 one knows $H^0(\mathbb{P}^1, E(-a)) = H^1(\mathbb{P}^1, E^*(a - 2))^* = 0$ for $a \gg 0$.

Thus, we have a short exact sequence

$$0 \longrightarrow \mathcal{O}(a_1) \longrightarrow E \longrightarrow E_1 \longrightarrow 0, \qquad (5.5)$$

where the quotient E_1 is *a priori* only a sheaf. It is locally free if the map $s : \mathcal{O}(a_1) \to E$ has constant rank, i.e. is nowhere trivial. But if s vanished in some point $x \in \mathbb{P}^1$, then we could divide by the equation $s_x \in H^0(\mathbb{P}^1, \mathcal{O}(1))$ of x in order to obtain a section $\mathcal{O}(a_1 + 1) \to E$, which would contradict the maximality of a_1. Hence, s is an inclusion of vector bundles and hence E_1 is a holomorphic vector bundle.

By induction hypothesis we may assume that E_1 is split, i.e. $E_1 = \bigoplus_{i>1} \mathcal{O}(a_i)$. It remains to show that the sequence (5.5) splits, which would yield $E = \mathcal{O}(a_1) \oplus \bigoplus_{i>1} \mathcal{O}(a_i)$.

In order to see this, we first show that $a_i \leq a_1$. Indeed, if we had $a_i > a_1$ for one i, then $H^0(\mathbb{P}^1, E_1(-a_1 - 1)) \neq 0$. This combined with $H^1(\mathbb{P}^1, \mathcal{O}(-1)) = 0$ and the long exact cohomology sequence associated with

$$0 \longrightarrow \mathcal{O}(-1) \longrightarrow E(-a_1 - 1) \longrightarrow E_1(-a_1 - 1) \longrightarrow 0$$

would yield $H^0(\mathbb{P}^1, E(-a_1 - 1)) \neq 0$, which contradicts the maximality of a_1. Secondly, the splitting of (5.5) is equivalent to the splitting of its dual sequence twisted by $\mathcal{O}(a_1)$:

$$0 \longrightarrow E_1^*(a_1) \longrightarrow E^*(a_1) \longrightarrow \mathcal{O} \longrightarrow 0.$$

The splitting of the latter follows from $H^1(\mathbb{P}^1, E_1^*(a_1)) = H^1(\mathbb{P}^1, \bigoplus_{i>1} \mathcal{O}(a_1 - a_i)) = 0$, as $a_1 - a_i \geq 0$, which implies the surjectivity of $H^0(E^*(a_1)) \to$

$H^0(\mathcal{O})$. The lift of $1 \in H^0(\mathbb{P}^1, \mathcal{O})$ considered as a homomorphism $\mathcal{O} \to E^*(a_1)$ splits the sequence. □

Very roughly, the corollary says that there are no interesting vector bundles others than line bundles on the projective line $\mathbb{P}^1$. The situation differs drastically on curves of positive genus and, as well, on higher dimensional projective spaces. Already an explicit classification of rank two vector bundles on the projective plane $\mathbb{P}^2$ is impossible, although the situation is expected to become easier on projective spaces $\mathbb{P}^n$ with $n \geq 5$.

Exercises

5.2.1 Let (E, h) be an hermitian holomorphic vector bundle on a compact Kähler manifold X. Suppose that the curvature F_∇ of the Chern connection is trivial, i.e. the Chern connection is flat. Prove that the Lefschetz operator Λ preserves the harmonicity of forms and thus defines a map $\Lambda : \mathcal{H}^{p,q}(X, E) \to \mathcal{H}^{p-1,q-1}(X, E)$. Deduce from this the existence of a Lefschetz decomposition on $H^{*,*}(X, E)$.

5.2.2 Let C be an elliptic curve. Show that $H^1(C, \mathcal{O}_C) = \mathbb{C}$ and use this to construct a non-splitting extension $0 \to \mathcal{O} \to E \to \mathcal{O} \to 0$. Prove that E cannot be written as a direct sum of two holomorphic line bundles.

(There is an algebraic argument using $H^1(C, \mathcal{O}_C) = \text{Ext}^1(\mathcal{O}_C, \mathcal{O}_C)$, but one might as well try to construct a new $\bar{\partial}$-operator on the trivial bundle of rank two by means of a representative of a non-trivial class in $H^1(C, \mathcal{O}_C) = H^{0,1}(C)$.)

5.2.3 Show that on $\mathbb{P}^2$ there exists a rank two vector bundle which is not isomorphic to the direct sum of holomorphic line bundles.

5.2.4 (**The degree-genus formula**) Let $C \subset \mathbb{P}^2$ be a smooth curve defined by a homogeneous polynomial of degree d. Show that the genus $g(C) = \dim H^0(C, K_C)$ is given by the formula

$$g(C) = \frac{1}{2}(d-1)(d-2).$$

Use this to show that there are curves which are not plane, i.e. not isomorphic to a smooth curve in $\mathbb{P}^2$. Prove that for a smooth curve $C \subset X$ in a K3 surface X one has $g(C) = ([C]^2 + 2)/2$.

5.2.5 Show that hypersurfaces in $\mathbb{P}^n$ with $n \geq 3$ do not admit non-trivial holomorphic one-forms. In particular, the Albanese of any such hypersurface is trivial.

5.2.6 Which complex tori could possibly be realized as complete intersections in $\mathbb{P}^n$?

5.2.7 Let L be an ample line bundle on a K3 surface X. Show that $h^0(X, L) = 2 + (1/2) \int_X c_1(L)^2$. Study ample line bundles on complex tori.

5.2.8 Use Serre duality to give a direct algebraic proof of the Kodaira vanishing theorem for curves.

5.2.9   Prove that $H^q(\mathbb{P}^n, \Omega^p(m)) = 0$ for $p + q > n$, $m > 0$ and for $p + q < n$, $m < 0$.

5.2.10   Let Y be a hypersurface of a compact complex manifold X with $\mathcal{O}(Y)$ positive. Suppose that $H^2(X, \mathbb{Z})$ and $H^2(Y, \mathbb{Z})$ are torsion free. Prove that the restriction induces an isomorphism $\mathrm{Pic}(X) \to \mathrm{Pic}(Y)$ if $\dim(X) \geq 4$. (Use the exponential sequence and the weak Lefschetz theorem. The assumption on $H^2(\,\,, \mathbb{Z})$ is superfluous, see the comments at the end of this section.)

5.2.11   Let X be a projective manifold of dimension n and let $L \in \mathrm{Pic}(X)$ be an ample line bundle.
   i) Show that $m \mapsto h^0(X, L^{\otimes m})$ for $m \gg 0$ is a polynomial of degree n with positive leading coefficient.
   ii) Deduce from this that $a(X) = \dim(X) = n$, i.e. X is Moishezon. Use the arguments of Section 2.2.

Comments: - The proof of the Kodaira vanishing follows closely the classical argument, see [35, 100]. An algebraic proof had long been missing. In fact, the Kodaira vanishing does not hold for arbitrary smooth projective varieties in positive characteristic. An algebraic proof was eventually found by Deligne and Illusie in [32] (see also [42]).
   - There are various versions of the Kodaira vanishing theorem for positive vector bundles, e.g. the so called Le Potier vanishing theorem [85]. Roughly, the positivity of a vector bundle E is translated into positivity of the relative $\mathcal{O}_\pi(1)$ on $\pi : \mathbb{P}(E) \to X$. Using Kodaira vanishing for $\mathcal{O}_\pi(1)$ on $\mathbb{P}(E)$ and the formula $\pi_*\mathcal{O}(1) \cong E^*$, one can deduce the vanishing of certain cohomology groups.
   - Another far-reaching generalization of the Kodaira vanishing theorem is the Kawamata–Viehweg vanishing which predicts the same sort of vanishing but this time for line bundles which are not quite positive, but only big and nef. The result is often used in Mori theory.
   - The Weak Lefschetz theorem actually holds for the integral cohomology groups. Moreover, there are versions of it for the homotopy groups. E.g. if $\dim(X) \geq 3$ then $\pi_1(X) \cong \pi_1(Y)$ for any positive hypersurface $Y \subset X$. See [50] for Lefschetz theorems for subvarieties of higher codimensions.
   - For the algebraic version of Serre's theorem 5.2.7 see [66, Ch.III, Thm.5.2]. In fact, E could be an arbitrary coherent sheaf.
   - A proof of Grothendieck's lemma can be found in many standard text books. The original source is [63]. Indecomposable vector bundles on elliptic curves have been investigated by Atiyah. The results of [4] are central for ongoing research in mathematical physics (F-theory).

5.3 Kodaira Embedding Theorem

Not every compact complex manifold is Kähler and not every compact Kähler manifold is projective. Of course, one would like to have a criterion that decides whether a Kähler manifold is projective. Such a criterion is provided by the Kodaira embedding theorem which will be proved now. Roughly, it suffices to be able to describe the Kähler cone inside $H^2(X, \mathbb{R})$ in order to decide whether an ample line bundle exists. The analogous question, which complex manifolds are in fact Kähler, is essentially open. In particular, it is not known whether being Kähler is a purely topological property.

Let L be a holomorphic line bundle on a compact complex manifold X. For any choice of a basis $s_0, \ldots, s_N \in H^0(X, L)$ there exists a natural rational map $\varphi_L : X \dashrightarrow \mathbb{P}^N$ given by $x \mapsto (s_0(x) : \ldots : s_N(x))$ (cf. Proposition 2.3.26). When does this map define a closed embedding of X? Clearly, the answer to this question does not depend on the choice of the basis.

i) The rational map φ_L is a morphism, i.e. everywhere defined on X, if and only if for any $x \in X$ there exists at least one section $s \in H^0(X, L)$ with $s(x) \neq 0$ or, equivalently, if and only if $\text{Bs}(L) = \varnothing$ (see Definition 2.3.25).

In other words, φ_L is a morphism if and only if for any $x \in X$ the natural restriction map

$$H^0(X, L) \longrightarrow L(x)$$

is surjective. Notice that this map sits in the long exact cohomology sequence associated to

$$0 \longrightarrow L \otimes \mathcal{I}_{\{x\}} \longrightarrow L \longrightarrow L(x) \longrightarrow 0,$$

where $\mathcal{I}_{\{x\}}$ denotes the ideal sheaf of the point $x \in X$.

ii) Suppose that φ_L is defined everywhere. Then φ_L is injective if and only if for two two arbitrary distinct points $x_1 \neq x_2 \in X$ there exists a section $s \in H^0(X, L)$ with $s(x_1) = 0$ and $s(x_2) \neq 0$. One says that φ_L (or L) *separates points*.

Together with the criterion i) we find that the complete linear system $|L|$ defines an injective morphism if and only if for any two distinct points $x_1, x_2 \in X$ the restriction map

$$H^0(X, L) \longrightarrow L(x_1) \oplus L(x_2)$$

is surjective. Again, this map is induced by a short exact sequence:

$$0 \longrightarrow L \otimes \mathcal{I}_{\{x_1, x_2\}} \longrightarrow L \longrightarrow L(x_1) \oplus L(x_2) \longrightarrow 0.$$

iii) Assume that $\varphi_L : X \to \mathbb{P}^N$ is an injective morphism. In order to ensure that φ_L is a closed embedding, one has to check that for any $x \in X$

the differential $d\varphi_{L,x} : T_x X \to T_{\varphi(x)}\mathbb{P}^N$ is injective. Choose a section $s_0 \in H^0(X, L)$ with $s_0(x) \neq 0$. Then we may find a basis $s_0, \ldots, s_N$ of $H^0(X, L)$ such that $s_i(x) = 0$ for $i > 0$. Thus, locally around $x \in X$ the map φ_L is given by $X \to \mathbb{C}^N$, $y \mapsto (t_1(y), \ldots, t_N(y))$ with $t_1(x) = \ldots = t_N(x) = 0$. Here $t_i = s_i/s_0$. Hence, $d\varphi_{L,x}$ is injective if and only if the one-forms $dt_1, \ldots, dt_N$ span the cotangent space $\bigwedge^1_x X$ in x.

Let us reformulate this as follows. The sections $s_1, \ldots, s_N$ form a basis of the subspace $H^0(X, L \otimes \mathcal{I}_{\{x\}}) \subset H^0(X, L)$ of all global sections of L that vanish in x. On this subspace there exists a natural map

$$d_x : H^0(X, L \otimes \mathcal{I}_{\{x\}}) \longrightarrow L(x) \otimes \textstyle\bigwedge^1_x X,$$

which can be defined in terms of a local trivialization $\psi : L|_U \cong U \times \mathbb{C}$ as $s \mapsto d(\psi s)_x$. If we change the local trivialization ψ to $\lambda\psi$, then $d(\lambda\psi s)_x = \lambda(x)d(\psi s)_x$ for any s vanishing in x. Thus, d_x is independent of ψ.

Next, it is easy to see that $d(t_i)_x = (\psi s_0)^{-1}d(\psi s_i)_x$ for $i = 1, \ldots, n$. Hence, $d\varphi_L$ is injective in $x \in X$ if and only if

$$d_x : H^0(X, L \otimes \mathcal{I}_{\{x\}}) \longrightarrow L(x) \otimes \textstyle\bigwedge^1_x X$$

is surjective. As before, the map d_x is induced by a short exact sequence which in this case takes the form

$$0 \longrightarrow L \otimes \mathcal{I}^2_{\{x\}} \longrightarrow L \otimes \mathcal{I}_{\{x\}} \longrightarrow L(x) \otimes \textstyle\bigwedge^1_x X \longrightarrow 0.$$

We leave it to the reader to verify that $\mathcal{I}_{\{x\}}/\mathcal{I}^2_{\{x\}}$ is canonically isomorphic to $\bigwedge^1_x X$.

Summarizing the above discussion, we find that the complete linear system $|L|$ induces a closed embedding $\varphi : X \hookrightarrow \mathbb{P}^N$ if and only if the global sections of L separate points $x_1 \neq x_2 \in X$ and tangent directions $v \in T_x X$. These two properties can be rephrased as has been done in i) and ii) respectively iii).

Recall that a line bundle L on a compact complex manifold X is called ample if and only if L^k for some $k > 0$ defines a closed embedding $\varphi_L : X \hookrightarrow \mathbb{P}^N$.

Proposition 5.3.1 (Kodaira embedding theorem) *Let X be a compact Kähler manifold. A line bundle L on X is positive if and only if L is ample. In this case, the manifold X is projective.*

Before actually proving the proposition we need to study positivity of line bundles under blow-ups. Let X be a complex manifold and let L be a positive line bundle on X. Denote by $\sigma : \hat{X} \to X$ the blow-up of X in a finite number of distinct points $x_1, \ldots, x_\ell \in X$ and by E_j the exceptional divisors $\sigma^{-1}(x_j)$, $j = 1, \ldots, \ell$.

Lemma 5.3.2 *For any line bundle M on X and integers $n_1, \ldots, n_\ell > 0$ the line bundle $\sigma^*(L^k \otimes M) \otimes \mathcal{O}(-\sum n_j E_j)$ on $\hat{X}$ is positive for $k \gg 0$.*

Proof. In a neighbourhood $x_j \in U_j \subset X$ of each point x_j the blow-up can be seen as the incidence variety $\hat{U}_j = \mathcal{O}(-1) \subset U_j \times \mathbb{P}^{n-1}$. Moreover, $\mathcal{O}(E_j)$ is isomorphic to to $p_j^* \mathcal{O}(-1)$, where $p_j : \hat{U}_j \to \mathbb{P}^{n-1}$ is the second projection (see the proof of Proposition 2.5.6). Thus, we can endow $\mathcal{O}(-E_j)$ with the pullback of the natural Fubini–Study hermitian structure on $\mathcal{O}(1)$ (cf. Examples 4.1.2 and 4.3.12).

Gluing these hermitian structures (or rather their n_j-th powers) by means of a partition of unity yields an hermitian metric on $\mathcal{O}(-\sum n_j E_j)$. Locally near any E_j the curvature F_∇ of the Chern connection on $\mathcal{O}(-\sum n_j E_j)$ with respect to this hermitian structure is $-n_j(2\pi i)p_j^* \omega_{\mathrm{FS}}$, where ω_{FS} is the Fubini–Study Kähler form on $\mathbb{P}^{n-1}$.

Thus, F_∇ semi-positive locally around each E_j and strictly positive for all tangent directions of E_j itself. Indeed, the curvature is compatible with pull-back and the curvature of the Chern connection of the Fubini–Study hermitian product on $\mathcal{O}(1)$ is $-(2\pi i)\omega_{\mathrm{FS}}$. (Note that $p_j^* \omega_{\mathrm{FS}}$ depends on the chosen local coordinates on U_j used to realize $\hat{U}_j$ inside $U_j \times \mathbb{P}^{n-1}$. But all positivity considerations are not affected by this ambiguity.)

For any real $(1,1)$-forms α and β on X with α positive the positivity property of F_∇ immediately implies that the form $\sigma^*(k \cdot \alpha + \beta) + (i/2\pi)F_\nabla$ is a positive form on $\hat{X}$ for $k \gg 0$.

To conclude, one chooses α and β such that $[\alpha] = c_1(L)$ and $[\beta] = c_1(M)$. $\square$

Let us now turn to the proof of the Kodaira embedding theorem.

Proof. If L is ample then for some $k \gg 0$ the line bundle L^k is isomorphic to the restriction of $\mathcal{O}(1)$ under an embedding $X \subset \mathbb{P}^N$. Thus, $c_1(L)$ is, up to a positive scalar, given by the restriction of the Fubini–Study Kähler form and thus positive. In other words, any ample line bundle is positive.

For the converse, we will show that a high power L^k of a positive line bundle L defines a closed embedding $\varphi_{L^k} : X \hookrightarrow \mathbb{P}^N$.

Let us first prove the injectivity φ_{L^k}. If $\sigma : \hat{X} \to X$ is the blow-up of X in $x \in X$ we denote the exceptional divisor $\sigma^{-1}(x)$ by E. Consider the commutative diagram:

$$
\begin{array}{ccc}
H^0(X, L^k) & \longrightarrow & L^k(x) \\
\downarrow & & \downarrow \wr \\
H^0(\hat{X}, \sigma^* L^k) & \longrightarrow & H^0(E, \mathcal{O}_E) \otimes L^k(x)
\end{array}
$$

The vertical map on the left is given by pulling back sections of L^k to $\sigma^* L^k$ on $\hat{X}$. Since the blow-up map is surjective, this map is injective. We shall show that it is in fact bijective.

If X is one-dimensional, $\sigma : \hat{X} \to X$ is an isomorphism and $H^0(X, L^k) \to H^0(\hat{X}, L^k)$ is clearly bijective.

If $\dim(X) \geq 2$ then, due to Hartogs' theorem (see Exercise 2.2.6), any section $s \in H^0(\hat{X}, \sigma^* L^k)$ can first be restricted to $\hat{X} \setminus E = X \setminus x$ and then extended to a global section of L^k. Thus, also in this case the pull-back $H^0(X, L^k) \to H^0(\hat{X}, L^k)$ is bijective.

The cokernel of $H^0(\hat{X}, \sigma^* L^k) \to H^0(E, \mathcal{O}_E) \otimes L^k(x)$ is contained in $H^1(\hat{X}, \sigma^* L^k \otimes \mathcal{O}(-E))$, as it is induced by the short exact sequence

$$0 \longrightarrow \sigma^* L^k \otimes \mathcal{O}(-E) \longrightarrow \sigma^* L^k \longrightarrow \sigma^* L^k|_E \longrightarrow 0$$
$$\downarrow \wr$$
$$L^k(x) \otimes \mathcal{O}_E$$

If $n = \dim(X)$, Proposition 2.5.3 shows $K_{\hat{X}} \cong \sigma^* K_X \otimes \mathcal{O}((n-1)E)$. Hence, by Lemma 5.3.2 the line bundle $L' := \sigma^* L^k \otimes K_{\hat{X}}^* \otimes \mathcal{O}(-E) = \sigma^*(L^k \otimes K_X^*) \otimes \mathcal{O}(-nE)$ is positive for $k \gg 0$. Thus, by Kodaira vanishing 5.2.2 one finds for $k \gg 0$:

$$H^1(\hat{X}, \sigma^* L^k \otimes \mathcal{O}(-E)) = H^1(\hat{X}, K_{\hat{X}} \otimes L') = 0.$$

Therefore, $x \in X$ is not a base-point of L^k for $k \gg 0$.

Using the map $H^0(X, L^{2^\ell}) \to H^0(X, L^{2^{\ell+1}})$, $s \mapsto s^2$ one finds $\mathrm{Bs}(L) \supset \ldots \supset \mathrm{Bs}(L^{2^\ell}) \supset \mathrm{Bs}(L^{2^{\ell+1}}) \supset \ldots$, which is a decreasing sequence of compact subsets with empty intersection. Hence, the line bundle L^k is base-point free for $k = 2^\ell \gg 0$. (In fact, with a bit more work one can even show that this holds true for any $k \gg 0$, but we won't need this.)

A similar argument, using the blow-up $\hat{X} \to X$ in two distinct points $x_1 \neq x_2 \in X$ and working with $\sigma^* L^k \otimes \mathcal{O}(-E_1 - E_2)$, shows that also $H^0(X, L^k) \to L^k(x_1) \oplus L^k(x_2)$ is surjective for all $k \gg 0$ and $x_1 \neq x_2 \in X$.

It thus remains to check that L^k separates tangent directions for $k \gg 0$. In order to show this, we compare the two exact sequences

$$0 \longrightarrow \mathcal{I}_{\{x\}}^2 \longrightarrow \mathcal{I}_{\{x\}} \longrightarrow \bigwedge_x^1 X \longrightarrow 0$$

and

$$0 \longrightarrow \mathcal{O}(-2E) \longrightarrow \mathcal{O}(-E) \longrightarrow \mathcal{O}_E(-E) \longrightarrow 0.$$

Clearly, pulling-back functions that vanish of order one respectively two in x yields a commutative diagram

$$\begin{array}{ccc} \sigma^* \mathcal{I}_{\{x\}}^2 & \longrightarrow & \sigma^* \mathcal{I}_{\{x\}} \\ \downarrow & & \downarrow \\ \mathcal{O}(-2E) & \longrightarrow & \mathcal{O}(-E) \end{array}$$

Twisting by L^k and passing to the quotients we obtain the commutative diagram

$$
\begin{array}{ccc}
H^0(X, L^k \otimes \mathcal{I}_{\{x\}}) & \longrightarrow & L^k(x) \otimes \bigwedge_x^1 X \\
\downarrow & & \downarrow \\
H^0(\hat{X}, \sigma^* L^k(-E)) & \longrightarrow & L^k(x) \otimes H^0(E, \mathcal{O}_E(-E))
\end{array}
$$

As before, the vertical arrow on the left is an isomorphism. In order to see that also the other one is bijective we recall that $E \cong \mathbb{P}(\mathcal{N}_{\{x\}/X}) = \mathbb{P}(T_x X)$ and $\mathcal{O}_E(-E) = \mathcal{O}(1)$ (cf. Corollary 2.5.6).

Hence, $H^0(E, \mathcal{O}_E(-E)) = H^0(\mathbb{P}(T_x X), \mathcal{O}(1)) = \bigwedge_x^1 X$. The map $\bigwedge_x^1 X \otimes \mathcal{O}_E \to \mathcal{O}_E(-E)$, which yields the vertical arrow on the right, is nothing but the evaluation map $\mathcal{O}^{\oplus n} \twoheadrightarrow \mathcal{O}_{\mathbb{P}^{n-1}}(1)$ and thus also surjective on the level of global sections.

Once the surjectivity of the vertical maps is shown, it suffices to prove that $H^1(\hat{X}, \sigma^* L^k \otimes \mathcal{O}(-2E)) = 0$ for $k \gg 0$, which again follows from Kodaira vanishing by applying Lemma 5.3.2 as before. $\square$

The projectivity of a compact Kähler manifold can now be read off the position of the Kähler cone $\mathcal{K}_X \subset H^2(X, \mathbb{R})$ relative to the integral lattice $\mathrm{Im}(H^2(X, \mathbb{Z}) \subset H^2(X, \mathbb{R}))$.

Corollary 5.3.3 *A compact Kähler manifold X is projective if and only if $\mathcal{K}_X \cap H^2(X, \mathbb{Z}) \neq \emptyset$.*

Proof. By abuse of notation we write $\mathcal{K}_X \cap H^2(X, \mathbb{Z})$ instead of $\mathcal{K}_X \cap \mathrm{Im}(H^2(X, \mathbb{Z}) \to H^2(X, \mathbb{R}))$. The Kähler cone $\mathcal{K}_X$ is by definition the cone of all Kähler classes on X and hence contained in $H^{1,1}(X)$. Thus, a class $\alpha \in \mathcal{K}_X \cap H^2(X, \mathbb{Z})$ is in particular of type $(1,1)$. Hence by Proposition 3.3.2 one has $\alpha = c_1(L)$ for some line bundle L. This line bundle L is positive and, therefore, by the Kodaira embedding theorem also ample. Hence, X is projective.

Conversely, if X is projective then $c_1(\mathcal{O}(1)|_X)$ for any projective embedding $X \subset \mathbb{P}^N$ yields a class in $\mathcal{K}_X \cap H^2(X, \mathbb{Z})$. $\square$

Definition 5.3.4 A class in $\mathcal{K}_X \cap H^2(X, \mathbb{Z})$ is a *Hodge class*.

For complex tori the projectivity criterion takes the following very precise form (compare with Proposition 3.C.11).

Corollary 5.3.5 *Let $X = V/\Gamma$ be a complex torus. Then X is projective if and only if there exists a Riemann form, i.e. an alternating bilinear form $\omega : V \times V \to \mathbb{R}$ such that*
i) $\omega(iu, iv) = \omega(u, v)$,
ii) $\omega(\ , i(\ ))$ *is positive definite, and*
iii) $\omega(u, v) \in \mathbb{Z}$ *whenever $u, v \in \Gamma$.*

Proof. Clearly, the Riemann form ω can be considered as a constant two-form. Recall that $\bigwedge^2 V^* \cong H^2(X, \mathbb{R})$. Conditions i) and ii) ensure that ω is in fact a Kähler form and iii) is equivalent to $\omega \in H^2(X, \mathbb{Z})$. $\square$

Examples 5.3.6 i) Any compact complex curve is projective. Indeed, any curve is Kähler and since $H^{1,1}(X) = H^2(X, \mathbb{C})$ the open subset $\mathcal{K}_X \subset H^2(X, \mathbb{R})$ contains integral classes. Of course, a direct proof without using Kodaira's theorems can also be given.

ii) Every compact Kähler manifold X with $H^{0,2}(X) = 0$ is projective. This is in fact a very useful statement which applies to many Calabi–Yau manifolds.

Corollary 5.3.7 *If X is projective, the natural homomorphism* $\mathrm{Div}(X) \to \mathrm{Pic}(X)$ *(see Section 2.3) is surjective.*

Proof. Let $L \in \mathrm{Pic}(X)$ be an ample line bundle. In particular, L is positive and we may apply Serre vanishing 5.2.7. If $M \in \mathrm{Pic}(X)$ is any line bundle then $\chi(X, M \otimes L^k) = h^0(X, M \otimes L^k)$ for $k \gg 0$.

On the other hand, the Hirzebruch–Riemann–Roch formula (5.2) shows that $\chi(X, M \otimes L^k)$ is a polynomial in k of degree $n := \dim(X)$ with leading coefficient $(1/n!) \int_X c_1(L)^n$. The latter is positive, as L is positive, and thus one obtains $h^0(X, M \otimes L^k) \neq 0$ for $k \gg 0$. Applied to $M = \mathcal{O}$ we find in particular that $H^0(X, L^k) \neq 0$ for $k \gg 0$ (which is actually clear from the definition of 'ample'). Following the discussion in Section 2.3 one sees that $M \otimes L^k$ and L^k are both contained in the image of the homomorphism $\mathrm{Div}(X) \to \mathrm{Pic}(X)$ and so is M.

More explicitly, if $0 \neq s_1 \in H^0(X, M \otimes L^k)$ and $0 \neq s_2 \in H^0(X, L^k)$ then $\mathcal{O}(Z(s_1) - Z(s_2)) \cong M$. $\square$

The reader may notice that we actually only used the easy direction of the Kodaira embedding theorem, namely that an ample line bundle is positive and thus satisfies the assumption of the Kodaira vanishing theorem.

The argument in the proof shows as well that any line bundle on a projective manifold admits non-trivial meromorphic sections, namely s_1/s_2 with the above notation.

The corollary also shows that the Neron–Severi group $\mathrm{NS}(X) = H^{1,1}(X, \mathbb{Z})$ of a projective manifold is indeed spanned by the fundamental classes of divisors. See Sections 2.3 and 3.3.

Exercises

5.3.1   Show that a complex torus $\mathbb{C}^2/\Gamma$ is abelian, i.e. projective, if and only if there exists a line bundle L with $\int c_1^2(L) > 0$. (In fact, this criterion is valid for any compact complex surface, but more difficult to prove in general [8].)

5.3.2 Show that any vector bundle E on a projective manifold X can be written as a quotient $(L^k)^{\oplus \ell} \twoheadrightarrow E$ with L an ample line bundle, $k \ll 0$ and $\ell \gg 0$. (Imitate the first step of the proof of Proposition 5.3.1, which shows the assertion for $E = \mathcal{O}$. You will also need to revisit Proposition 5.2.7.)

5.3.3 Let $\sigma : \hat{X} \to X$ be the blow-up of X in $x \in X$ with the exceptional divisor E and let L be an ample line bundle on X. Show that $\sigma^* L^k \otimes \mathcal{O}(-E)$ is ample for $k \gg 0$.

(This is obtained by a revision of the proof of the Kodaira embedding theorem. In particular, the blow-up $\hat{X}$ of a projective manifold X is again projective.)

5.3.4 Continue Exercise 5.1.6 and show, by using the techniques of this section, that any line bundle L of degree $\deg(L) > 2g$ on a compact curve C of genus g is very ample, i.e. the linear system associated with L embeds C.

Conclude that any elliptic curve is isomorphic to a plane curve, i.e. to a hypersurface in $\mathbb{P}^2$.

5.3.5 Show that there exists a complex torus X of dimension n such that X is projective and, therefore, $a(X) = n$.

Comments: - The Kodaira embedding theorem is complemented by a theorem of Chow saying that any complex submanifold $X \subset \mathbb{P}^N$ can be described as the zero set of finitely many homogeneous polynomials. This, eventually, establishes the precise relation between our notion of projective complex manifolds and the notion of smooth projective varieties (over $\mathbb{C}$) used in algebraic geometry.

The proof of Chow's theorem is not difficult, see e.g. [59]. It works more generally for analytic subvarieties and uses, besides some elementary linear projections etc., the proper mapping theorem. The proof for analytic hypersurfaces can easily be deduced from $\mathrm{Pic}(\mathbb{P}^N) = \mathbb{Z} \cdot \mathcal{O}(1)$ and the description of all global holomorphic sections of $\mathcal{O}(1)$.

We also mention the deep result of Moishezon showing that a compact complex manifold X is projective if and only if X is Kähler and Moishezon (cf. Exercise 5.2.11). A proof using singular metrics can be found in [35].

- There is yet another relevant result dealing with the embedded manifold $X \subset \mathbb{P}^N$. Namely, Bertini's theorem asserting that the generic hyperplane section of X, i.e. the generic divisor in the linear system defined by $\mathcal{O}(1)|_X$, is smooth. For its algebro-geometric proof see [66] or [99].

- It might be worth emphasizing that in the proof of the Kodaira embedding theorem blow-ups are used in order to pass from a high codimension situation (e.g. a point x in X) and the non-locally free sheaf that comes with it (the ideal sheaf $\mathcal{I}_x$) to a codimension one situation to which cohomological methods can be applied by viewing the ideal sheaf of the exceptional divisor as a holomorphic line bundle.

6

Deformations of Complex Structures

We have seen many examples of complex manifolds. Some of them, like the projective space $\mathbb{P}^n$, seemed rather unique and 'rigid', others, like complex tori, never come alone, as other complex manifolds seem to exist close by.

Classifying all complex structures, e.g. on a real torus $(S^1 \times S^1)^n$, eventually leads to the theory of moduli spaces. Riemann himself brought to light the most fascinating aspect: The set of all complex structures on a given differentiable manifold comes itself with a natural differentiable and in fact complex structure.

This chapter has the modest aim to acquaint the reader with certain local aspects of this theory. The two sections treat the deformation theory of complex manifolds, but each from a slightly different angle.

Section 6.1 considers all almost complex structures on a given differentiable manifold. A deformation of a certain integrable almost complex structure is studied by a power series expansion of this linear operator. It turns out, that the recursive equations for the coefficients imposed by the integrability condition cannot always be solved. For the important class of Calabi–Yau manifolds however a (formal) solution exists due to the Tian–Todorov lemma 6.1.9. Appendix 6.A presents a formalized approach to this circle of ideas using the language of differential algebras. This has recently been promoted by Kontsevich, Manin and others. The underlying philosophy to study deformations in terms of differential graded Lie algebras goes back to Deligne. Section 6.2 surveys a number of rather deep theorems mostly due to Kodaira and collaborators. Here, a deformation is viewed as a complex manifold fibred over a base space, the fibres being deformations of a distinguished fibre isomorphic to a given complex manifold.

6.1 The Maurer–Cartan Equation

A complex manifold is by definition a differentiable manifold M endowed with an atlas whose transition functions are holomorphic. As we have seen, a complex structure defined by such an atlas can also be encoded by an integrable almost complex structure I, which is just an endomorphism of the tangent bundle satisfying a certain non-linear integrability condition. In this section, we will work with the latter description of complex structures. We will explain how to view the various complex structures I on M or, more precisely, the (small) deformations of a given one.

In order to study all complex structures, one might naively consider just the subset of the endomorphisms of the tangent bundle that define integrable almost complex structures, but many of them will define isomorphic complex manifolds. Recall that two complex manifolds (M, I) and (M', I') are isomorphic if there exists a diffeomorphism $F : M \to M'$ such that $dF \circ I = I' \circ dF$. Thus, the set of diffeomorphism classes of complex structures I on a fixed differentiable manifold M is the quotient of the set

$$\mathcal{A}_c(M) := \{I \mid I = \text{integrable almost complex structure}\}$$

of all complex structures by the action of the diffeomorphism group

$$\text{Diff}(M) \times \mathcal{A}_c(M) \longrightarrow \mathcal{A}_c(M), \quad (F, I) \longmapsto dF \circ I \circ (dF)^{-1}.$$

The description of the space of isomorphism classes of complex structures comprises therefore two things:

i) Describe the subset of integrable almost complex structures inside the space of all almost complex structures on M.

ii) Divide out by the action of the diffeomorphism group $\text{Diff}(M)$.

In this section we will deal with these questions on an infinitesimal level, i.e. we shall consider the power series expansion of any deformation of a complex structure and study the coefficients of it.

We start out with the set

$$\mathcal{A}_{ac}(M) := \{I \mid I^2 = -\text{id}\} \subset \text{End}(TM)$$

of all almost complex structures on M. Although we will not discuss this in any detail here, the space $\mathcal{A}_{ac}(M)$ is a nice space, i.e. (after completion) it is an infinite dimensional manifold. This, in general, is no longer true for the subspace $\mathcal{A}_c(M) \subset \mathcal{A}_{ac}(M)$ of integrable almost complex structures.

Let us first consider an arbitrary almost complex structure $I \in \mathcal{A}_{ac}(M)$. Recall that an almost complex structure I is uniquely determined by a decomposition of the tangent bundle $T_{\mathbb{C}}M = T^{1,0} \oplus T^{0,1}$ with $I = i \cdot \text{id}$ on $T^{1,0}$ and $I = -i \cdot \text{id}$ on $T^{0,1}$. In fact, giving $T^{0,1} \subset T_{\mathbb{C}}M$ is enough, for $T^{1,0} = \overline{T^{0,1}}$.

If $I(t)$ is a continuous family of almost complex structures with $I(0) = I$, one has a continuous family of such decompositions $T_{\mathbb{C}}M = T_t^{1,0} \oplus T_t^{0,1}$ or, equivalently, of subspaces $T_t^{0,1} \subset T_{\mathbb{C}}M$.

Thus, for small t the deformation $I(t)$ of I can be encoded by a map

$$\phi(t) : T^{0,1} \longrightarrow T^{1,0} \text{ with } v + \phi(t)(v) \in T_t^{0,1}.$$

We write $T^{1,0}$ and $T^{0,1}$ for the subbundles defined by I. Explicitly, one has

$$\phi(t) = -\mathrm{pr}_{T_t^{1,0}} \circ j,$$

where $j : T^{0,1} \subset T_{\mathbb{C}}$ and $\mathrm{pr}_{T_t^{1,0}} : T_{\mathbb{C}} \to T_t^{1,0}$ are the natural inclusion respectively projection.

Conversely, if $\phi(t)$ is given, then one defines for small t

$$T_t^{0,1} := (\mathrm{id} + \phi(t))(T^{0,1}).$$

The condition t be 'small' has to be imposed in order to ensure that with this definition $T_t^{0,1}$ has the right dimension. More precisely, t small means that $T_t^{0,1} \subset T_{\mathbb{C}} \to T^{0,1}$ is an isomorphism, which is certainly an open condition, at least on a compact manifold. To avoid problems caused by non-compactness we will henceforth assume that M is compact.

Let us now consider the power series expansion

$$\phi(t) = \phi_0 + \phi_1 t + \phi_2 t^2 + \dots$$

of a given ϕ. Then $\phi_0 = 0$ and the higher order coefficients $\phi_{i>0}$ will be regarded as sections of $\bigwedge^{0,1} \otimes T^{1,0}$.

Next, we shall study the integrability of the deformed almost complex structures $I(t)$. From now on we will assume that I is integrable and we will denote the complex manifold (M, I) by X. In particular, the higher order coefficients ϕ_i correspond to elements in $\mathcal{A}^{0,1}(T_X)$.

The integrability condition for $I(t)$ can be expressed as (cf. Proposition 2.6.17)

$$[T_t^{0,1}, T_t^{0,1}] \subset T_t^{0,1}.$$

Let us rephrase this in terms of $\phi(t)$. First note that there exists the $\bar{\partial}$-operator on the holomorphic tangent bundle T_X of $X = (M, I)$, which can in particular be applied to $\phi(t) \in \mathcal{A}^{0,1}(T_X)$. Secondly, one defines a 'Lie bracket'

$$[\ , \ ] : \mathcal{A}^{0,p}(T_X) \times \mathcal{A}^{0,q}(T_X) \longrightarrow \mathcal{A}^{0,p+q}(T_X)$$

by taking the Lie bracket in T_X and the exterior product in the form part in $\mathcal{A}^{0,*}$. Spelled out in local coordinates this gives for $\alpha = \sum d\bar{z}_I \otimes v_I$ and $\beta = \sum d\bar{z}_J \otimes w_J$ with $v_I = \sum \alpha_{Ij} \frac{\partial}{\partial z_j}$ and $w_J = \sum \beta_{J\ell} \frac{\partial}{\partial z_\ell}$ the following formula

$$[\alpha, \beta] = \sum (d\bar{z}_I \wedge d\bar{z}_J)[v_I, w_J] = \sum_{I,j,J,\ell} (d\bar{z}_I \wedge d\bar{z}_J) \left[\alpha_{Ij} \frac{\partial}{\partial z_j}, \beta_{J\ell} \frac{\partial}{\partial z_\ell} \right]. \quad (6.1)$$

This is indeed well-defined, as any holomorphic coordinate change would change the tangent vector fields v_I, w_J by antiholomorphic factors $\bar{f}$ and $\frac{\partial \bar{f}}{\partial z_m} = 0$.

Remark 6.1.1 Using the local description, one easily checks that $\bar\partial[\alpha, \beta] = [\bar\partial\alpha, \beta] \pm [\alpha, \bar\partial\beta]$. In particular, the Lie bracket of two $\bar\partial$-closed sections in $\mathcal{A}^{0,*}(\mathcal{T}_X)$ is again closed and the bracket of a closed one and an exact one is exact. Thus, the bracket induces a map

$$H^p(X, \mathcal{T}_X) \times H^q(X, \mathcal{T}_X) \longrightarrow H^{p+q}(X, \mathcal{T}_X).$$

Lemma 6.1.2 *The integrability equation* $[T_t^{0,1}, T_t^{0,1}] \subset T_t^{0,1}$ *is equivalent to the* Maurer–Cartan *equation*

$$\bar\partial\phi(t) + [\phi(t), \phi(t)] = 0. \tag{6.2}$$

Proof. The assertion can be proved locally. We will omit the parameter t in the notation and shall write

$$\phi = \sum_{i,j} \phi_{ij} d\bar z_i \otimes \frac{\partial}{\partial z_j}.$$

If the integrability condition holds true, then we have in particular

$$\left[\frac{\partial}{\partial\bar z_i} + \phi\left(\frac{\partial}{\partial\bar z_i}\right), \frac{\partial}{\partial\bar z_k} + \phi\left(\frac{\partial}{\partial\bar z_k}\right)\right] \in T_t^{0,1}.$$

Since $[\frac{\partial}{\partial\bar z_i}, \frac{\partial}{\partial\bar z_k}] = 0$, this is equivalent to

$$\left(\sum_\ell \left[\frac{\partial}{\partial\bar z_i}, \phi_{k\ell}\frac{\partial}{\partial z_\ell}\right] + \sum_j \left[\phi_{ij}\frac{\partial}{\partial z_j}, \frac{\partial}{\partial\bar z_k}\right] + \sum_{j,\ell} \left[\phi_{ij}\frac{\partial}{\partial z_j}, \phi_{k\ell}\frac{\partial}{\partial z_\ell}\right]\right) \in T_t^{0,1}.$$

Now,

$$\left[\frac{\partial}{\partial\bar z_i}, \phi_{k\ell}\frac{\partial}{\partial z_\ell}\right] = \frac{\partial\phi_{k\ell}}{\partial\bar z_i}\frac{\partial}{\partial z_\ell} \quad\text{and}\quad \left[\phi_{ij}\frac{\partial}{\partial z_j}, \frac{\partial}{\partial\bar z_k}\right] = -\frac{\partial\phi_{ij}}{\partial\bar z_k}\frac{\partial}{\partial z_j}$$

and, hence,

$$\sum_\ell \left[\frac{\partial}{\partial\bar z_i}, \phi_{k\ell}\frac{\partial}{\partial z_\ell}\right] + \sum_j \left[\phi_{ij}\frac{\partial}{\partial z_j}, \frac{\partial}{\partial\bar z_k}\right]$$

$$= \sum_j \left(\frac{\partial\phi_{kj}}{\partial\bar z_i} - \frac{\partial\phi_{ij}}{\partial\bar z_k}\right)\frac{\partial}{\partial z_j}$$

$$= (\bar\partial\phi)\left(\frac{\partial}{\partial\bar z_i}, \frac{\partial}{\partial\bar z_k}\right).$$

On the other hand,

$$\sum_{j,\ell} \left[\phi_{ij}\frac{\partial}{\partial z_j}, \phi_{k\ell}\frac{\partial}{\partial z_\ell}\right] = [\phi, \phi]\left(\frac{\partial}{\partial\bar z_i}, \frac{\partial}{\partial\bar z_k}\right).$$

Therefore, the integrability condition implies that $\bar\partial\phi + [\phi,\phi] \in \mathcal{A}^{0,2}(T^{1,0}X)$ satisfies

$$\left(\bar\partial\phi + [\phi,\phi]\right)\left(\frac{\partial}{\partial\bar z_i}, \frac{\partial}{\partial\bar z_k}\right) \in T_t^{0,1}.$$

Hence, $\bar\partial\phi + [\phi,\phi] \in \mathcal{A}^{0,2}(T^{1,0}X \cap T_t^{0,1})$. But for small t one has $T^{1,0}X \cap T_t^{0,1} = 0$ and, therefore, $\bar\partial\phi + [\phi,\phi] = 0$.

Conversely, if the Maurer–Cartan equation (6.2) holds true, then the integrability condition is satisfied for a local frame of $T_t^{0,1}$. This immediately shows that it is true for all sections of $T_t^{0,1}$. □

Let us now consider the power series expansion of the Maurer–Cartan equation, i.e. we replace ϕ in (6.2) by $\phi = \sum_{i=1}^{\infty} \phi_i t^i$:

$$\bar\partial\left(\sum_{i=1}^{\infty} \phi_i t^i\right) + \sum_{i,j=1}^{\infty} [\phi_i, \phi_j] \, t^{i+j} = 0.$$

This yields a recursive system of equations:

$$0 = \bar\partial\phi_1 \qquad\qquad\qquad (6.3)$$
$$0 = \bar\partial\phi_2 + [\phi_1, \phi_1]$$
$$\vdots$$
$$0 = \bar\partial\phi_k + \sum_{0<i<k} [\phi_i, \phi_{k-i}].$$

We shall first study the first-order equation $\bar\partial\phi_1 = 0$ in detail. The Maurer–Cartan equation is, in particular, saying that the first-order deformation of the complex structure I is described by a $\bar\partial$-closed $(0,1)$-form ϕ_1 with values in the holomorphic tangent bundle $\mathcal{T}_X$. Thus, it defines an element $[\phi_1] \in H^1(X, \mathcal{T}_X)$.

Definition 6.1.3 The *Kodaira–Spencer class* of a one-parameter deformation I_t of the complex structure I is the induced cohomology class $[\phi_1] \in H^1(X, \mathcal{T}_X)$.

In order to identify isomorphic (first order) deformations, the (infinitesimal) action of the diffeomorphism group $\mathrm{Diff}(M)$ has to be taken into account. Let F_t be a one-parameter family of diffeomorphisms of the manifold M. With respect to local coordinate functions x_i this can be expanded into a power series $\tilde x_i = x_i + tF_i(x) + t^2 \ldots$. The first order term $\sum_i F_i \frac{\partial}{\partial x_i}$ defines a global vector field

$$\frac{dF_t}{dt}\Big|_{t=0} \in \mathcal{A}^0(TM).$$

Moreover, any global vector field is obtained in this way. If the diffeomorphism group $\mathrm{Diff}(M)$ is considered as an infinite-dimensional Lie group, this calculation identifies the tangent space of it with $\mathcal{A}^0(TM)$. (It is intuitively clear how to view $\mathrm{Diff}(M)$ a s a Lie group, but special attention has to be paid to the fact that everything is infinite-dimensional.)

We can rewrite this identification on the complex manifold $X = (M, I)$ with respect to holomorphic coordinates z_i as $\tilde{z}_i = z_i + t f_i(z, \bar{z}) + t^2 \dots$. For the antiholomorphic coordinates $\bar{z}_i$ we then have $\bar{\tilde{z}}_i = \bar{z}_i + t \bar{f}_i(z, \bar{z}) + t^2 \dots$. Thus, $\frac{dF}{dt}|_{t=0} \in \mathcal{A}^0(TX)$ in local coordinates (up to the factor 2) is given as

$$\sum f_i \frac{\partial}{\partial z_i} + \sum \bar{f}_i \frac{\partial}{\partial \bar{z}_i}.$$

Any diffeomorphism can be used to construct new complex structures out of the given one I by pushing-forward I via the differential of the diffeomorphism. More explicitly, for the diffeomorphism F_t the new complex structure is given by $dF_t \circ I \circ (dF_t)^{-1}$ (see the formula on page 256). The antiholomorphic tangent bundle of this complex structure is thus

$$T_t^{0,1} = dF_t(T^{0,1}X).$$

In local coordinates the first order deformation of $T^{0,1}X$ is described as the image of $T^{0,1}X$ under

$$\mathrm{id} + \sum_{i,j} \underbrace{\frac{\partial f_i}{\partial z_j} dz_j \otimes \frac{\partial}{\partial z_i}}_{=0 \text{ on } T^{0,1}X} + \frac{\partial f_i}{\partial \bar{z}_j} d\bar{z}_j \otimes \frac{\partial}{\partial z_i} + \underbrace{\frac{\partial \bar{f}_i}{\partial z_j} dz_j \otimes \frac{\partial}{\partial \bar{z}_i}}_{=0 \text{ on } T^{0,1}X} + \frac{\partial \bar{f}_i}{\partial \bar{z}_j} d\bar{z}_j \otimes \frac{\partial}{\partial \bar{z}_i}.$$

In this sum, only the third and the fifth terms are non-trivial on $T^{0,1}X$. Moreover, the fifth term is an endomorphism of $T^{0,1}$.

Thus, the first order deformation of $T^{0,1}X$ induced by a one-parameter family of diffeomorphisms F_t of M is described by the homomorphism

$$T^{0,1}X \longrightarrow T^{1,0}X \;, \quad \frac{\partial}{\partial \bar{z}_j} \longmapsto \sum_i \frac{\partial f_i}{\partial \bar{z}_j} \frac{\partial}{\partial z_i} \;.$$

This yields

Lemma 6.1.4 *The first-order deformation of the complex structure I on the manifold M (defining the complex manifold $X = (M, I)$) induced by a one-parameter family F_t of diffeomorphism of M is determined by the map*

$$\phi_1 = \bar{\partial}\left(\left(\tfrac{dF_t}{dt}|_{t=0}\right)^{1,0}\right) : T^{0,1}X \longrightarrow T^{1,0}X \;.$$

$\square$

As a consequence of this, we derive a description of all first-order deformations of isomorphism classes of complex structures of $X = (M, I)$. We will present another point of view on this result in the next section.

Proposition 6.1.5 *Let X be a complex manifold. There is a natural bijection between all first-order deformations of X and elements of $H^1(X, T_X)$.*

Proof. Write $X = (M, I)$. Then the first order deformations of I correspond to $\bar{\partial}$-closed elements in $\mathcal{A}^{0,1}(T_X)$ and by the previous lemma those that define isomorphic ones differ by elements of the image of $\bar{\partial} : \mathcal{A}^0(T_X) \to \mathcal{A}^{0,1}(T_X)$, i.e. $\bar{\partial}$-exact ones. As the Dolbeault cohomology just computes the sheaf cohomology of T_X, this proves the assertion $\square$

In other words, the Kodaira–Spencer class determines the first-order deformations of X and any class in $H^1(X, T_X)$ occurs as a Kodaira–Spencer class.

The principal task in deformation theory is to integrate given first-order deformations $v \in H^1(X, T_X)$, i.e. to find a one-parameter family I_t such that its Kodaira–Spencer class is v. In general this is not possible. Obstructions may occur at any order. Thus, it frequently happens that we find $\phi_1 t + \ldots + \phi_k t^k$, but a ϕ_{k+1} does not exist.

Moreover, since we have to divide by the action of the diffeomorphism group at any step, we have to study all possible ϕ_i carefully. This is needed in order to ensure the convergence of the power series and to solve the recursive equations.

That there in general exist obstructions can already be guessed at order two: Once a lift ϕ_1 of $v \in H^1(X, T_X)$ is chosen, we have to find ϕ_2 such that $\bar{\partial}\phi_2 = -[\phi_1, \phi_1]$. But it may very well happen that the form $[\phi_1, \phi_1]$ is never $\bar{\partial}$-exact, no matter how ϕ_1 is chosen.

Using Remark 6.1.1 we can associate to any first order deformation $v \in H^1(X, T_X)$ a class $[v, v] \in H^2(X, T_X)$.

Corollary 6.1.6 *A first-order deformation $v \in H^1(X, T_X)$ cannot be integrated if $[v, v] \in H^2(X, T_X)$ does not vanish.* $\square$

Next, we will describe how the recursive system of equations (6.3) given by the Maurer–Cartan equation can be solved when the underlying manifold X admits a holomorphic volume form, i.e. when X is a *Calabi–Yau manifold*. More precisely, by a Calabi–Yau manifold we mean here a compact Kähler manifold X of dimension n with trivial canonical bundle $\bigwedge^n \Omega_X = K_X \cong \mathcal{O}_X$. Moreover, we fix a holomorpic volume form, i.e. a trivializing section $\Omega \in H^0(X, K_X)$.

The form Ω can be used to define a natural isomorphism

$$\eta : \bigwedge^p T_X \cong \Omega_X^{n-p}.$$

As different sign conventions are possible here, we will try to be very concrete. One defines

$$\eta(v_1 \wedge \ldots \wedge v_p) = i_{v_1} \ldots i_{v_p}(\Omega),$$

where i_v is the contraction defined by $i_v(\alpha)(w_1, \ldots, w_i) = \alpha(v, w_1, \ldots, w_i)$. If Ω is locally written in the form $f dz_1 \wedge \ldots \wedge dz_n$, then

$$\eta \left(\frac{\partial}{\partial z_{i_1}} \wedge \ldots \wedge \frac{\partial}{\partial z_{i_p}} \right) = (-1)^{(\sum i_j)-p} f dz_1 \wedge \ldots \wedge \widehat{dz_{i_1}} \wedge \ldots \wedge \widehat{dz_{i_p}} \wedge \ldots dz_n,$$

for $i_1 < \ldots < i_p$.

Moreover, the isomorphism η induces canonical isomorphisms

$$\eta : \mathcal{A}^{0,q}(\bigwedge^p \mathcal{T}_X) \cong \mathcal{A}^{n-p,q}(X).$$

Definition 6.1.7 The operator $\Delta : \mathcal{A}^{0,q}(\bigwedge^p \mathcal{T}_X) \to \mathcal{A}^{0,q}(\bigwedge^{p-1} \mathcal{T}_X)$ is defined as

$$\Delta : \mathcal{A}^{0,q}(\bigwedge^p \mathcal{T}_X) \xrightarrow{\eta} \mathcal{A}^{n-p,q}(X) \xrightarrow{\partial} \mathcal{A}^{n-p+1,q}(X) \xrightarrow{\eta^{-1}} \mathcal{A}^{0,q}(\bigwedge^{p-1} \mathcal{T}_X) .$$

Note that ∂ itself is not well-defined on $\mathcal{A}^{0,q}(\bigwedge^p \mathcal{T}_X)$.

Warning: The operator Δ has nothing to do with the Laplacian. The notation is a bit unfortunate here, but historical. In fact, we have not (yet) even chosen a metric on X, so strictly speaking there is no Laplacian.

Lemma 6.1.8 *The operator Δ anti-commutes with $\bar{\partial}$, i.e. $\Delta \circ \bar{\partial} = -\bar{\partial} \circ \Delta$.*

Proof. We use the fact that $\Omega \in \mathcal{A}^{n,0}(X)$ is holomorphic, i.e. $\bar{\partial}$-closed. If we write in local coordinates $\Omega = f dz_1 \wedge \ldots \wedge dz_n$, then f is a holomorphic function. Then

$$\eta \left(g \cdot d\bar{z}_I \otimes \frac{\partial}{\partial z_J} \right) = \pm (fg) \cdot d\bar{z}_I \otimes dz_{\{1,\ldots,n\} \setminus J}.$$

Hence, $\bar{\partial} \circ \eta = \eta \circ \bar{\partial}$ and, therefore,

$$\begin{aligned} \bar{\partial}(\Delta(\alpha)) &= \eta^{-1}(\bar{\partial}\partial\eta(\alpha)) = -\eta^{-1}(\partial\bar{\partial}\eta(\alpha)) \\ &= -\eta^{-1}(\partial\eta(\bar{\partial}\alpha)) = -\Delta(\bar{\partial}(\alpha)). \end{aligned}$$

$\square$

Also note that $\Delta^2 = (\eta^{-1}\partial\eta)(\eta^{-1}\partial\eta) = 0$. Thus, Δ can be used as a differential, but it is not well-behaved with respect to the exterior product $\mathcal{A}^{0,p}(\bigwedge^r \mathcal{T}_X) \times \mathcal{A}^{0,q}(\bigwedge^s \mathcal{T}_X) \to \mathcal{A}^{0,p+q}(\bigwedge^{r+s} \mathcal{T}_X)$. This is made more precise by the following key result due to Tian and Todorov ([107, 108]).

Lemma 6.1.9 (Tian–Todorov lemma) *If $\alpha \in \mathcal{A}^{0,p}(\mathcal{T}_X)$ and $\beta \in \mathcal{A}^{0,q}(\mathcal{T}_X)$, then*

$$(-1)^p [\alpha, \beta] = \Delta(\alpha \wedge \beta) - \Delta(\alpha) \wedge \beta - (-1)^{p+1} \alpha \wedge \Delta(\beta).$$

Proof. As the claim is local and additive in α and β, it suffices to prove it for $\alpha = a d\bar{z}_I \otimes \frac{\partial}{\partial z_i}$ and $\beta = b d\bar{z}_J \otimes \frac{\partial}{\partial z_j}$.

Then, by definition, $[\alpha, \beta] = d\bar{z}_I \wedge d\bar{z}_J [a \frac{\partial}{\partial z_i}, b \frac{\partial}{\partial z_j}]$.

In a first step we shall reduce the assertion to the case $p = q = 0$. Using the short hand

$$G(\alpha, \beta) := \Delta(\alpha \wedge \beta) - \Delta(\alpha) \wedge \beta - (-1)^{p+1} \alpha \wedge \Delta(\beta),$$

one does this by proving

$$G(\alpha, \beta) = (-1)^p (d\bar{z}_I \wedge d\bar{z}_J) \otimes G\left(a\frac{\partial}{\partial z_i}, b\frac{\partial}{\partial z_j}\right).$$

Using the standard sign rules, one indeed computes

$$\Delta(\alpha\beta) = \eta^{-1}\partial\eta(\alpha \wedge \beta) = (-1)^q \eta^{-1}\partial\eta\left((d\bar{z}_I \wedge d\bar{z}_J) \otimes (a\frac{\partial}{\partial z_i} \wedge b\frac{\partial}{\partial z_j})\right)$$

$$= (-1)^q \eta^{-1}\partial\left((d\bar{z}_I \wedge d\bar{z}_J) \otimes \eta\left(a\frac{\partial}{\partial z_i} \wedge b\frac{\partial}{\partial z_j}\right)\right)$$

$$= (-1)^q \cdot (-1)^{p+q}\eta^{-1}\left((d\bar{z}_I \wedge d\bar{z}_J) \otimes \partial\eta\left(a\frac{\partial}{\partial z_i} \wedge b\frac{\partial}{\partial z_j}\right)\right)$$

$$= (-1)^p (d\bar{z}_I \wedge d\bar{z}_J) \otimes \Delta\left(a\frac{\partial}{\partial z_i} \wedge b\frac{\partial}{\partial z_j}\right)$$

Similarly,

$$\Delta(\alpha) \wedge \beta = \eta^{-1}\partial\eta\left(d\bar{z}_I \otimes \eta\left(a\frac{\partial}{\partial z_i}\right)\right) \wedge \beta = (-1)^p \left(d\bar{z}_I \otimes \Delta\left(a\frac{\partial}{\partial z_i}\right)\right) \wedge \beta$$

$$= (-1)^p (d\bar{z}_I \wedge d\bar{z}_J) \otimes \left(\Delta\left(a\frac{\partial}{\partial z_i}\right) \wedge b\frac{\partial}{\partial z_j}\right)$$

and

$$(-1)^{p+1}\alpha \wedge \Delta(\beta) = (-1)^{p+1}\alpha \wedge \left((-1)^q d\bar{z}_J \otimes \Delta\left(b\frac{\partial}{\partial z_j}\right)\right)$$

$$= (-1)^p (-1)(d\bar{z}_I \wedge d\bar{z}_J) \otimes \left(a\frac{\partial}{\partial z_i} \wedge \Delta\left(b\frac{\partial}{\partial z_j}\right)\right)$$

Thus,

$$G(\alpha, \beta) = (-1)^p (d\bar{z}_I \wedge d\bar{z}_J) \otimes \left(\Delta\left(a\frac{\partial}{\partial z_i} \wedge b\frac{\partial}{\partial z_j}\right)\right.$$

$$\left. - \Delta\left(a\frac{\partial}{\partial z_i}\right) \wedge b\frac{\partial}{\partial z_j} - (-1)a\frac{\partial}{\partial z_i} \wedge \Delta\left(b\frac{\partial}{\partial z_j}\right)\right)$$

$$= (-1)^p (d\bar{z}_I \wedge d\bar{z}_J) \otimes G\left(a\frac{\partial}{\partial z_i}, b\frac{\partial}{\partial z_j}\right)$$

In order to treat the case $p = q = 0$, let us first mention explicitly

$$\eta\left(\frac{\partial}{\partial z_i}\right) = (-1)^i f\, dz_1 \wedge \ldots \wedge \widehat{dz_i} \wedge \ldots \wedge dz_n$$

and

$$\eta\left(\frac{\partial}{\partial z_i} \wedge \frac{\partial}{\partial z_j}\right) = (-1)^{i+j-2} f dz_1 \wedge \ldots \wedge \widehat{dz_i} \wedge \ldots \wedge \widehat{dz_j} \wedge \ldots \wedge dz_n.$$

This also shows $\eta^{-1}(dz_1 \wedge \ldots \wedge \widehat{dz_i} \wedge \ldots \wedge dz_n) = f^{-1}\frac{\partial}{\partial z_i}$ and $\eta(dz_1 \wedge \ldots \wedge dz_n) = f^{-1}$.

Hence,

$$\Delta\left(a\frac{\partial}{\partial z_i}\right) = \eta^{-1}\partial\left((-1)^{i-1}(af)dz_1 \wedge \ldots \wedge \widehat{dz_i} \wedge \ldots \wedge dz_n\right)$$

$$= \eta^{-1}\frac{\partial(af)}{\partial z_i}dz_1 \wedge \ldots \wedge dz_n = \frac{\partial(af)}{\partial z_i}f^{-1}$$

and similarly

$$\Delta\left(b\frac{\partial}{\partial z_j}\right) = \frac{\partial(bf)}{\partial z_j}f^{-1}.$$

Furthermore,

$$\Delta\left(a\frac{\partial}{\partial z_i}, b\frac{\partial}{\partial z_j}\right) = \eta^{-1}\partial\left((-1)^{i+j}(abf)dz_1 \wedge \ldots \wedge \widehat{dz_i} \wedge \ldots \wedge \widehat{dz_j} \wedge \ldots \wedge dz_n\right)$$

$$= \eta^{-1}\left((-1)^{j-1}\frac{\partial(abf)}{\partial z_i}dz_1 \wedge \ldots \wedge \widehat{dz_j} \wedge \ldots \wedge dz_n\right.$$

$$\left.+ (-1)^i\frac{\partial(abf)}{\partial z_j}dz_1 \wedge \ldots \wedge \widehat{dz_i} \wedge \ldots \wedge dz_n\right)$$

$$= \frac{\partial(abf)}{\partial z_i}f^{-1}\frac{\partial}{\partial z_j} - \frac{\partial(abf)}{\partial z_j}f^{-1}\frac{\partial}{\partial z_i}$$

$$= \left[a\frac{\partial}{\partial z_i}, b\frac{\partial}{\partial z_j}\right] + \Delta\left(a\frac{\partial}{\partial z_i}\right)b\frac{\partial}{\partial z_j} - a\frac{\partial}{\partial z_i}\Delta\left(b\frac{\partial}{\partial z_j}\right)$$

This proves the assertion for $p = q = 0$.   □

Corollary 6.1.10 *Let α and β be Δ-closed elements of $\mathcal{A}^{0,*}(\mathcal{T}_X)$. Then the bracket $[\alpha, \beta]$ is Δ-exact.*   □

The Tian–Todorov lemma will be applied to the coefficients $\phi_i \in \mathcal{A}^{0,1}(\mathcal{T}_X)$ of the power series expansion of a deformation of the complex structure. Before proving the general result, we shall spell out the first two steps.

Let $v \in H^1(X, \mathcal{T}_X)$ be given. We first look for a good representative $\phi_1 \in \mathcal{A}^{0,1}(\mathcal{T}_X)$. In order to solve the Maurer–Cartan equation $\bar{\partial}\phi_2 = -[\phi_1, \phi_1]$, we have to ensure that

   i) $[\phi_1, \phi_1]$ is $\bar{\partial}$–closed   and   ii) $[\phi_1, \phi_1]$ has no harmonic part.

Here we use the Hodge decomposition $\mathcal{A}^{0,1}(\mathcal{T}_X) = \bar{\partial}\mathcal{A}^0(\mathcal{T}_X) \oplus \mathcal{H}^{0,1}(\mathcal{T}_X) \oplus \bar{\partial}^*\mathcal{A}^{0,2}(\mathcal{T}_X)$ with respect to an hermitian metric on X and thus on $\mathcal{T}_X$ (see Theorem 4.1.13).

The first condition is always satisfied:
- If $\alpha_i \in \mathcal{A}^{0,q_i}(\mathcal{T}_X)$, $i = 1, 2$, are $\bar{\partial}$-closed, then so is $[\alpha_1, \alpha_2]$.
This follows from Remark 6.1.1, but you may also use the Tian–Todorov lemma and Lemma 6.1.8 to deduce directly that $\bar{\partial}[\alpha_1, \alpha_2] = 0$.

The Tian–Todorov lemma also yields:
- If $\alpha_i \in \mathcal{A}^{0,q_i}(\mathcal{T}_X)$, $i = 1, 2$ are chosen such that $\eta(\alpha_i) \in \mathcal{A}^{n-1,q_i}(X)$ are ∂-closed, then $\eta[\alpha_1, \alpha_2]$ is ∂-exact. More explicitly, $\eta[\alpha_1, \alpha_2] = (-1)^{q_1} \cdot \partial\eta(\alpha_1 \wedge \alpha_2)$

Eventually, we find:
- If $\phi_1 \in \mathcal{A}^{0,1}(\mathcal{T}_X)$ is chosen such that $\eta(\phi_1) \in \mathcal{A}^{n-1,1}(X)$ is harmonic with respect to a Kähler metric on X, then $\bar{\partial}\phi_2 = -[\phi_1, \phi_1]$ admits a solution $\phi_2 \in \mathcal{A}^{0,1}(\mathcal{T}_X)$ which can, moreover, be chosen such that $\eta(\phi_2)$ is ∂-exact.

Indeed, by the above arguments $\eta[\phi_1, \phi_1]$ is ∂-exact and $\bar{\partial}$-closed. Since by Hodge theory on a compact Kähler manifold a ∂-exact form cannot have a non-trivial harmonic part, this proves that $[\phi_1, \phi_1]$ is $\bar{\partial}$-exact. But using the $\partial\bar{\partial}$-lemma (cf. Corollary 3.2.10), we can, moreover, say that there exists a form $\gamma \in \mathcal{A}^{n-2,0}(X)$ with $\bar{\partial}\partial\gamma = -\eta[\phi_1, \phi_1]$. Then set $\phi_2 := \eta^{-1}(\partial\gamma)$. Since η commutes with $\bar{\partial}$, this proves the claim.

We wish to emphasize that not only have we shown that $[\phi_1, \phi_1]$ is $\bar{\partial}$-exact for any choice of ϕ_1, but also that for the harmonic choice for $\eta(\phi_1)$ the form $\eta[\phi_1, \phi_1]$ is ∂-exact. In fact, for the latter it suffices to assume that $\eta(\phi_1)$ is ∂-closed.

All these arguments can now easily be generalized to prove the general result, which is also due to Tian and Todorov.

Proposition 6.1.11 *Let X be a Calabi–Yau manifold and let $v \in H^1(X, \mathcal{T}_X)$. Then there exists a formal power series $\phi_1 t + \phi_2 t^2 + \ldots$ with $\phi_i \in \mathcal{A}^{0,1}(\mathcal{T}_X)$ satisfying the Maurer–Cartan equations*

$$\bar{\partial}\phi_1 = 0 \quad \text{and} \quad \bar{\partial}\phi_k = -\sum_{0<i<k} [\phi_i, \phi_{k-i}],$$

with $[\phi_1] = v$ and such that

$$\eta(\phi_i) \in \mathcal{A}^{n-1,1}(X) \text{ is } \partial-\text{exact}$$

for all $i > 1$.

Proof. We begin with $\phi_1 \in \mathcal{A}^{0,1}(\mathcal{T}_X)$ representing v such that $\eta(\phi_1) \in \mathcal{A}^{n-1,1}(X)$ is harmonic (∂-closed would be enough). Then one finds ϕ_2 as above. Let us suppose that we have found $\phi_2, \ldots, \phi_{k-1}$ as claimed by the

Proposition. Then $\eta[\phi_i, \phi_{k-i}]$ is ∂-exact for $0 < i < k$ by the Tian–Todorov lemma. Thus, in order to find ϕ_k satisfying the assertion, we have to ensure that the sum $\sum_{0<i<k}[\phi_i, \phi_{k-i}]$ is $\bar{\partial}$-closed. As we have used implicitly before, one has

$$\bar{\partial}\left(\sum_{0<i<k}[\phi_i, \phi_{k-i}]\right) = \sum_{0<i<k}\left([\bar{\partial}\phi_i, \phi_{k-i}] + [\phi_i, \bar{\partial}\phi_{k-i}]\right)$$

and by induction hypothesis this equals

$$- \sum_{0<i<k}\left(\sum_{0<j<i}[[\phi_j, \phi_{i-j}], \phi_{k-i}] + \sum_{0<\ell<k-i}[\phi_i, [\phi_\ell, \phi_{k-i-\ell}]]\right)$$
$$= - \sum_{0<i<k}\sum_{0<j<i}[[\phi_j, \phi_{i-j}], \phi_{k-i}] + \sum_{0<i<k}\sum_{0<\ell<i}[\phi_{k-i}, [\phi_\ell, \phi_{i-\ell}]] = 0,$$

where we use $[\alpha, \beta] = -[\beta, \alpha]$ for $\alpha \in \mathcal{A}^{0,2}(\mathcal{T}_X)$ and arbitrary β.

The $\partial\bar{\partial}$-lemma eventually ensures that we can choose ϕ_k of the form $\eta^{-1}(\partial\gamma)$, i.e. $\eta(\phi_k)$ is ∂-exact.   □

Remarks 6.1.12 i) There might be very well other solutions of the Maurer–Cartan equation which do not satisfy the extra condition that $\eta(\phi_i)$ is ∂-exact for $i > 1$. In fact, the $\eta(\phi_i)$ need not even be ∂-closed.

ii) Even with the extra condition that $\eta(\phi_1)$ is harmonic the solution given by the proposition is not unique. At any step of the recursion we may change ϕ_k (or rather $\eta(\phi_k)$) by a $\partial\bar{\partial}$-exact form. This then affects all higher order equations.

By evoking the Green operator G for the chosen Kähler metric it is possible to single out a distinguished solution. More precisely, one can recursively define the ϕ_k by

$$\eta(\phi_k) := -\bar{\partial}^*\left(G\left(\eta\sum_{0<i<k}[\phi_i, \phi_{k-i}]\right)\right).$$

This is due to the fundamental property of the Green operator, that $\Delta_{\bar{\partial}} \circ G = $ id on the space of all $\bar{\partial}$-exact forms. In this way, one thus kills all ambiguity in the power series expansion $\phi_1 t + \phi_2 t^2 + \ldots$, which only depends on the chosen Kähler structure on X.

iii) So far, we have not said anything about the existence of a convergent solution. In fact, there is a standard procedure to turn any formal solution into a convergent one. At this point, a good deal of analysis enters the game. For the details we refer to [80, Sect. 5.3].

The principal idea is to consider the operator $\bar{\partial}$ on $\mathcal{A}^{0,q}(\mathcal{T}_X)$ and its adjoint operator with respect to a chosen Kähler (or just hermitian) metric. For clarity, let us write $\bar{\partial}_{\mathcal{T}_X}$ respectively $\bar{\partial}^*_{\mathcal{T}_X}$ for these two operators. Then one shows that a formal solution converges if the coefficients ϕ_k are $\bar\partial_{T_X}^*$-exact. It is then not too difficult to transform any formal solution into one that satisfies this additional condition, as the mere existence of a formal solution ensures that all possible obstructions are trivial.

iv) It is noteworthy that so far we have not used the existence of a Kähler–Einstein metric on X (cf. Corollary 4.B.22), but have worked with a completely arbitrary Kähler metric. In fact, once the formal solution in the proposition has been chosen as proposed in ii), i.e. $\eta(\phi)$ harmonic and $\eta(\phi_k) := -\bar\partial^* \left(G \left(\eta \sum_{0<i<k} [\phi_i, \phi_{k-i}] \right) \right)$, then it is automatically convergent provided the chosen Kähler metric is indeed Kähler–Einstein. In other words, the existence of the Kähler–Einstein metric ensures that a convergent formal solution can be constructed directly.

The argument, why the Kähler–Einstein metric works better than any other Kähler metric goes as follows. In iii) we have quoted the fact that a formal solution converges if the $\eta(\phi_k)$ are $\bar\partial_{T_X}^*$-exact. But now, with respect to a Kähler–Einstein metric the isomorphism $\eta : \mathcal{A}^{0,q}(T_X) \cong \mathcal{A}^{1,q}(X)$ is compatible with the natural hermitian structures on T_X and Ω_X (up to a constant scalar). In particular, $\bar\partial^* \circ \eta = \eta \circ \bar\partial_{T_X}^*$. The compatibility of the hermitian metrics under η follows from Corollary 4.B.23. A priori, under η the two metrics might pointwise differ by a scalar. But, due to the description of Kähler–Einstein metrics in terms of the holomorphic volume form provided by Corollary 4.B.23, this scalar is a constant.

Exercises

6.1.1 Compute the dimensions of $H^1(X, T_X)$ in the following cases: i) $X = \mathbb{P}^n$, ii) X a compact complex torus, and iii) X a curve.

6.1.2 Let X be a compact complex manifold (not necessarily Kähler) with a everywhere non-degenerate holomorphic two-form $\sigma \in H^0(X, \Omega_X^2)$, i.e. the induced homomorphism $T_X \to \Omega_X$ is an isomorphism. Show that also in this case there exists a formal solution $\sum \phi_i t^i$ satisfying the Maurer–Cartan equation, such that $[\phi_1] = v \in H^1(X, T_X)$ is a given class.

6.1.3 Let X be a compact complex manifold with $H^2(X, T_X) = 0$. Show that any $v \in H^1(X, T_X)$ can formally be integrated. This in particular applies to complex curves.

Comments: - The original proof of the Tian–Todorov lemma, which strikes one as indeed very clever, has been reproduced at various places. Our proof is very close to the original one.

- The convergence seems more difficult than the mere existence of a formal solution, but this kind of problem had been studied and well-understood before.

- For results on the structure of the diffeomorphism group of a manifold we refer to [6].

6.2 General Results

In this section we collect standard results from deformation theory of compact complex manifolds, but essentially no proofs are given. In this sense the section is closer in spirit to the various appendices, but we think that such a central topic deserves its own section. So, the following can be read as a survey which should be enough for the understanding of many applications, but for a thorough understanding the reader is advised to consult other sources.

As stressed earlier, deformation theory is difficult to learn and to teach. The analysis required is far beyond the scope of this book. In fact, [80] is a whole book on the subject. Moreover, the appropriate language uses complex spaces which we have avoided so far. Our approach to deformation theory and the use of complex spaces will thus appear very ad hoc to the more expert reader.

Let $\mathcal{X}$ and S be connected complex manifolds and let $\pi : \mathcal{X} \to S$ be a smooth proper morphism. Then the fibres $\mathcal{X}_t := \pi^{-1}(t)$ are compact complex submanifolds of $\mathcal{X}$ and we say that $\pi : \mathcal{X} \to S$ is a *smooth family* of complex manifolds parametrized by S.

Examples 6.2.1 i) Let E be a vector bundle on S. Then $\mathcal{X} := \mathbb{P}(E) \to S$ is a family. All fibres are isomorphic to $\mathbb{P}^{r-1}$ with $r = \mathrm{rk}(E)$. However, the isomorphism $\mathcal{X}_t \cong \mathbb{P}^{r-1}$ is not canonical.

Furthermore, $\pi : \mathcal{X} = \mathbb{P}(E) \to S$ is a *locally trivial family*, i.e. if $S = \bigcup U_i$ is a trivializing covering for the vector bundle E then $\pi^{-1}(U_i)$ is isomorphic as a complex manifold to $U_i \times \mathbb{P}^{r-1}$. Often, however, the family is not trivial, i.e. $\mathcal{X}$ is not isomorphic to $S \times \mathbb{P}^{r-1}$. As an example one may take $E = \mathcal{O} \oplus \mathcal{O}(1)$ on $S = \mathbb{P}^1$.

ii) Consider the linear system $|\mathcal{O}(k)|$ on $\mathbb{P}^n$ of hypersurfaces of degree k and let $\mathcal{X} \subset |\mathcal{O}(k)|_\mathrm{o} \times \mathbb{P}^n$ be the *universal hypersurface*. Here, $|\mathcal{O}(k)|_\mathrm{o} \subset |\mathcal{O}(k)|$ is the open subset of all smooth hypersurfaces. In coordinates, $\mathcal{X}$ is defined by a single equation $\sum a_j f_j$, where $f_1, \ldots, f_N$ is a basis of $H^0(\mathbb{P}^n, \mathcal{O}(k))$ and $a_1, \ldots, a_N$ are linear coordinates on $|\mathcal{O}(k)|$.

The projection yields a family $\mathcal{X} \to S := |\mathcal{O}(k)|_\mathrm{o}$. Note that there are fibres which might be isomorphic to each other. Indeed, $\mathrm{PGl}(n+1)$ acts on $\mathbb{P}^n$ and thus on S. The orbits of this action parametrize isomorphic hypersurfaces. On the other hand, if $k \geq 2$ the family is not locally trivial.

iii) Consider $\mathcal{X} := (\mathbb{C} \times \mathbb{H})/\sim$ with $(z, \tau) \sim (z', \tau)$ if and only if $z - z' \in \mathbb{Z} + \mathbb{Z}\tau$. Then $\mathcal{X} \to S := \mathbb{H}$ is the universal elliptic curve over the upper half-plane. As was mentioned before, two fibres $\mathcal{X}_\tau$ and $\mathcal{X}_{\tau'}$ are isomorphic if and only if τ and τ' are in the same orbit of the natural $\mathrm{Sl}(2, \mathbb{Z})$-action on $\mathbb{H}$ (see page 57).

iv) Let us consider the manifold $\mathcal{X}$ of all tuples $(y, x, t) \in \mathbb{P}^1 \times \mathbb{P}^2 \times \mathbb{C}$ satisfying $y_0^2 x_0 - y_1^2 x_1 - t y_0 y_1 x_2 = 0$. Then the projection $\mathcal{X} \to \mathbb{C}$ defines a family of smooth surfaces. It turns out that for $t \neq 0$ the fibre $\mathcal{X}_t$ is isomorphic to $\mathbb{P}^1 \times \mathbb{P}^1$ (e.g. using the map $\mathbb{P}^1 \times \mathbb{P}^1 \to \mathcal{X}_t$, $(y, z) \mapsto (y, ((y_1 z_0 + y_1 z_1)^2 :$

$(y_0z_0 - y_0z_1)^2, 4t^{-1}z_0z_1y_0y_1), t))$. However, the special fibre $\mathcal{X}_0$ is not at all isomorphic to $\mathbb{P}^1 \times \mathbb{P}^1$, it is the Hirzebruch surface $\mathbb{P}(\mathcal{O} \oplus \mathcal{O}(2))$ over $\mathbb{P}^1$.

In these examples we observe many different phenomena:

- The fibres of a family might be all isomorphic without the family being isomorphic to a product (see i)).
- The fibres might be very different and the family parametrizes them all in a more or less effective way (see iii)).
- The general fibres might be all isomorphic, but still converging to a non-isomorphic central fibre (see iv)).

In the differentiable category life is much easier due to the following

Proposition 6.2.2 (Ehresmann) *Let* $\pi : \mathcal{X} \to S$ *be a proper family of differentiable manifolds. If S is connected, then all fibres are diffeomorphic.*

Proof. By connecting two points of the base by an arc, we may assume that the base S is an interval $(-\varepsilon, 1 + \varepsilon)$. Then one has to show that the fibres $\mathcal{X}_0$ and $\mathcal{X}_1$ are diffeomorphic. Locally in $\mathcal{X}$, the morphism π looks like the projection $\mathbb{R}^{m+1} \to \mathbb{R}$ and we may thus lift the vector field $\partial/\partial t$ on S to $\mathbb{R}^{m+1}$.

More precisely, there exist finitely many open subsets $U_i \subset \mathcal{X}$ covering the fibre $\mathcal{X}_0$ such that $\partial/\partial t$ can be lifted to a vector field v_i on U_i. Using a partition of unity, one constructs in this way a vector field v on $\bigcup U_i$ that projects to $\partial/\partial t$ on some open neighbourhood of $0 \in (-\varepsilon, 1 + \varepsilon)$.

Since the family is proper, there exists a point $t_0 > 0$ such that $\mathcal{X}_{t_0}$ is contained in $\bigcup U_i$. Using the compactness of $[0, 1]$ it suffices to show that $\mathcal{X}_0$ and $\mathcal{X}_{t_0}$ are diffeomorphic. A diffeomorphism $\mathcal{X}_0 \to \mathcal{X}_{t_0}$ is provided by the flow associated to the vector field v.   □

The arguments in the proof can be used to show more:

Corollary 6.2.3 *Every proper family of differentiable manifolds is locally diffeomorphic to a product.*   □

It should also be clear from the proof that the arguments do not work in the complex setting. But once we know that the fibres are all isomorphic as complex manifolds, the family is in fact locally trivial. This result is due to Fischer and Grauert [45]:

Theorem 6.2.4 *Let $\mathcal{X} \to S$ be proper family of complex manifolds such that all fibres are isomorphic as complex manifolds. Then $\mathcal{X} \to S$ is locally trivial.*□

We will not deal with the more global aspects of families of complex manifolds. In particular, the construction of moduli spaces and universal families, i.e. families parametrizing all complex manifolds of a certain type effectively, will not be touched upon. Let us rather turn to the local aspects of deforming a complex manifold X.

Let $\pi : \mathcal{X} \to S$ be a smooth family and $0 \in S$. Then $\pi : \mathcal{X} \to S$ can be considered as a deformation of the fibre $X := \mathcal{X}_0$. Usually, one restricts to a germ of $0 \in S$, i.e. the family π is only studied over arbitrary small open neighbourhoods of $0 \in S$. In this context, $\mathcal{X} \to S$ will be called a *deformation of X*.

In particular, we may trivialize $\pi : \mathcal{X} \to S$ as a differentiable family. More precisely, we may choose a diffeomorphism $\mathcal{X} \cong X \times S$ (passing to a small open neighbourhood of $0 \in S$ is tacitly assumed). In this way, the family of complex manifolds $\pi : \mathcal{X} \to S$ can be viewed as a family of complex structures on the differentiable manifold M underlying the complex fibre X. This explains the relation between our approach in this section and the discussion of Section 6.1.

The first result we want to quote deals with compact Kähler manifolds that have interested us throughout the whole course.

Theorem 6.2.5 (Kodaira) *Let X be a compact Kähler manifold. If $\mathcal{X} \to S$ is a deformation of $X = \mathcal{X}_0$, then any fibre $\mathcal{X}_t$ is again Kähler.* □

We stress that this is a local result. By definition the fibres of a deformation $\mathcal{X} \to S$ are all close to the central fibre X. So, more precisely, the above result says that if $\mathcal{X} \to S$ is a smooth family and the fibre $\mathcal{X}_0$ over $0 \in S$ is Kähler, then there exists an open neighbourhood $0 \in U \subset S$ such that for any $t \in U$ the fibre $\mathcal{X}_t$ is again Kähler. There are indeed examples where for $t \in S$ 'far' away from $0 \in S$ the fibre $\mathcal{X}_t$ is not Kähler anymore.

Let us next describe an alternative approach to the definition of the Kodaira–Spencer class of an infinitesimal deformation. Let $\pi : \mathcal{X} \to S$ be a smooth proper family as before and consider the fibre $X := \mathcal{X}_0$ over a distinguished point $0 \in S$. Since π is smooth, we obtain a surjection of the tangent spaces $\pi : T_x\mathcal{X} \to T_0 S$ for any point $x \in X$. The kernel of this surjection is the tangent space $T_x X$ of the fibre. More globally, we obtain a short exact sequence of the form

$$0 \longrightarrow \mathcal{T}_X \longrightarrow \mathcal{T}_\mathcal{X}|_X \longrightarrow T_0 S \otimes \mathcal{O}_X \longrightarrow 0 \qquad (6.4)$$

Definition 6.2.6 The boundary map

$$\kappa : T_0 S \longrightarrow H^1(X, \mathcal{T}_X)$$

of the associated long exact cohomology sequence is called the *Kodaira–Spencer map*. The class $\kappa(v) \in H^1(X, \mathcal{T}_X)$ is the *Kodaira–Spencer* class associated with the infinitesimal deformation $v \in T_0 S$.

Remark 6.2.7 It is not too difficult to check that this definition of the Kodaira–Spencer class is compatible with the one given in Definition 6.1.3. In order to see this, one has to pass from the Čech description of the boundary map $\kappa : T_0 S \to H^1(X, \mathcal{T}_X)$ to the Dolbeault interpretation of $H^1(X, \mathcal{T}_X)$.

Often, our notion of families and deformation used so far is not flexible enough. In the language of the previous section, it might happen that the Maurer–Cartan equation has solutions up to a certain order, but a complete power series expansion (we are not even talking about convergence yet) does not exist. In our context, this amounts to allow as the base space S of a family or a deformation not only complex manifolds but more singular and even non-reduced structures. The crucial concept here is that of a complex space which in many respects parallels the notion of a scheme in algebraic geometry.

Definition 6.2.8 A *complex space* consist of a Hausdorff topological space X and a sheaf of rings $\mathcal{O}_X$ such that locally $(X, \mathcal{O}_X)$ is isomorphic to an analytic subset $Z \subset U \subset \mathbb{C}^n$ (see Definition 1.1.23) endowed with the sheaf $\mathcal{O}_U/\mathcal{I}$ where $\mathcal{I}$ is a sheaf of holomorphic functions with $Z = Z(\mathcal{I})$.

A complex space may have various properties. The topological ones, like connectedness, properness, etc., only concern the underlying topological space X, whereas the ring theoretic ones, like reduced, integral, etc., deal with the properties of the structure sheaf $\mathcal{O}_X$ and often just its germs $\mathcal{O}_{X,x}$ which are local rings. The maximal ideal of $\mathcal{O}_{X,x}$ will be denoted $\mathfrak{m}_x$.

Examples 6.2.9 i) Clearly, any complex manifold provides an example of a complex space which is locally modeled on $Z = U \subset \mathbb{C}^n$ and $\mathcal{I} = (0)$.

ii) Also the equation $z_1 \cdot z_2 = 0$ defines a complex space X within $\mathbb{C}^2$. Its structure sheaf is $\mathcal{O}_{\mathbb{C}^2}/(z_1 \cdot z_2)$. Here, $0 \in X$ is the only singular point, i.e. $X \setminus \{x\}$ is a complex manifold..

iii) The easiest non-reduced complex space is the *double point* $\mathrm{Spec}(\mathbb{C}[\varepsilon])$. By definition, $\mathrm{Spec}(\mathbb{C}[\varepsilon])$ consist of one point 0 and the value of the structure sheaf over the only non-empty open subset $\{0\}$ is $\mathbb{C}[x]/x^2$.

A morphism $(X, \mathcal{O}_X) \to (Y, \mathcal{O}_Y)$ between complex spaces consists of a continuous map $\varphi : X \to Y$ and a ring homomorphism $\tilde{\varphi} : \varphi^{-1}\mathcal{O}_Y \to \mathcal{O}_X$. The *fibre* of a morphism $\varphi : (X, \mathcal{O}_X) \to (Y, \mathcal{O}_Y)$ over a point $y \in Y$ is by definition the complex space $(\varphi^{-1}(y), \mathcal{O}_X/\tilde{\varphi}^{-1}(\mathfrak{m}_y))$, where $\mathfrak{m}_y$ is the maximal ideal of the local ring $\mathcal{O}_{Y,y}$. Note that this sometimes deviates from our use of the word. E.g. if one considers the map $\mathbb{C} \to \mathbb{C}$, $z \mapsto z^2$ the fibre over $0 \in \mathbb{C}$ in the new sense would be the double point and not the complex submanifold $\{0\} \subset \mathbb{C}$.

A morphism $(X, \mathcal{O}_X) \to (Y, \mathcal{O}_Y)$ is *flat* if the stalk $\mathcal{O}_{X,x}$ is a flat $\mathcal{O}_{Y,\varphi(x)}$-module for any $x \in X$ via the natural ring homomorphism $\mathcal{O}_{Y,\varphi(x)} \to \mathcal{O}_{X,x}$. In the following, the mentioning of the structure sheaf $\mathcal{O}_X$ of a complex space will often be suppressed.

Using these more general spaces, one defines a smooth *family of complex manifolds* as a flat proper morphism $\mathcal{X} \to S$ of complex spaces such that all fibres $\mathcal{X}_t$ are smooth. Similarly, one defines a *deformation* of a complex manifold X as a complex space $\mathcal{X}$ that is flat over a base space S, which could be an even non-reduced complex space, and such that all the fibres are smooth.

The *tangent space* of a complex space $(S, \mathcal{O}_S)$ at a point $0 \in S$ is by definition the vector space

$$T_0 S := \operatorname{Hom}_{\mathbb{C}}(\mathfrak{m}_0/\mathfrak{m}_0^2, \mathbb{C}).$$

Analogously to the smooth situation (cf. (6.4)), every deformation $\mathcal{X} \to S$ induces a short exact sequence

$$0 \longrightarrow \mathcal{T}_X \longrightarrow \Theta_{\mathcal{X}}|_X \longrightarrow T_0 S \otimes \mathcal{O}_X \longrightarrow 0.$$

As soon as the spaces involved are no longer necessarily smooth, one changes the notation from $\mathcal{T}_{\mathcal{X}}$ to $\Theta_{\mathcal{X}}$ or sometimes $\operatorname{Der}(\mathcal{O}_{\mathcal{X}})$. We are not going into the precise definition of the tangent sheaf(!) Θ of a complex space here, but one obtains, as before, the Kodaira–Spencer map $T_0 S \to H^1(X, \mathcal{T}_X)$ as the boundary map of the associated long exact cohomology sequence.

From here it is not difficult to prove the following (cf. Proposition 6.1.5)

Proposition 6.2.10 *Let X be a compact complex manifold. Then there is a natural bijection between classes in $H^1(X, \mathcal{T}_X)$ and isomorphism classes of deformations $\mathcal{X} \to \operatorname{Spec}(\mathbb{C}[\varepsilon])$ of X.* □

A deformation over the double point $\operatorname{Spec}(\mathbb{C}[\varepsilon])$ is called an infinitesimal deformation (of first order). It corresponds to the first coefficient in the power series expansion in Section 6.1. In this sense, infinitesimal deformations are well understood. A central question in deformation theory is whether any infinitesimal deformation can be integrated, i.e. given a class $v \in H^1(X, \mathcal{T}_X)$ can one find a deformation over a smooth base S such that v is contained in the image of the Kodaira–Spencer map $T_0 S \to H^1(X, \mathcal{T}_X)$. Since this is not always possible, one has to deal with arbitrary complex spaces and not only the double point.

Let $\pi : \mathcal{X} \to S$ be a deformation of X and $f : S' \to S$ a morphism of germs (in particular, f maps the distinguished point $0' \in S'$ to $0 \in S$). Then the pull-back $f^*\mathcal{X} = \mathcal{X} \times_S S'$ defines a deformation of X over S'.

Definition 6.2.11 A deformation $\pi : \mathcal{X} \to S$ of the compact complex manifold $X = \mathcal{X}_0$ is called *complete* if any other deformation $\pi' : \mathcal{X}' \to S'$ of X is obtained by pull-back under some $f : S' \to S$.

If in addition f is always unique, then $\pi : \mathcal{X} \to S$ is called *universal*. If only its differential $T_{S'}(0') \to T_S(0)$ is unique, then the deformation is called *versal*.

The existence of a complete (let alone versal) deformation for any compact complex manifold is a deep theorem. Let us begin with the following criterion for completeness.

Theorem 6.2.12 (Kodaira–Spencer) *A deformation $\pi : \mathcal{X} \to S$ of $X = \mathcal{X}_0$ over a reduced base S is complete if and only if the Kodaira–Spencer map $T_0 S \to H^1(X, \mathcal{T}_X)$ is surjective.*

In the examples, a complete deformation, which roughly is a deformation that parametrizes, in a possibly non-effective way, all small deformations of a given complex manifold, can often be found. It is often much harder to find a deformation that parametrizes all possible deformations in an effective way. In fact, in many situations a universal deformation does not exist. But at least one has:

Theorem 6.2.13 (Kuranishi) *Any compact complex manifold admits a versal deformation.*

Note that for a versal deformation $\mathcal{X} \to S$ the Kodaira–Spencer map $T_0 S \to H^1(X, \mathcal{T}_X)$ is bijective.

Using the existence of the versal deformation, one might try to develop criteria for various additional properties. None of the following results is easy. We refer to the literature for more details.

- A versal deformation is always complete for any of its fibres.

- If $H^0(X, \mathcal{T}_X) = 0$, then any versal deformation is universal and hence unique.

- If the dimension of $H^1(\mathcal{X}_t, \mathcal{T}_{\mathcal{X}_t})$ stays constant in a versal deformation of $\mathcal{X} \to S$ of S, then it is versal for any of its fibres.

- One says that X has *unobstructed deformations* if X admits a smooth versal deformation, i.e. the base S is an honest complex manifold. This is the case if $H^2(X, \mathcal{T}_X) = 0$.

In fact, the results explained in the last section (cf. Proposition 6.1.11) show that any Calabi–Yau manifold X, i.e. a compact Kähler manifold with trivial canonical bundle K_X, has unobstructed deformations. An algebraic approach was developed by Ran and Kawamata [74], using the so called T^1-lifting property.

The idea why Calabi–Yau manifolds should have unobstructed deformations is the following: One observes that for any deformation $\mathcal{X}_t$ of a Calabi–Yau manifold X of dimension n one has $H^1(\mathcal{X}_t, \mathcal{T}_{\mathcal{X}_t}) = H^1(\mathcal{X}_t, \Omega_{\mathcal{X}_t}^{n-1})$. The dimension of the latter space does not depend on t, as it occurs as a direct summand of the Hodge decomposition of $H^n(\mathcal{X}_t, \mathbb{C})$ which in turn only depends on the underlying topological manifold. By iii) this shows that the family is versal and that all tangent spaces of S are of the same dimension. Now, if S is reduced this would be enough to conclude that S is in fact smooth, i.e. that X has unobstructed deformations. Of course, the fact that S could *a priori* be non-reduced makes the problem much harder (see e.g. [61, Sect. 14] for more details).

Comments: - For the general deformation theory of complex manifolds see Kodaira's original papers or his book [80]. We also recommend Douady's survey [39].

- For the theory of complex spaces one might consult [58].

- Note that a small deformation of a projective manifold need not be projective again. Thus, if one wants to work exclusively in the algebraic context, one has to modify the techniques accordingly. In particular, the first-order deformation of a polarized manifold (X, L) are in general not parametrized by $H^1(X, \mathcal{T}_X)$.

Appendix to Chapter 6

6.A dGBV-Algebras

In this appendix we will give a first introduction to differential Gerstenhaber–Batalin–Vilkovisky algebras and their deformations. It will turn out that the Tian–Todorov lemma, and thus the unobstructedness of Calabi–Yau manifolds, can be interpreted in this more general framework. This more abstract approach makes very clear which algebraic structures are really needed to make the Tian–Todorov deformations work.

Let us begin with the definition of an abstract dGBV algebra. In the following we let $(\mathcal{A}, d)$ be a *differential supercommutative* $\mathbb{C}$-*algebra*. This is a slightly weaker notion than that of a dga used in Appendix 3.A. Here, only a $\mathbb{Z}/2\mathbb{Z}$-grading of $\mathcal{A}$ is required. In particular, there is a decomposition $\mathcal{A} = \mathcal{A}^0 \oplus \mathcal{A}^1$, where $\mathcal{A}^0$ (resp. $\mathcal{A}^1$) is the set of even (resp. odd) elements. If $\alpha \in \mathcal{A}$ is a homogeneous element then $\tilde{\alpha} \in \mathbb{Z}/2\mathbb{Z}$ is determined by $\alpha \in \mathcal{A}^{\tilde{\alpha}}$.

The differential d is a $\mathbb{C}$-linear odd derivative. More precisely, $d(\mathcal{A}^0) \subset \mathcal{A}^1$ and $d(\mathcal{A}^1) \subset \mathcal{A}^0$, $d(\alpha \cdot \beta) = d(\alpha) \cdot \beta + (-1)^{\tilde{\alpha}} \alpha \cdot d(\beta)$ for all homogeneous α, and $d^2 = 0$.

Definition 6.A.1 A supercommutative $\mathbb{C}$-algebra $\mathcal{A}$ (no differential yet) together with an odd $\mathbb{C}$-linear map $\Delta : \mathcal{A} \to \mathcal{A}$ is called a *Batalin–Vilkovisky algebra* (BV algebra, for short) if for any $\alpha \in \mathcal{A}$ the map

$$\delta_\alpha : \mathcal{A} \longrightarrow \mathcal{A}, \quad \beta \longmapsto (-1)^{\tilde{\alpha}} \Delta(\alpha \cdot \beta) - (-1)^{\tilde{\alpha}} \Delta(\alpha) \cdot \beta - \alpha \cdot \Delta(\beta)$$

is a derivation of parity $\tilde{\alpha} + 1$. If in addition $\Delta^2 = 0$, then $(\mathcal{A}, \Delta)$ is a *Gerstenhaber–Batalin–Vilkovisky algebra* (GBV algebra for short).

One speaks of a differential (Gerstenhaber–)Batalin–Vilkovisky algebra if $\mathcal{A}$ is also endowed with a differential d satisfying $[d, \Delta] = d \circ \Delta + \Delta \circ d = 0$.

Let now $(\mathcal{A}, d, \Delta)$ be a dGBV algebra. then one introduces the following bracket. For any $\alpha, \beta \in \mathcal{A}$ one sets

$$[\alpha \bullet \beta] := \delta_\alpha(\beta).$$

At first sight, this is just a new notation for δ_α, but it turns out that $[\ \bullet\ ]$ shares indeed many properties of a Lie bracket. In fact, it is what one calls an *odd Lie bracket*. What this means, is expressed by

Proposition 6.A.2 *Let $(\mathcal{A}, d, \Delta)$ be a dGBV algebra. Then*

i) $[\alpha \bullet \beta] = -(-1)^{(\tilde{\alpha}+1)(\tilde{\beta}+1)}[\beta \bullet \alpha]$.

ii) $[\alpha \bullet [\beta \bullet \gamma]] = [[\alpha \bullet \beta] \bullet \gamma] + (-1)^{(\tilde{\alpha}+1)(\tilde{\beta}+1)}[\beta \bullet [\alpha \bullet \gamma]]$.

iii) $[d, \delta_\alpha] = \delta_{d(\alpha)} = [d(\alpha) \bullet\ ]$.

Proof. All these properties can be proved by a rather straightforward computation. As an example let us show iii) for an even α. By definition and using $d \circ \Delta = -\Delta \circ d$ one computes

$$
\begin{aligned}
[d, \delta_\alpha](\beta) &= d(\delta_\alpha(\beta)) + \delta_\alpha(d(\beta)) \\
&= d\left(\Delta(\alpha \cdot \beta) - \Delta(\alpha) \cdot \beta - \alpha \cdot \Delta(\beta)\right) \\
&\quad + \Delta(\alpha \cdot d\beta) - \Delta(\alpha) \cdot d\beta - \alpha \cdot \Delta(d\beta) \\
&= -\Delta((d\alpha) \cdot \beta) + \Delta(d\alpha) \cdot \beta - d(\alpha) \cdot \Delta(\beta) \\
&= \delta_{d\alpha}(\beta).
\end{aligned}
$$

$\square$

Proposition 6.A.3 *Let $(\mathcal{A}, d, \Delta)$ be a dGBV algebra and let $\phi \in \mathcal{A}^0$. Then $d_\phi := d + \delta_\phi$ is an odd derivation. Furthermore, $d_\phi^2 = 0$, i.e. d_ϕ is a differential, if and only if the master equation*

$$
\left[\left(d(\phi) + \frac{1}{2}\delta_\phi(\phi)\right) \bullet \quad \right] = 0 \tag{6.5}
$$

holds true.

Proof. Since d and δ_φ are both odd derivations, their sum d_ϕ is a derivation as well.

Clearly, $d_\phi^2 = [d, \delta_\phi] + [\phi \bullet [\phi \bullet \quad]]$. By iii) of Proposition 6.A.2 we know that $[d, \delta_\phi] = [d(\phi) \bullet \quad]$ and using ii) one finds $[\phi \bullet [\phi \bullet \quad]] = \frac{1}{2}[[\phi \bullet \phi] \bullet \quad]$. This shows that $d_\phi^2 = 0$ if and only if $[(d(\phi) + \frac{1}{2}[\phi \bullet \phi]) \bullet \quad] = 0$. $\square$

Clearly, the master equation can also be written as

$$
\left[\left(d(\phi) + \frac{1}{2}[\phi \bullet \phi]\right) \bullet \quad \right] = 0.
$$

We now come back to the example that interests us most: The case of a complex manifold and more specifically of a Calabi–Yau manifold X. It will turn out that deforming a complex structure can also be viewed as a deformation of a differential algebra as above.

Let X be a complex manifold. One defines

$$
\mathcal{A}_X := \bigoplus \mathcal{A}^{0,q}(X, \textstyle\bigwedge^p T_X).
$$

The $\mathbb{Z}/2\mathbb{Z}$-grading given by $p + q$ and the exterior product determine the structure of a supercommutative algebra on $\mathcal{A}_X$. Note that, we require $d\bar{z}_i \wedge (\partial/\partial z_j) = -(\partial/\partial z_j) \wedge d\bar{z}_i$ (otherwise the product wouldn't be $\mathbb{Z}/2$-commutative.)

Since $\mathcal{T}_X$ and all exterior powers $\bigwedge^p \mathcal{T}_X$ are holomorphic bundles, $\bar{\partial}$ endows $\mathcal{A}_X$ with the structure of a differential graded algebra. (Verify the signs!)

This example shows an additional feature, namely the existence of a Lie bracket

$$[\ , \ ] : \mathcal{A}^{0,q}(\textstyle\bigwedge^p \mathcal{T}_X) \times \mathcal{A}^{0,q'}(\textstyle\bigwedge^{p'} \mathcal{T}_X) \longrightarrow \mathcal{A}^{0,q+q'}(\textstyle\bigwedge^{p+p'-1} \mathcal{T}_X) \, ,$$

called the *Schouten–Nijenhuis bracket*. In local coordinates it is given by

$$\left[\alpha_{IJ} d\bar{z}_I \frac{\partial}{\partial z_J}, \beta_{KL} d\bar{z}_K \frac{\partial}{\partial z_L} \right] = d\bar{z}_I \wedge d\bar{z}_K \left[\alpha_{IJ} \frac{\partial}{\partial z_J}, \beta_{KL} \frac{\partial}{\partial z_L} \right]$$

with $[v_1 \wedge \ldots \wedge v_p, w_1 \wedge \ldots \wedge w_{p'}] = \sum (-1)^{i+j} [v_i, w_j] \wedge v_1 \wedge \ldots \wedge \widehat{v_i} \wedge \ldots \wedge v_p \wedge w_1 \wedge \ldots \wedge \widehat{w_j} \wedge \ldots \wedge w_{p'}$ and $[v_1 \wedge \ldots \wedge v_p, f] = \sum (-1)^{p-i} v_i(f) v_1 \wedge \ldots \wedge \widehat{v_i} \wedge \ldots \wedge v_p$ for $p' = 0$. In particular, it generalizes the Lie bracket $[\ , \ ]$ on $\mathcal{A}^{0,*}(\mathcal{T}_X)$ introduced in Section 6.1. The sign convention used here yields

$$[\alpha, \beta] = -(-1)^{(p+q+1)(p'+q'+1)} [\beta, \alpha].$$

In fact, the Schouten–Nijenhuis bracket is the analogue of the odd Lie algebra bracket $[\ \bullet \ ]$ on a dGBV algebra. The difference is that for an arbitrary complex manifold X the algebra $\mathcal{A}_X$ is not endowed with the additional structure Δ. This is only the case for a Calabi–Yau manifold, as we will see shortly, but for the moment we stick to the case of a general complex manifold. Although, the Schouten–Nijenhuis bracket is not induced by an additional differential Δ, it still satisfies most of the properties of $[\ \bullet \ ]$. In particular, we will again write $\delta_\alpha := [\alpha, \ ]$ and with the local description at hand, one easily verifies:

Lemma 6.A.4 *Let $\alpha \in \mathcal{A}^{0,q}(\bigwedge^p \mathcal{T}_X)$. Then:*
i) $[\alpha, \ ] : \mathcal{A} \to \mathcal{A}$ *is a derivation.*
ii) $[\alpha, [\alpha, \ ]] = \frac{1}{2}[[\alpha, \alpha], \ ]$.
iii) $[\bar{\partial}, \delta_\alpha] = \delta_{\bar{\partial}(\alpha)} = [\bar{\partial}(\alpha), \ ]$.

This yields immediately the analogue of Proposition 6.A.3 for $\mathcal{A}_X$ with X an arbitrary complex manifold:

Corollary 6.A.5 *Let X be a complex manifold and let $\phi \in \mathcal{A}_X^0$.*
Then $\bar{\partial}_\phi := \bar{\partial} + \delta_\phi$ is an odd derivation on $\mathcal{A}_X$. Furthermore, $\bar{\partial}_\phi^2 = 0$, i.e. $\bar{\partial}_\phi$ is a differential, if and only if the master equation

$$\bar{\partial}(\phi) + \frac{1}{2}[\phi, \phi] = 0 \tag{6.6}$$

holds true. □

Remark 6.A.6 There is an obvious similarity between the Maurer–Cartan equation (6.2) and the master equation (6.6). More precisely, if $\phi(t) \in \mathcal{A}^{0,1}(T_X)$ describes the deformation $I(t)$ of the complex structure defining X then $I(t)$ is integrable, which is equivalent to $\phi(t)$ satisfying the Maurer–Cartan equation (6.2), if and only if $\frac{1}{2}\phi(t) \in \mathcal{A}^{0,1}(T_X) \subset \mathcal{A}_X^0$ satisfies the master equation.

In other words, we can study deformations of the complex structure I in terms of deformations of the differential $\bar{\partial}$ on $\mathcal{A}^{0,*}(\bigwedge^* T_X)$. (This should not be confused with the differential $\bar{\partial}(t)$ associated with the complex structure $I(t)$, which a priori lives on a different algebra.) The factor $1/2$ in the transition from the Maurer–Cartan equation to the master equation is harmless.

In this sense, solutions of the master equations can be viewed as generalized deformations of the given complex structure. However, a satisfactory geometric interpretation of those is still missing for the time being.

For an arbitrary manifold X the differential algebra $(\mathcal{A}_X, d)$ has no reason to carry the additional structure of a BV algebra. This is only ensured for Calabi–Yau manifolds, as we shall explain next.

Let X be a complex manifold with trivial canonical bundle $K_X \cong \mathcal{O}_X$ and fix a trivializing section $\Omega \in H^0(X, K_X)$. Then, due to Definition 6.1.7, one obtains an odd operator

$$\Delta : \mathcal{A}^{0,q}(\textstyle\bigwedge^p T_X) \longrightarrow \mathcal{A}^{0,q}(\textstyle\bigwedge^{p-1} T_X) .$$

As in the general context, one then defines the operators δ_α for any $\alpha \in \mathcal{A}^{0,q}(\bigwedge^p T_X)$ in terms of Δ.

Proposition 6.A.7 (Generalized Tian–Todorov lemma) *The above assumption implies*

$$-[\alpha, \beta] = (-1)^{\tilde{\alpha}}\Delta(\alpha \cdot \beta) - (-1)^{\tilde{\alpha}}\Delta(\alpha) \cdot \beta - \alpha \cdot \Delta(\beta)$$

for any $\alpha, \beta \in \mathcal{A}_X$.

Proof. The proof is slightly more involved than the one of Lemma 6.1.9, but the idea is of course the same.   □

This immediately yields:

Corollary 6.A.8 *Let X be a Calabi–Yau manifold X. Then $\bar{\partial}$ and Δ define the structure of a dGBV algebra on $\mathcal{A}^{0,*}(X, \bigwedge^* T_X)$.*

Proof. Since $[\alpha, \beta] = (-1)^{\tilde{\alpha}}\Delta(\alpha \cdot \beta) - (-1)^{\tilde{\alpha}}\Delta(\alpha) \cdot \beta - \alpha \cdot \Delta(\beta)$ by Proposition 6.A.7 and since $[\alpha, \ ]$ is a derivation by Lemma 6.A.4, $(\mathcal{A}^{0,*}(X, \bigwedge^* T_X), \Delta)$ is a BV algebra. It is in fact GBV, as $\Delta^2 = \eta^{-1} \circ \partial^2 \circ \eta = 0$.

Moreover, since the holomorphic volume form Ω is holomorphic, one has $\bar{\partial}\Delta(\alpha) = \bar{\partial}\eta^{-1}\partial\eta(\alpha) = \eta^{-1}\bar{\partial}\partial\eta(\alpha) = -\eta^{-1}\partial\bar{\partial}\eta(\alpha) = -\Delta(\bar{\partial}\alpha)$. Hence $[\bar{\partial}, \Delta] = 0$.   □

From now on, for a Calabi–Yau manifold X, we will use the two notations $[\ ,\ ]$ and $[\ \bullet\ ]$ interchangeably, the difference in the sign being of no importance and often suppressed altogether.

Remark 6.A.9 In the general context of a dGBV algebra $(\mathcal{A}, d, \Delta)$, the deformation d_ϕ of the differential d yields again a dGBV algebra $(\mathcal{A}, d_\phi, \Delta)$ if one assumes in addition $\Delta(\phi) = 0$, because this implies immediately $[d_\phi, \Delta] = 0$. In the geometric situation of a Calabi–Yau manifold X, the condition $\Delta(\phi) = 0$ amounts to $\eta(\phi)$ be ∂-closed. This condition has been encountered before in the proof of Proposition 6.1.11.

Once the analogy between the Maurer–Cartan equation for a Calabi–Yau manifold X and the master equation for $\mathcal{A}_X$ has been understood, one can go one and construct (formal) solutions of the master equation for a general dGBV algebra following the Tian–Todorov approach. In the case of a Calabi–Yau manifold X, we have seen that any first-order deformation, which are parametrized by elements in $H^1(X, \mathcal{T}_X)$, can be extended to a formal solution. In other words, we have seen how to construct a formal solution of the Maurer–Cartan equation as an element in $\mathcal{A}^{0,1}(X, \mathcal{T}_X) \otimes \mathbb{C}[H^1(X, \mathcal{T}_X)^*]$. We will proceed similarly in the case of a dGBV algebra.

So, let $(\mathcal{A}, d, \Delta)$ again be a general dGBV algebra.

Definition 6.A.10 The cohomology $H(\mathcal{A})$ of a $(\mathcal{A}, d, \Delta)$ is the cohomology with respect to the differential d, i.e.

$$H(\mathcal{A}) = \frac{\mathrm{Ker}(d : \mathcal{A} \to \mathcal{A})}{\mathrm{Im}(d : \mathcal{A} \to \mathcal{A})}.$$

Clearly, the cohomology $H(\mathcal{A})$ is again a $\mathbb{Z}/2\mathbb{Z}$-graded algebra. To simplify the discussion, we will assume however that $H^1(\mathcal{A}) = 0$, i.e. only even elements give rise to non-trivial cohomology classes. The aim is to construct a formal solution of the master equation as an element in $\mathcal{A} \otimes \mathbb{C}[H(\mathcal{A})^*]$. In order to achieve this one has to add two further conditions on our dGBV algebra $(\mathcal{A}, d, \Delta)$:

(1) The cohomology $H(\mathcal{A})$ is finite-dimensional.
(2) $\mathrm{Im}(d \circ \Delta) = (\mathrm{Ker}(d) \cap \mathrm{Ker}(\Delta)) \cap (\mathrm{Im}(d) + \mathrm{Im}(\Delta))$

Proposition 6.A.11 *Let $\mathcal{A}_X$ be the dGBV algebra associated to a compact Kähler manifold X with trivial canonical bundle. Then the conditions (1) and (2) hold true.*

Proof. Clearly, (1) is satisfied, as the cohomology $H(\mathcal{A}_X)$ is just the direct sum $\bigoplus H^q(X, \bigwedge^p \mathcal{T}_X)$ of Dolbault cohomology groups, which is finite-dimensional as soon as X is compact.

The assertion for (2) follows from formality, i.e. from the $\partial\bar{\partial}$-lemma, for compact Kähler manifolds. Indeed, the inclusion '$\subset$' is obvious and to verify

'$\supset$' one first writes any element $\gamma \in (\mathrm{Im}(d) + \mathrm{Im}(\Delta))$ as $\eta^{-1}(\bar\partial\alpha + \partial\eta\beta)$. Then one shows that $\gamma \in (\mathrm{Ker}(d) \cap \mathrm{Ker}(\Delta))$ implies that one can write $\bar\partial\alpha = \bar\partial\partial\alpha'$ and $\partial\eta(\beta) = \partial\bar\partial(\eta(\beta'))$, which is an immediate consequence of the $\partial\bar\partial$-lemma. $\square$

Proposition 6.A.12 *Let $(\mathcal{A}, d, \Delta)$ be a dGBV algebra satisfying the additional hypotheses* **(1)** *and* **(2)**.

Then there exists an element $\Phi \in \mathcal{A} \otimes \mathbb{C}[H(\mathcal{A})^*]$ whose homogeneous part of degree k is denoted Φ_k, such that

i) $d(\Phi) + \frac{1}{2}[\Phi \bullet \Phi] = 0$ *(master equation)*.

ii) $\Phi_0 = 0$.

iii) $\Phi_1 \in \mathcal{A} \otimes H(\mathcal{A})^*$ *defines a split of the surjection* $\mathrm{Ker}(d : \mathcal{A} \to \mathcal{A}) \to H(\mathcal{A})$. *Moreover, one may assume that* $\Phi_1 \in (\mathrm{Ker}(d) \cap \mathrm{Ker}(\Delta)) \otimes H(\mathcal{A})^*$.

iv) $\Phi_k \in \mathrm{Ker}(\Delta) \otimes \mathbb{C}[H(\mathcal{A})^*]$ *for $k \geq 2$*.

Proof. We leave the proof to the reader, as it is identical to the original proof of Proposition 6.1.11. The reader is advised to again construct Φ_1 and Φ_2 first, in order to get a feeling for how the recursion works. The recursion for the higher order term is analogous. $\square$

Comments: - All results of this section can be found in [7], where Barannikov and Kontsevich used dGBV algebras in order to construct Frobenius manifolds. Our presentation is based on Manin's book [89] which also contains a detailed account of the relation between dGBV algebras and Frobenius manifolds.

- There is also a symplectic version of the Barannikov–Kontsevich approach due to Merkulov, see [90].

A

Hodge Theory on Differentiable Manifolds

This appendix is meant to remind the reader of a few basic definitions and facts from differential geometry, but it cannot replace an introduction to the subject. We use the opportunity to introduce the related notations used throughout the text. No proofs are given, the material is far from being complete and the reader is advised to go back to any of the standard textbooks for details.

Definition A.0.1 An m-dimensional $\mathcal{C}^k$-manifold is a topological space M together with an open covering $M = \bigcup U_i$ and homeomorphisms $\varphi_i : U_i \cong V_i$ onto open subsets $V_i \subset \mathbb{R}^m$ such that
  i) M is Hausdorff.
  ii) The topology of M admits a countable basis.
  iii) The transition functions $\varphi_j \circ \varphi_i^{-1} : \varphi_i(U_i \cap U_j) \to \varphi_j(U_i \cap U_j)$ are $\mathcal{C}^k$-maps.

A *differentiable manifold* is a $\mathcal{C}^\infty$-manifold and only those will be considered. The datum $\{(U_i, \varphi_i)\}$ is called an *atlas* and each tuple (U_i, φ_i) is a *chart*. We say that two atlases define the same manifold if the transition functions $\varphi'_j \circ \varphi_i^{-1}$ are differentiable.

If M is a differentiable manifold, then one can introduce differentiable functions on M. By $\mathcal{C}_M$ we denote the sheaf of differentiable functions, i.e. for any open subset $U \subset M$ the value of $\mathcal{C}_M$ on U is the space of differentiable functions $f : U \to \mathbb{R}$, i.e. functions such that $f \circ \varphi_i^{-1} : \varphi_i(U \cap U_i) \to \mathbb{R}$ is differentiable for any chart (U_i, φ_i). Analogously, one introduces differentiable maps between differentiable manifolds.

In particular, there is the stalk $\mathcal{C}_{M,x}$ of the sheaf of differentiable functions at every point $x \in M$. The *tangent space* $T_x M$ of M at the point $x \in M$ can be defined as
$$T_x M := \mathrm{Der}_{\mathbb{R}}(\mathcal{C}_{M,x}, \mathbb{R}),$$
the vector space of derivations $D : \mathcal{C}_{M,x} \to \mathbb{R}$, i.e. of $\mathbb{R}$-linear maps satisfying $D(f \cdot g) = f(x) \cdot D(g) + D(f) \cdot g(x)$. E.g. any curve $\gamma : (-\varepsilon, \varepsilon) \to M$ with $\gamma(0) = x$ defines a tangent vector D_γ by $D_\gamma(f) = (d(f \circ \gamma)/dt)(0)$.

All the tangent spaces $T_x M$ glue to the *tangent bundle* $TM = \bigcup_{x \in M} T_x M$ which is an example of a differentiable real vector bundle on M.

Definition A.0.2 Let M be a differentiable manifold. A differentiable *vector bundle* of rank r on M consists of a differentiable manifold E, a differentiable map $\pi : E \to M$, and the structure of a real vector space on any fibre $E(x) := \pi^{-1}(x)$, such that there exists an open covering $M = \bigcup U_i$ and diffeomorphisms $\psi_i : \pi^{-1}(U_i) \to U_i \times \mathbb{R}^r$ with $\mathrm{pr}_{U_i} \circ \psi_i = \pi$ and such that for all $x \in U_i$ the map $\psi_i(x) : E(x) \to \mathbb{R}^r$ is an isomorphism of real vector spaces.

A real vector bundle can also be described in terms of the cocycle $\psi_{ij} : U_i \cap U_j \to \mathrm{Gl}(r, \mathbb{R})$. For line bundles, i.e. vector bundles of rank one, this leads to a complete parametrization of isomorphism classes by the sheaf cohomology $H^1(M, C_M^*)$ of the multiplicative sheaf of nowhere vanishing differentiable functions. Similarly, isomorphism classes of complex line bundles are in bijection with the elements of $H^1(M, C_{M,\mathbb{C}}^*)$, where $C_{M,\mathbb{C}}$ is the sheaf of complex valued differentiable functions. See also Appendix B.

In fact, all this works also for higher rank vector bundles, which are parametrized by $H^1(M, \mathrm{Gl}(r, C_M))$ with the difference that the latter cohomology group needs an extra definition, for the sheaf of functions with values in the group of invertible matrices is not abelian for $r > 1$.

In the cocycle language the tangent bundle TM corresponds to $\{D(\varphi_i \circ \varphi_j^{-1}) \circ \varphi_j\}$, where $D(\varphi_j \circ \varphi_j^{-1})$ is the total differential of the transition function.

To any vector bundle $\pi : E \to M$ one associates its sheaf of differentiable sections, also denoted E, by

$$E(U) := \Gamma(U, E) := \{ s : U \to E \mid \pi \circ s = \mathrm{id}_U \}.$$

(This might lead to confusion: In the main body of the text we often speak about holomorphic vector bundles E, which in particular are differentiable vector bundles. Thus, there are two sheaves associated with it: the sheaf of holomorphic sections and the sheaf of differentiable sections, both denoted E. It should be clear from the context which one is meant.)

Recall that vector bundles E, F on a manifold M give rise to new vector bundles (still on M) by taking direct sums $E \oplus F$, tensor products $E \otimes F$, homomorphisms $\mathrm{Hom}(E, F)$, etc. Moreover, any sub-bundle $F \subset E$ is a direct summand of E by writing $E = F \oplus F^\perp$, where $F^\perp$ is the orthogonal complement of F with respect to a metric on E.

In this vain, one defines the *cotangent bundle* $\bigwedge M$ as the dual $(TM)^* = \mathrm{Hom}(TM, M \times \mathbb{R})$ and the bundles of k-forms

$$\bigwedge^k M := \bigwedge^k (\bigwedge M).$$

Their sheaves of sections are given special names: By $\mathcal{A}_M^k$ one denotes the sheaf of sections of $\bigwedge^k M$, the *sheaf of k-forms*. In particular, $\mathcal{A}_M^0 = C_M$, which is a sheaf of real algebras, and all the higher $\mathcal{A}_M^k$ are sheaves of modules over $\mathcal{A}_M^0$.

If $f : M \to N$ is a differentiable map, then there exists a natural pull-back map $f^{-1}\mathcal{A}_N^k \to \mathcal{A}_M^k$ for any k.

Using the local description of $\mathcal{A}^k(U)$ for an open subset $U \subset \mathbb{R}^m$ as the space of differential forms $\sum f_{i_1\ldots i_k} dx_{i_1} \wedge \ldots \wedge dx_{i_k}$ with $f_{i_1\ldots i_k} \in \mathcal{C}(U)$, one defines the *exterior product*

$$\mathcal{A}_M^k \times \mathcal{A}_M^\ell \longrightarrow \mathcal{A}_M^{k+\ell}, \quad (\alpha, \beta) \longmapsto \alpha \wedge \beta$$

which is a map of $\mathcal{C}_M$ sheaves, and the *exterior differential* $d : \mathcal{A}_M^k \to \mathcal{A}_M^{k+1}$ which is a map of sheaves of $\mathbb{R}$-vector spaces only. Using vector fields, i.e. sections of TM, one can invariantly define the exterior differential by

$$(d\alpha)(v_1, \ldots, v_{k+1}) := \sum_{i=1}^{k+1} (-1)^{i+1} v_i \left(\alpha(v_1, \ldots, \hat{v}_i, \ldots, v_{k+1}) \right)$$

$$+ \sum_{1 \leq i < j \leq k+1} (-1)^{i+j} \alpha \left([v_i, v_j], v_1, \ldots, \hat{v}_i, \ldots, \hat{v}_j, \ldots v_{k+1} \right).$$

Here, the Lie bracket is understood as the Lie bracket of derivations.

Since $d^2 = 0$ locally, one obtains a complex of sheaves, the *de Rham complex*

$$\mathcal{C}_M \longrightarrow \mathcal{A}_M^1 \longrightarrow \mathcal{A}_M^2 \longrightarrow \cdots$$

Proposition A.0.3 (Poincaré lemma) *The de Rham complex of sheaves is a resolution of the sheaf of locally constant functions* $\mathbb{R} \subset \mathcal{C}_M$.

Definition A.0.4 The *de Rham cohomology* of a differentiable manifold M is defined as

$$H^k(M, \mathbb{R}) = \frac{\text{Ker} \left(d : \mathcal{A}^k(M) \to \mathcal{A}^{k+1}(M) \right)}{\text{Im} \left(d : \mathcal{A}^{k-1}(M) \to \mathcal{A}^k(M) \right)}.$$

Due to the Poincaré lemma and the fact that the sheaves $\mathcal{A}_M^k$ are acyclic, because they are soft (cf. Definition B.0.38), the de Rham cohomology co-incides with the sheaf cohomology of the sheaf $\mathbb{R}$ of locally constant real functions.

Definition A.0.5 The k-th *Betti number* of M is $b_k(M) := \dim_{\mathbb{R}} H^k(M, \mathbb{R})$. The *Euler number* is $e(M) := \sum (-1)^k b_k(M)$.

Of course, this definition makes only sense if the cohomology is finite dimensional, which is often the case in particular for compact manifolds.

The exterior product yields a multiplicative structure on the de Rham cohomology $H^*(M, \mathbb{R}) = \bigoplus H^k(M, \mathbb{R})$:

$$H^k(M, \mathbb{R}) \times H^\ell(M, \mathbb{R}) \longrightarrow H^{k+\ell}(M, \mathbb{R}) \,.$$

If M is compact and oriented, i.e. $\bigwedge^m M$ is trivial and a trivializing section has been chosen up to scaling by positive functions, then integration yields a linear map

$$\int_M : H^m(M, \mathbb{R}) \longrightarrow \mathbb{R}, \quad [\alpha] \longmapsto \int_M \alpha \,.$$

If M is in addition connected, then the integral is an isomorphism and one obtains a pairing

$$H^k(M, \mathbb{R}) \times H^{m-k}(M, \mathbb{R}) \longrightarrow \mathbb{R}, \quad ([\alpha], [\beta]) \longmapsto \int_M \alpha \wedge \beta.$$

Proposition A.0.6 (Poincaré duality) *Under the above assumptions the pairing is non-degenerate.*

In particular, $b_k(M) = b_{m-k}(M)$. One possible approach to prove the proposition is via harmonic forms (see Corollary A.0.15).

Examples A.0.7 Using the Mayer–Vietoris sequence one can easily compute the Betti-numbers of some of the basic examples:

i) $H^k(S^m, \mathbb{R}) = \mathbb{R}$ for $k = 0, m$ and trivial otherwise.

ii) $H^*(\mathbb{P}^n, \mathbb{R}) = H^{2*}(\mathbb{P}^n, \mathbb{R}) = \mathbb{R}[t]/(t^{n+1})$ with $\deg(t) = 2$. (Here, as always for us, $\mathbb{P}^n$ is the complex projective space.)

Let us now turn to *Riemannian manifolds*. A Riemannian manifold is a differentiable manifold M with the additional structure of a Riemannian metric g, i.e. a section g of $\bigwedge M \otimes \bigwedge M$ inducing a positive definite symmetric bilinear form, i.e. a scalar product, on each $T_x M$.

By standard linear algebra, the metric g also endows $\bigwedge_x M$ and all $\bigwedge_x^k M$ with a natural scalar product.

If M is in addition oriented, then one has a unique m-form, the *volume form*, $\mathrm{vol} = \mathrm{vol}_{(M,g)}$ which is of norm one and positive oriented at every point $x \in M$. The volume of the Riemannian manifold (M, g) is $\int_X \mathrm{vol}_{(M,g)}$.

Let (M, g) be an oriented Riemannian manifold of dimension m. Using the metric and the orientation one introduces the Hodge $*$-operator

$$* : \mathcal{A}^k(M) \longrightarrow \mathcal{A}^{m-k}(M)$$

(cf. Section 1.2). The form $*1$ is the volume form $\mathrm{vol}_{(M,g)}$. The adjoint d^* of the exterior differential d is given by

$$d^* := (-1)^{m(k+1)+1} * d * \quad \text{on} \quad \mathcal{A}^k(M)$$

and the *Laplace operator* is

$$\Delta := d^* d + dd^*.$$

Clearly, d^* is of degree -1, whereas Δ is of degree zero, i.e. Δ induces an endomorphism of each $\mathcal{A}^k(M)$.

Since the metric g induces a natural scalar product on any fibre $\bigwedge_x^k M$ for all $x \in M$, one can introduce a scalar product on the space of global k-forms whenever M is compact.

Definition A.0.8 If (M, g) is a compact oriented Riemannian manifold then for $\alpha, \beta \in \mathcal{A}^k(M)$ one defines

$$(\alpha, \beta) := \int_M g(\alpha, \beta) \cdot \mathrm{vol}_{(M,g)} = \int_M \alpha \wedge *\beta.$$

Lemma A.0.9 *If M is compact, then*

$$(d\alpha, \beta) = (\alpha, d^*\beta) \quad \text{and} \quad (\Delta\alpha, \beta) = (\alpha, \Delta\beta),$$

i.e. d^* is the adjoint operator of d with respect to $(\, , \,)$ and Δ is self-adjoint.

Remark A.0.10 Note that for the standard metric on $\mathbb{R}^m$ the Laplacian Δ as defined above applied to a function $f : \mathbb{R}^m \to \mathbb{R}$ is $\Delta(f) = -\sum \partial^2 f/\partial x_i^2$, i.e. Δ differs from the standard definition by a sign.

Definition A.0.11 The form $\alpha \in \mathcal{A}^k(M)$ is *harmonic* if $\Delta(\alpha) = 0$. The space of all harmonic k-forms is denoted by $\mathcal{H}^k(M, g)$ or simply $\mathcal{H}^k(M)$ if the metric is understood.

Lemma A.0.12 *On a compact oriented manifold (M, g) a form α is harmonic, i.e. $\Delta(\alpha) = 0$, if and only if $d\alpha = d^*\alpha = 0$.*

Corollary A.0.13 *The natural map $\mathcal{H}^k(M) \to H^k(M, \mathbb{R})$ that associates to any harmonic form its cohomology class is injective. (In fact, it is bijective, as will be explained shortly.)*

Lemma A.0.14 *For any Riemannian manifold (M, g), not necessarily compact, one has $*\Delta = \Delta*$ and $* : \mathcal{H}^k(M, g) \cong \mathcal{H}^{m-k}(M, g)$.*

Corollary A.0.15 (Poincaré duality) *Let (M, g) be a compact oriented connected Riemannian manifold. Then the pairing*

$$\mathcal{H}^k(M, g) \times \mathcal{H}^{m-k}(M, g) \longrightarrow \mathbb{R}, \quad (\alpha, \beta) \longmapsto \int_M \alpha \wedge \beta$$

is non-degenerate.

It is not difficult to check that on a compact manifold the three subspaces $d(\mathcal{A}^{k-1}(M))$, $d^*(\mathcal{A}^{k+1}(M))$ and $\mathcal{H}^k(M, g)$ of $\mathcal{A}^k(M)$ are pairwise orthogonal. All results mentioned so far are more or less elementary. The following however requires some hard, but by now standard, analysis.

Theorem A.0.16 (Hodge decomposition) *Let (M, g) be a compact oriented Riemannian manifold. Then, with respect to the scalar product $(\ ,\ )$, there exists an orthogonal sum decomposition with respect to the scalar product $(\ ,\ )$:*

$$\mathcal{A}^k(M) = d(\mathcal{A}^{k-1}(M)) \oplus \mathcal{H}^k(M, g) \oplus d^*(\mathcal{A}^{k+1}(M)).$$

Moreover, the space of harmonic forms $\mathcal{H}^k(M, g)$ is finite-dimensional.

Corollary A.0.17 *The projection in Corollary A.0.13 yields an isomorphism*

$$\mathcal{H}^k(M, g) \cong H^k(M, \mathbb{R}).$$

Equivalently, every cohomology class has a unique harmonic representative.

The following characterization of harmonic forms is often useful.

Lemma A.0.18 *Let (M, g) be a compact oriented Riemannian manifold and let $\alpha \in \mathcal{A}^k(M)$ be closed. Then α is harmonic if and only if $\|\alpha\|$ is minimal among all forms representing the cohomology class $[\alpha]$ of α.*

Comments: - Any textbook on global differential geometry will cover most of what we need, e.g. [12, 10, 79, 102].

- The approach to cohomology via differential forms is stressed in [16].

B

Sheaf Cohomology

In the bulk of the book we make frequent use of the theory of sheaves. It could often be avoided, but it is the appropriate language whenever one studies the interaction of local and global properties of a topological space, e.g. a manifold. The most relevant aspect for our purpose is sheaf cohomology and it would be very unnatural to avoid using this powerful machinery.

The appendix reminds the reader of the basic definitions (sheaves, resolutions, cohomology) and how to use them in some easy situations. With the appendix at hand, the parts of the book where sheaf cohomology is used should be accessible also to the reader with little or no prior knowledge.

In the following, M will be a topological space. In most of the examples it is the topological space underlying a differentiable or even complex manifold.

Definition B.0.19 A *pre-sheaf* $\mathcal{F}$ of abelian groups (or vector spaces, rings, etc.) on M consists of an abelian group (resp. a vector space, a ring, etc.) $\Gamma(U,\mathcal{F}) = \mathcal{F}(U)$ for every open subset $U \subset M$ and a group homomorphism (resp. linear map, ring homomorphism, etc.) $r_{U,V} : \mathcal{F}(V) \to \mathcal{F}(U)$ for any two nested open subsets $U \subset V$ satisfying the following two conditions:

i) $r_{U,U} = \mathrm{id}_{\mathcal{F}(U)}$.
ii) For open subsets $U \subset V \subset W$ one has $r_{U,V} \circ r_{V,W} = r_{U,W}$.

Sometimes, one additionally requires $\mathcal{F}(\emptyset) = 0$. In order to lighten the notation a bit, one may also write $s|_U$ instead of $r_{U,V}(s)$.

Example B.0.20 The basic example is $\mathcal{F} = \mathcal{C}_M^o$, the pre-sheaf of continuous functions on M. More precisely, $\mathcal{C}_M^o(U)$ is the ring of all continuous maps $f : U \to \mathbb{R}$.

For $\mathcal{F} = \mathcal{C}_M^o$ one easily verifies the following additional conditions, which do not hold for arbitrary pre-sheaves. We let $U = \bigcup U_i$ be the union of open subsets $U_i \subset M$. Then:

iii) If $f, g \in \mathcal{C}_M^o(\bigcup U_i)$ with $r_{U_i,U}(f) = r_{U_i,U}(g)$ for all i, then $f = g$.

iv) If functions $f_i \in \mathcal{C}_M^o(U_i)$ are given for all i such that $r_{U_i \cap U_j, U_i}(f_i) = r_{U_i \cap U_j, U_j}(f_j)$ for any j, then there exists a continuous function $f \in \mathcal{C}_M^o(U)$ with $r_{U_i, U}(f) = f_i$ for all i.

This leads to the definition of a sheaf.

Definition B.0.21 A pre-sheaf $\mathcal{F}$ is called a *sheaf* if iii) and iv) are satisfied.

Examples B.0.22 i) The constant pre-sheaf $\mathbb{R}$, for which $\mathcal{F}(U) = \mathbb{R}$ for all open subsets $\varnothing \neq U \subset M$ is not a sheaf as soon as M contains a disconnected open subset.

Usually one works rather with the constant sheaf(!) $\underline{\mathbb{R}}$, which, on an open set $U \subset M$, yields the set of all continuous functions $f : U \to \mathbb{R}$, where $\mathbb{R}$ is endowed with the discrete topology.

Of course, one defines in the same manner constant sheaves associated to other vector spaces, groups, rings, etc. E.g., we frequently use $\underline{\mathbb{Z}}$, which is the constant sheaf associated with $\mathbb{Z}$. Often, the notation is simplified by writing $\mathbb{R}$ and $\mathbb{Z}$ instead of $\underline{\mathbb{R}}$ respectively $\underline{\mathbb{Z}}$.

ii) Another important example is the sheaf $\mathcal{E}$ of sections of a (topological) vector bundle $\pi : E \to M$. By definition $\mathcal{E}(U)$ is the set of all (continuous) maps $s : U \to E$ with $\pi \circ s = \mathrm{id}_U$ (cf. Definition A.0.2).

In fact, $\mathcal{E}$ is a sheaf of $\mathcal{C}_M^o$-modules, i.e. each $\mathcal{E}(U)$ is a $\mathcal{C}_M^o(U)$-module and the restriction maps are compatible with the module structures on the different open subsets.

Since the vector bundle E can be recovered from its sheaf of sections $\mathcal{E}$, one often uses the same notation E for both.

Definition B.0.23 Let $\mathcal{F}$ and $\mathcal{G}$ be two (pre-)sheaves. A *(pre-)sheaf homomorphism* $\varphi : \mathcal{F} \to \mathcal{G}$ is given by group homomorphisms (linear maps, ring homomorphisms, etc.) $\varphi_U : \mathcal{F}(U) \to \mathcal{G}(U)$ for any open subset $U \subset M$ satisfying $r_{U,V}^{\mathcal{G}} \circ \varphi_V = \varphi_U \circ r_{U,V}^{\mathcal{F}}$ for any $U \subset V$.

Once a homomorphism $\varphi : \mathcal{F} \to \mathcal{G}$ of (pre-)sheaves of abelian groups is given, one constructs the associated pre-sheaves $\mathrm{Ker}(\varphi)$, $\mathrm{Im}(\varphi)$, and $\mathrm{Coker}(\varphi)$ which are defined in the obvious way, e.g. $\mathrm{Coker}(\varphi)(U) = \mathrm{Coker}(\varphi_U : \mathcal{F}(U) \to \mathcal{G}(U))$.

There is an important subtlety here. If φ is a sheaf homomorphism then $\mathrm{Ker}(\varphi)$ is a sheaf itself, but $\mathrm{Im}(\varphi)$ and $\mathrm{Coker}(\varphi)$, in general, are just presheaves.

In order to define the cokernel and the image of a sheaf homomorphism as honest sheaves, one needs to introduce the notion of a stalk.

Definition B.0.24 Let $\mathcal{F}$ be a (pre-)sheaf on M and $x \in M$. Then the *stalk* of $\mathcal{F}$ at x is

$$\mathcal{F}_x := \{(U, s) \mid x \in U \subset M, \ s \in \mathcal{F}(U)\}/ \sim .$$

Here, for two open subsets U_i, $i = 1, 2$ and sections $s_i \in \mathcal{F}(U_i)$, $i = 1, 2$, one sets $(U_1, s_1) \sim (U_2, s_2)$ if there exists an open subset $x \in U \subset U_1 \cap U_2$ such that $r_{U, U_1}(s_1) = r_{U, U_2}(s_2)$.

Equivalently, one could introduce the stalk $\mathcal{F}_x$ as the direct limit $\mathcal{F}_x = \lim_{x \in U} \mathcal{F}(U)$.

One immediately finds that any section $s \in \mathcal{F}(U)$ induces an element $s_x \in \mathcal{F}_x$ for any point $x \in U$. Furthermore, any (pre-)sheaf homomorphism $\varphi : \mathcal{F} \to \mathcal{G}$ induces homomorphisms $\mathcal{F}_x \to \mathcal{G}_x$ for any $x \in M$.

Definition B.0.25 The *sheaf* $\mathcal{F}^+$ *associated to a pre-sheaf* $\mathcal{F}$ is the sheaf for which $\mathcal{F}^+(U)$ of an open subset $U \subset M$ is the set of all maps $s : U \to \bigcup_{x \in U} \mathcal{F}_x$ with $s(x) \in \mathcal{F}_x$ and such that for all $x \in U$ there exists an open subset $x \in V \subset U$ and a section $t \in \mathcal{F}(V)$ with $s(y) = t(y)$ for all $y \in V$.

With this definition, $\mathcal{F}^+$ is a sheaf and the natural inclusion $\mathcal{F} \subset \mathcal{F}^+$ is an isomorphism if the pre-sheaf $\mathcal{F}$ was already a sheaf. For many constructions one needs to pass from a naturally defined pre-sheaf to its *sheafification*. E.g. the tensor product $\mathcal{F} \otimes_\mathcal{R} \mathcal{G}$ of two $\mathcal{R}$-modules $\mathcal{F}$ and $\mathcal{G}$ is defined as the sheafification of $U \mapsto \mathcal{F}(U) \otimes_{\mathcal{R}(U)} \mathcal{G}(U)$.

Definition B.0.26 Let $\varphi : \mathcal{F} \to \mathcal{G}$ be a homomorphism of sheaves. Then the *image sheaf* $\text{Im}(\varphi)$ is the sheaf associated with the image pre-sheaf $U \mapsto \text{Im}(\varphi_U)$. Analogously, one defines the *cokernel sheaf* $\text{Coker}(\varphi)$.

The sheaf homomorphism φ is injective if and only if $\text{Ker}(\varphi)$ is trivial. Similarly, one says that φ is surjective if its cokernel sheaf $\text{Coker}(\varphi)$ is trivial. The essential difference between these two properties is that φ is injective if and only if φ_U is injective for any open subset U. On the other hand, φ might be surjective without φ_U being surjective for all/any open subset. However, both properties can be detected by their stalks. More precisely, φ is injective or surjective if and only if $\varphi_x : \mathcal{F}_x \to \mathcal{G}_x$ is injective respectively surjective for any point $x \in M$.

Definition B.0.27 A sequence $\mathcal{F}^\bullet$ of sheaf homomorphisms

$$\cdots \longrightarrow \mathcal{F}^i \xrightarrow{\varphi^i} \mathcal{F}^{i+1} \xrightarrow{\varphi^{i+1}} \mathcal{F}^{i+2} \xrightarrow{\varphi^{i+2}} \cdots$$

is a *complex* if $\varphi^{i+1} \circ \varphi^i = 0$ for all i. It is an *exact complex* if $\text{Ker}(\varphi^{i+1}) = \text{Im}(\varphi^i)$ for all i.

An exact complex of the form $0 \longrightarrow \mathcal{F}^0 \longrightarrow \mathcal{F}^1 \longrightarrow \mathcal{F}^2 \longrightarrow 0$ is called *short exact sequence*.

Corollary B.0.28 *A complex of the form*

$$0 \longrightarrow \mathcal{F}^0 \longrightarrow \mathcal{F}^1 \longrightarrow \mathcal{F}^2 \longrightarrow 0$$

is exact if and only if the induced complex of stalks

$$0 \longrightarrow \mathcal{F}^0_x \longrightarrow \mathcal{F}^1_x \longrightarrow \mathcal{F}^2_x \longrightarrow 0$$

is exact for any $x \in M$.

Since surjectivity does not mean surjectivity for any open subset, a short exact sequence as above does not necessarily define short exact sequences

$$0 \longrightarrow \mathcal{F}^0(U) \longrightarrow \mathcal{F}^1(U) \longrightarrow \mathcal{F}^2(U)(\longrightarrow 0)$$

for any open subset $U \subset M$ and in particular not for M. This is where cohomology comes in. It turns out that the failure of surjectivity of $\mathcal{F}^1(M) \to \mathcal{F}^2(M)$ is measured by the cohomology of $\mathcal{F}^0$.

In order to introduce sheaf cohomology, one has to make a choice. There is the theoretically superior but rather abstract approach via derived categories or the more ad hoc one using acyclic resolutions. We outline the second one.

One first has to single out special sheaves with no cohomology in order to define cohomology for all other ones by resolving them.

Definition B.0.29 A *resolution* of a sheaf $\mathcal{F}$ is a complex $0 \to \mathcal{F}^0 \to \mathcal{F}^1 \to \ldots$ together with a homomorphism $\mathcal{F} \to \mathcal{F}^0$ such that

$$0 \longrightarrow \mathcal{F} \longrightarrow \mathcal{F}^0 \longrightarrow \mathcal{F}^1 \longrightarrow \mathcal{F}^2 \longrightarrow \cdots$$

is an exact complex of sheaves.

One possible choice for sheaves without cohomology is provided by flasque sheaves.

Definition B.0.30 A sheaf $\mathcal{F}$ is called *flasque* if for any open subset $U \subset M$ the restriction map $r_{U,M} : \mathcal{F}(M) \to \mathcal{F}(U)$ is surjective.

Why flasque sheaves are the right ones is explained by the following

Lemma B.0.31 *If*

$$0 \longrightarrow \mathcal{F}^0 \longrightarrow \mathcal{F}^1 \longrightarrow \mathcal{F}^2 \longrightarrow 0$$

is a short exact sequence and $\mathcal{F}^0$ is flasque, then the induced sequence

$$0 \longrightarrow \mathcal{F}^0(U) \longrightarrow \mathcal{F}^1(U) \longrightarrow \mathcal{F}^2(U) \longrightarrow 0$$

is exact for any open subset $U \subset M$.

Next, one has to ensure that any sheaf can be resolved by flasque sheaves. This will allow to define the cohomology of any sheaf.

Proposition B.0.32 *Any sheaf $\mathcal{F}$ on M admits a resolution*

$$0 \longrightarrow \mathcal{F} \longrightarrow \mathcal{F}^0 \longrightarrow \mathcal{F}^1 \longrightarrow \mathcal{F}^2 \longrightarrow \cdots$$

such that all sheaves $\mathcal{F}^i$, $i = 0, 1, \ldots$ are flasque.

Definition B.0.33 The *i-th cohomology group* $H^i(M, \mathcal{F})$ of a sheaf $\mathcal{F}$ is the *i*-th cohomology of the complex

$$\mathcal{F}^0(M) \xrightarrow{\varphi^0} \mathcal{F}^1(M) \xrightarrow{\varphi^1} \mathcal{F}^2(M) \xrightarrow{\varphi^2} \cdots$$

induced by a flasque resolution $\mathcal{F} \to \mathcal{F}^\bullet$. Explicitly,

$$H^i(M, \mathcal{F}) = \frac{\mathrm{Ker}\left(\varphi^i_M : \mathcal{F}^i(M) \to \mathcal{F}^{i+1}(M)\right)}{\mathrm{Im}\left(\varphi^{i-1}_M : \mathcal{F}^{i-1}(M) \to \mathcal{F}^i(M)\right)}.$$

Clearly, with this definition any flasque sheaf $\mathcal{F}$ has vanishing cohomology $H^i(M, \mathcal{F}) = 0$ for $i > 0$. Moreover, for any sheaf $\mathcal{F}$ one has $H^0(M, \mathcal{F}) = \Gamma(M, \mathcal{F}) = \mathcal{F}(M)$. That this definition of cohomology is really independent of the chosen flasque resolution is due to

Proposition B.0.34 *If $\mathcal{F} \to \mathcal{F}^\bullet$ and $\mathcal{F} \to \mathcal{G}^\bullet$ are two flasque resolutions of a sheaf $\mathcal{F}$ then both define naturally isomorphic cohomology groups.*

The most striking feature of cohomology is that it explains fully the non-exactness of short exact sequences on the level of global sections.

Proposition B.0.35 *Let*

$$0 \longrightarrow \mathcal{F}^0 \longrightarrow \mathcal{F}^1 \longrightarrow \mathcal{F}^2 \longrightarrow 0$$

be a short exact sequence of sheaves on M. Then there exists a long exact cohomology *sequence*

$$0 \longrightarrow H^0(M, \mathcal{F}^0) \longrightarrow H^0(M, \mathcal{F}^1) \longrightarrow H^0(M, \mathcal{F}^2)$$

$$H^1(M, \mathcal{F}^0) \longrightarrow H^1(M, \mathcal{F}^1) \longrightarrow H^1(M, \mathcal{F}^2)$$

$$H^2(M, \mathcal{F}^0) \longrightarrow H^2(M, \mathcal{F}^1) \longrightarrow H^2(M, \mathcal{F}^2) \ldots$$

Definition B.0.36 Suppose $\mathcal{F}$ is a sheaf of vector spaces on M. Then $h^i(M, \mathcal{F})$ denotes the dimension of $H^i(M, \mathcal{F})$, which inherits a natural vector space structure. If all $h^i(M, \mathcal{F})$ are finite and only finitely many are non-trivial ($\mathcal{F}$ has 'finite cohomology'), one defines the *Euler–Poincaré characteristic* of $\mathcal{F}$ as

$$\chi(M, \mathcal{F}) := \sum (-1)^i h^i(M, \mathcal{F}).$$

An easy consequence of the existence of the long exact cohomology sequence is

Corollary B.0.37 *Let*

$$0 \longrightarrow \mathcal{F}^0 \longrightarrow \mathcal{F}^1 \longrightarrow \mathcal{F}^2 \longrightarrow 0$$

be a short exact sequences of sheaves of vector spaces with finite cohomology. Then

$$\chi(M, \mathcal{F}^1) = \chi(M, \mathcal{F}^0) + \chi(M, \mathcal{F}^2).$$

It might happen that flasque resolutions are difficult to find. But in order to compute the cohomology of a sheaf, any resolution by *acyclic sheaves*, i.e. sheaves with trivial higher cohomology groups, can be used. What kind of acyclic sheaves are convenient depends on the situation. For topological, differentiable, and complex manifolds, the following one is very useful. For the Zariski topology one has to use different ones.

Definition B.0.38 A sheaf $\mathcal{F}$ is called *soft* if the restriction $\Gamma(M, \mathcal{F}) \to \Gamma(K, \mathcal{F})$ is surjective for any closed subset $K \subset M$.

The space of sections $\Gamma(K, \mathcal{F})$ of $\mathcal{F}$ over the closed set K is defined as the direct limit of the spaces of sections over all open neighbourhoods of K.

Proposition B.0.39 *Soft sheaves are acyclic. Any sheaf of modules over a soft sheaf of commutative rings is soft and hence acyclic.*

This is frequently applied to the sheaf of continuous (or differentiable) functions on a manifold which is easily shown to be soft. Notice that the sheaf of holomorphic functions on a complex manifold is not soft.

Čech cohomology is another cohomology theory. It has the advantage to be defined without using any sheaf resolution. Since it often coincides with the cohomology defined above, we will sketch the main steps of its construction.

Let us first fix an open covering $M = \bigcup_i U_i$ with I an ordered set and consider the intersections $U_{i_0 \ldots i_p} := U_{i_0} \cap \ldots \cap U_{i_p}$. Then we set

$$C^p(\{U_i\}, \mathcal{F}) := \prod_{i_0 < \ldots < i_p} \Gamma(U_{i_0 \ldots i_p}, \mathcal{F}).$$

There is a natural differential

$$d : C^p(\{U_i\}, \mathcal{F}) \longrightarrow C^{p+1}(\{U_i\}, \mathcal{F}), \quad \alpha = \prod \alpha_{i_0 \ldots i_p} \longmapsto d\alpha$$

with

$$(d\alpha)_{i_0 \ldots i_{p+1}} = \sum_{k=0}^{p+1} (-1)^k \alpha_{i_0 \ldots \widehat{i_k} \ldots i_{p+1}} |_{U_{i_0 \ldots i_{p+1}}}$$

A calculation shows that $d^2 = 0$, i.e.

$$C^0(\{U_i\}, \mathcal{F}) \xrightarrow{\ d\ } C^1(\{U_i\}, \mathcal{F}) \xrightarrow{\ d\ } \cdots$$

is a complex. Since $\mathcal{F}$ is a sheaf, one finds $\Gamma(M, \mathcal{F}) = \mathrm{Ker}(d : C^0(\{U_i\}, \mathcal{F}) \to C^1(\{U_i\}, \mathcal{F}))$.

Definition B.0.40 The i-th *Čech cohomology group* with respect to the fixed open covering $M = \bigcup_i U_i$ is

$$\check{H}^i(\{U_i\}, \mathcal{F}) = \frac{\operatorname{Ker}\left(d : C^i(\{U_i\}, \mathcal{F}) \to C^{i+1}(\{U_i\}, \mathcal{F})\right)}{\operatorname{Im}\left(d : C^{i-1}(\{U_i\}, \mathcal{F}) \to C^i(\{U_i\}, \mathcal{F})\right)}.$$

In this way, we have defined cohomology groups without using acyclic sheaves, but which still depend on the open covering $M = \bigcup U_i$. This can be remedied by passing to the limit. More precisely, if $M = \bigcup_i U_i$ is refined by an open covering $M = \bigcup_j V_j$, then there exists a natural map

$$\check{H}^i(\{U_i\}, \mathcal{F}) \longrightarrow \check{H}^i(\{V_j\}, \mathcal{F})$$

and one defines the Čech cohomology of a sheaf without specifying an open covering as

$$\check{H}^i(M, \mathcal{F}) := \varinjlim \check{H}^i(\{U_i\}, \mathcal{F}).$$

Examples B.0.41 i) Let us compute the Čech cohomology of S^1. Take the standard open covering of S^1 by the two hemispheres U_1 and U_2, both homeomorphic to $\mathbb{R}^1$. Their intersection $U_1 \cap U_2$ consists of two disjoints open intervals. Hence, the Čech complex in this situation for $\mathcal{F} = \mathbb{Z}$ is

$$\mathbb{Z} \times \mathbb{Z} \longrightarrow \mathbb{Z} \times \mathbb{Z}, \quad (a, b) \longmapsto (a - b, a - b).$$

This determines the Čech cohomology of S^1 with respect to this open covering, but in fact also in the limit: $\check{H}^i(S^1, \mathbb{Z}) = \mathbb{Z}$ for $i = 0, 1$ and $= 0$ otherwise.

ii) This calculation can be generalized in two ways, depending on whether one views S^2 as a sphere or as the projective line $\mathbb{P}^1$.

For the higher dimensional spheres S^n one finds that $\check{H}^i(S^n, \mathbb{Z}) = \mathbb{Z}$ for $i = 0, n$ and trivial otherwise.

For the complex projective space $\mathbb{P}^n$ one considers the standard open covering by open subsets homeomorphic to $\mathbb{C}^n$. This yields $\check{H}^i(\mathbb{P}^n, \mathbb{Z}) = \mathbb{Z}$ for $i = 0, 2, \ldots, 2n$ and $\check{H}^i(\mathbb{P}^n, \mathbb{Z}) = 0$ otherwise.

Using a sheaf version of the Čech complex one obtains

Proposition B.0.42 *For any open covering $M = \bigcup_i U_i$ there exists a natural homomorphism*

$$\check{H}^i(\{U_i\}, \mathcal{F}) \longrightarrow H^i(M, \mathcal{F}).$$

Of course, these homomorphisms are in general not bijective, as the open covering might be very coarse. However, passing to the limit often results in isomorphisms with the true cohomology groups, i.e. when the topological space is reasonable (e.g. paracompact), the induced maps $\check{H}^i(M, \mathcal{F}) \to H^i(M, \mathcal{F})$ are indeed bijective. E.g. the examples in B.0.41 do compute the true cohomology groups $H^i(S^n, \mathbb{Z})$ and $H^i(\mathbb{P}^n, \mathbb{Z})$. The case that interests us most is:

Proposition B.0.43 *The natural map $\check{H}^1(M, \mathcal{F}) \to H^1(M, \mathcal{F})$ is always bijective.*

This opens the way to parametrize line bundles by cohomology classes. Let us sketch this for differentiable real line bundles.

Let $\pi : L \to M$ be a line bundle that can be trivialized over open subsets U_i of an open cover $M = \bigcup U_i$ by maps $\psi_i : L|_{U_i} \cong U_i \times \mathbb{R}$. Then $\{\psi_{ij} := \psi_i \circ \psi_j^{-1} \in \mathcal{C}_M^*(U_i \cap U_j)\}$ can be considered as a cocycle and thus gives rise to an element in $\check{H}^1(\{U_i\}, \mathcal{C}_M^*)$, where $\mathcal{C}_M^*$ is the sheaf of differentiable functions without zeros.

One verifies that this cohomology class does not depend on the choice of the trivializations ψ_i. Moreover, any class in $\check{H}^1(\{U_i\}, \mathcal{C}_M^*)$ can be interpreted as a cocycle of a line bundle trivialized over the open subsets U_i.

Thus, by definition every line bundle L gives rise to a cohomology class in the covering independent Čech cohomology group $\check{H}^1(M, \mathcal{C}_M^*)$ which does not depend neither on the open covering nor on the trivializations. The above proposition then yields

Corollary B.0.44 *There is a bijection between the set of isomorphism classes of real line bundles on a differentiable manifold M and the group $H^1(M, \mathcal{C}_M^*)$.*

There is also the algebraic topology way of computing cohomology. Let M be a manifold. A continuous map $\gamma : [0,1]^k \to M$ is called a k-chain. For any $t \in [0,1]$ and any $1 \leq i \leq k$ one obtains a $k-1$-chain $d(i,t)(\gamma) : [0,1]^{k-1} \to M$ by $d(i,t)(\gamma)(t_1, \ldots, t_{k-1}) = \gamma(t_1, \ldots, t_{i-1}, t, t_i, \ldots, t_{k-1})$. The k-chain γ is non-degenerate if for any i the map $t \mapsto d(i,t)(\gamma)$ is non-constant.

Let for any open subset $U \subset M$ the group $C_k(U)$ be the abelian group generated by k-chains in U. The differential $d : C_k(U) \to C_{k-1}(U)$ is by definition $d(\gamma) := \sum_{i=1}^{k} (-1)^i (d(i,0)(\gamma) - d(i,1)(\gamma))$. This yields a complex

$$\cdots \longrightarrow C_{k+1} \longrightarrow C_k(U) \longrightarrow C_{k-1}(U) \longrightarrow \cdots$$

We are more interested in its dual

$$\cdots \longrightarrow C^{k-1}(U) \longrightarrow C^k(U) \longrightarrow C^{k+1}(U) \longrightarrow \cdots$$

with $C^k(U) = \mathrm{Hom}(C_k(U), \mathbb{Z})$.

It turns out, that for the unit ball in $\mathbb{R}^n$ the sequence

$$\mathbb{Z} \longrightarrow C^0(U) \longrightarrow C^1(U) \longrightarrow C^2(U) \longrightarrow \cdots$$

is exact. It is not difficult to verify that the sheaves $\mathcal{C}^k$ are actually soft and one, therefore, obtains a soft resolution of the constant sheaf

$$\mathbb{Z} \longrightarrow \mathcal{C}^0 \longrightarrow \mathcal{C}^1 \longrightarrow \mathcal{C}^2 \longrightarrow \cdots$$

Thus, the sheaf cohomology $H^i(M, \mathbb{Z})$ of the constant sheaf $\mathbb{Z}$ on M can be computed as the cohomology of the complex

$$C^0(M) \longrightarrow C^1(M) \longrightarrow C^2(M) \longrightarrow \cdots.$$

Similarly, one can compute the cohomology $H^i(M, \mathbb{R})$ by using $\mathrm{Hom}(C_k, \mathbb{R})$.

The advantage of the approach is that it allows to define the fundamental class of any submanifold $N \subset M$ of an oriented compact manifold M.

Indeed, if $\gamma : [0,1]^k \to M$ is a k-chain, where k is the codimension of $N \subset M$, then one might define the value of $[N] \in H^k(M, \mathbb{Z})$ as the number of intersection points of N and γ. (One has to show that any γ is homologous to a chain transversal to N.) Passing to the real cohomology $H^k(M, \mathbb{R})$ one recovers the fundamental class defined by $\alpha \mapsto \int_N \alpha$.

Evidently, a good theory should also deal with morphisms. E.g. the *direct image sheaf* $f_* \mathcal{F}$ of a sheaf $\mathcal{F}$ on M under a continuous map $f : M \to N$ is the sheaf $U \mapsto \mathcal{F}(f^{-1}(U))$.

We rarely use direct images (let alone higher direct images), but we do use the pull-back of a sheaf. In fact, there are usually two of them.

Definition B.0.45 Let $f : M \to N$ be a continuous map and $\mathcal{F}$ a sheaf on N. Then the *inverse image* $f^{-1}\mathcal{F}$ is the sheaf on M such that for an open subset $U \subset M$ its space of sections $(f^{-1}\mathcal{F})(U)$ is the set of all maps $s : U \to \bigcup_{x \in U} \mathcal{F}_{f(x)}$ such that for any $x \in U$ there exist open subsets $x \in U_x \subset U$, $f(U_x) \subset V \subset N$ and an element $t \in \mathcal{F}(V)$ with $s(y) = t(f(y))$ for all $y \in U_x$.

Clearly, with this definition the stalks are given by $(f^{-1}\mathcal{F})_x = \mathcal{F}_{f(x)}$.

Often, one considers a space M together with a sheaf of 'functions' $\mathcal{R}_M$, e.g. $\mathcal{R}_M$ could be any subsheaf (of rings) of $\mathcal{C}_M$ (of differentiable, analytic, or holomorphic functions). One calls $(M, \mathcal{R}_M)$ a *ringed space*. A continuous map $f : (M, \mathcal{R}_M) \to (N, \mathcal{R}_N)$ between two ringed spaces consist of a continuous map $f : M \to N$ and a homomorphism of sheaves of rings $f^{-1}\mathcal{R}_N \to \mathcal{R}_M$. (Compare this with the definition of a complex space 6.2.8.)

Definition B.0.46 Let $f : (M, \mathcal{R}_M) \to (N, \mathcal{R}_N)$ be a continuous map of ringed spaces. Then the *pull-back* $f^*\mathcal{F}$ of any $\mathcal{R}_N$-module $\mathcal{F}$ on N is the $\mathcal{R}_M$-module given by $f^{-1}\mathcal{F} \otimes_{f^{-1}\mathcal{R}_N} \mathcal{R}_M$.

Remark B.0.47 If $(M, \mathcal{R}_M)$ is a locally ringed space, i.e. the stalks of $\mathcal{R}_M$ are all local, then one can define the *fibre* of any $\mathcal{R}_M$-module $\mathcal{F}$ at a point $x \in M$ as $\mathcal{F}(x) := \mathcal{F} \otimes_{\mathcal{R}_{M,x}} \mathcal{R}_{M,x}/\mathfrak{m}$, where $\mathfrak{m} \subset \mathcal{R}_{M,x}$ is the maximal ideal. One should be aware of the essential difference between the stalk $\mathcal{F}_x$ and the fibre $\mathcal{F}(x)$ of a sheaf.

If $\mathcal{F}$ is the sheaf of sections of a vector bundle E, then its fibre $\mathcal{F}(x)$ is nothing but the fibre $E(x) = \pi^{-1}(x)$ of the vector bundle. Its stalks are much bigger.

Comment: For a complete treatment of sheaf theory we recommend [71, 20] or the classic [55]. A survey of some basic aspects can also be found in [116]. Singular cohomology is defined in [51].

References

1. J. Amorós, M. Burger, K. Corlette, D. Kotschick, D. Toledo *Fundamental groups of compact Kähler manifolds.* Math. Surv. Mono. 44. AMS (1996).
2. M. Atiyah, I. MacDonald *Introduction to commutative algebra.* Addison-Wesley Publishing Co. (1969).
3. M. Atiyah *Complex analytic connections in fibre bundles.* Trans. Amer. Math. Soc. 85 (1957), 181–207.
4. M. Atiyah *Vector bundles over an elliptic curve.* Proc. London Math. Soc. (3) 7 (1957), 414–452.
5. *Première classe de Chern et courbure de Ricci : preuve de la conjecture de Calabi.* Astérisque 58 (1978).
6. A. Banyaga *The structure of classical diffeomorphism groups.* Math. and its appl. 400. Kluwer (1997).
7. S. Barannikov, M. Kontsevich *Frobenius manifolds and formality of Lie algebras of polyvector fields.* Internat. Math. Res. Notices no. 4 (1998), 201–215.
8. W. Barth, C. Peters, A. Van de Ven *Compact complex surfaces.* Erg. Math. Grenzgebiete (3), 4. Springer (1984).
9. A. Beauville *Riemannian Holonomy and Algebraic Geometry.* Preprint (1999).
10. M. Berger, B. Gostiaux *Differential geometry: manifolds, curves, and surfaces.* GTM 115. Springer (1988).
11. M. Berger, A. Lascoux *Variétés Kähleriennes compactes.* Lect. Notes Math. 154. Springer (1970).
12. A. Besse *Einstein manifolds.* Erg. Math. Grenzgebiete (3) 10. Springer (1987).
13. C. Birkenhake, H. Lange *Complex tori.* Progr. Math. 177. Birkhäuser (1999).
14. F. Bogomolov *Unstable vector bundles and curves on surfaces.* Proc. Int. Congr. Math. (Helsinki, 1978), 517–524.
15. B. Booss, D. Bleecker *Topology and analysis. The Atiyah-Singer index formula and gauge-theoretic physics.* Universitext. Springer (1985).
16. R. Bott, L. Tu *Differential forms in algebraic topology.* GTM 82. Springer (1982).
17. N. Bourbaki *Topologie Générale.* Hermann (1971).
18. J.-P. Bourguignon *Eugenio Calabi and Kähler Metrics.* Manifolds and geometry (Pisa, 1993), 61–85. Sympos. Math., XXXVI, Cambridge Univ. Press, Cambridge, (1996).

19. J.-P. Bourguignon *Métriques d'Einstein-Kähler sur les variétés de Fano: obstructions et existence (d'après Y. Matsushima, A. Futaki, S. T. Yau, A. Nadel et G. Tian).* Séminaire Bourbaki Exp. No. 830. Astérisque 245 (1997), 277–305.

20. G. Bredon *Sheaf theory.* GTM 170. Springer (1997).

21. J. Carlson, M. Green, P. Griffiths, J. Harris *Infinitesimal variations of Hodge structure. I.* Comp. Math. 50 (1983), 109–205.

22. J. Carlson, C. Peters, St. Müller-Stach *Period mappings and Period domains.* Cambridge Univ. Press (2003).

23. H.D. Cao, J. Zhou *Susy in CY geometry.* Preprint.

24. B. Chen, K. Ogiue *Some characterizations of complex space forms in terms of Chern classes.* Quart. J. Math. Oxford 26 (1975), 459–464.

25. S.-S. Chern *Complex manifolds without potential theory.* Second edition. Universitext. Springer (1979).

26. S. S. Chern *Characteristic classes of Hermitian Manifolds.* Ann. Math. 47 (1946), 85-121.

27. Clay Milleniums problems: www.claymath.org/Millennium_Prize_Problems.

28. O. Debarre *Tores et variétés abéliennes complexes.* Cours Spécialisés 6. SMF, (1999).

29. O. Debarre *Higher-dimensional algebraic geometry.* Universitext. Springer (2001).

30. J. de Jong *Smoothness, semi-stability and alterations.* Inst. Hautes Études Sci. Publ. Math. 83 (1996), 51–93.

31. P. Deligne *Équations différentielles à points singuliers réguliers.* Lect. Notes Math. 163. Springer (1970).

32. P. Deligne, L. Illusie *Relèvements modulo p^2 et décomposition du complexe de de Rham.* Invent. Math. 89 (1987), 247–270.

33. P. Deligne, P. Griffiths, J. Morgan, D. Sullivan *Real homotopy theory of Kähler manifolds.* Invent. Math. 29 (1975), 245–274.

34. P. Deligne, J. Morgan *Notes on supersymmetry (following J. Bernstein).* in Quantum Fields and Strings. A Course for Mathematicians. Vol. 1. AMS/IAS (1999).

35. J.-P. Demailly *Complex analytic and algebraic geometry.* http://www-fourier.ujf-grenoble.fr/~demailly/books.html.

36. S. Donaldson *Anti self-dual Yang–Mills connections over complex algebraic surfaces and stable vector bundles.* Proc. London Math. Soc. 50 (1985), 1–26.

37. S. Donaldson *Scalar curvature and projective embeddings. I.* J. Diff. Geom. 59 (2001), 479–522.

38. S. Donaldson, P. Kronheimer *The Geometry of four-manifolds.* Oxford University Press (1990).

39. A. Douady *Le problème des modules pour les variétés analytiques complexes.* Séminaires Bourbai 277 (1964/65).

40. J. Duistermaat, J. Kolk *Lie groups.* Universitext. Springer (2000).

41. F. El Zein *Introduction à la théorie de Hodge mixte.* Act. Math. Hermann, (1991).

42. H. Esnault, E. Viehweg *Lectures on vanishing theorems.* DMV Seminar, 20. Birkhäuser, (1992).

43. Y. Félix, St. Halperin, J.-C. Thomas *Rational homotopy theory.* GTM 205. Springer (2001).

44. J. Figueroa-O'Farrill, C. Köhl, B. Spence *Supersymmetry and the cohomology of (hyper)Kähler manifolds.* Nuclear Phys. B 503 (1997), 614–626.

45. W. Fischer, H. Grauert *Lokal triviale Familien kompakter komplexer Mannigfaltigkeiten.* Nachr. Akad. Wiss. Göttingen II (1965), 89–94.

46. J. Fröhlich, O. Grandjean, A. Recknagel *Supersymmetric quantum theory and differential geometry.* Comm. Math. Phys. 193 (1998), 527–594.

47. W. Fulton *Intersection theory.* Erg. Math. Grenzgebiete (2) 3. Springer (1998).

48. W. Fulton *Introduction to toric varieties.* Annals Math. Stud., 131. Princeton University Press (1993).

49. W. Fulton, S. Lang *Riemann–Roch algebra.* Grundlehren Math. Wiss. 277. Springer (1985).

50. W. Fulton, R. Lazarsfeld *On the connectedness of degeneracy loci and special divisors.* Acta Math. 146 (1981), 271–283.

51. W. Fulton *Algebraic topology. A first course.* GTM 153. Springer (1995).

52. A. Fujiki *On the de Rham cohomology group of a compact Kähler symplectic manifold.* Alg. geom., Sendai, 1985, 105–165, Adv. Stud. Pure Math., 10, (1987).

53. B. van Geemen *Kuga–Satake varieties and the Hodge conjecture.* The arithmetic and geometry of algebraic cycles. Kluwer (2000), 51–82.

54. P. Gilkey *The index theorem and the heat equation.* Math. Lecture Ser. 4. Publish or Perish, Inc. (1974).

55. R. Godement *Topologie algébrique et théorie des faisceaux.* Hermann (1958).

56. B. Gordon *A survey of the Hodge conjecture for abelian varieties.* Appendix to J. Lewis *A survey of the Hodge conjecture.* AMS, (1999).

57. H. Grauert, R. Remmert *Analytische Stellenalgebren.* Grundlehren math. Wiss. 176. Springer (1971).

58. H. Grauert, R. Remmert *Coherent Analytic Sheaves.* Grundlehren math. Wiss. 265. Springer (1984).

59. P. Griffith, J. Harris *Principles of Algebraic Geometry.* Wiley (1978).

60. P. Griffiths, J. Morgan *Rational homotopy theory and differential forms.* Progr. Math. 16. Birkhäuser (1981).

61. M. Gross, D. Huybrechts, D. Joyce *Calabi–Yau manifolds and Related Geometries.* Universitext. Springer (2003).

62. A. Grothendieck *Standard conjectures on algebraic cycles.* Algebraic Geometry (Bombay 1968) Oxford University Press London (1969), 193–199.

63. A. Grothendieck *Sur la classification des fibrés holomorphes sur la sphère de Riemann.* Amer. J. Math. 79 (1957), 121–138.

64. R. Gunning, H. Rossi *Analytic Functions of Several Complex Variables.* Prentice Hall (1965).

65. J. Harris *Algebraic geometry. A first course.* GTM 133. Springer (1992).

66. R. Hartshorne *Algebraic geometry.* GTM 52. Springer (1977).

67. H. Hironaka *Resolution of singularities of an algebraic variety over a field of characteristic zero. I, II.* Ann. Math. 79 (1964), 109–326.

68. F. Hirzebruch, D. Zagier *The Atiyah-Singer theorem and elementary number theory.* Math. Lecture Ser. 3. Publish or Perish (1974).

69. F. Hirzebruch *The signature theorem. Reminiscences and recreation.* Prospects in math. 3–31. Ann. Math. Stud. 70, Princeton Univ. Press (1971).

70. D. Huybrechts, M. Lehn *The geometry of the moduli spaces of shaves.* Asp. of Math. E 31. Vieweg (1997).

71. B. Iversen *Cohomology of sheaves*. Universitext. Springer (1986).
72. D. Joyce *Compact manifolds with special holonomy*. Oxford University Press (2000).
73. E. Kähler *Über eine bemerkenswerte Hermitesche Metrik*. Abh. Math. Sem. Hamburg Univ. 9 (1933), 173–186.
74. Y. Kawamata *Unobstructed deformations. A remark on a paper of Z. Ran: "Deformations of manifolds with torsion or negative canonical bundle"*. J. Alg. Geom. 1 (1992), 183–190.
75. M. Kapranov *Rozansky–Witten invariants via Atiyah classes*. Comp. Math. 115 (1999) 71–113.
76. G. Kempf *Complex abelian varieties and theta functions*. Universitext. Springer (1991).
77. F. Kirwan *Cohomology of quotients in symplectic and algebraic geometry*. Math. Notes 31. Princeton University Press (1984).
78. S. Kobayashi *Differential Geometry of Complex Vector Bundles*. Publ. Math. Soc. Japan 15, Princeton Univ. Press and Iwanami Shoten, (1987).
79. S. Kobayashi, K. Nomizu *Foundations of differential geometry*. Wiley.
80. K. Kodaira *Complex manifolds and deformation of complex structures*. Grundlehren math. Wiss. 283. Springer (1986).
81. J. Kollár, T. Matsusaka *Riemann–Roch type inequalities*. Amer. J. Math. 105 (1983), 229–252.
82. A. Lamari *Courants kählériens et surfaces compactes*. Ann. Inst. Fourier 49 (1999), 263–285.
83. R. Lazarsfeld *Positivity in Algebraic Geometry*. to appear. Springer (2004).
84. C. Le Brun *Topology versus Chern numbers for complex 3-folds*. Pac. J. 191 (1999), 132–131.
85. J. Le Potier *Annulation de la cohomolgie à valeurs dans un fibré vectoriel holomorphe positif de rang quelconque*. Math. Ann. 218 (1975), 35–53.
86. A. Lichnerowicz *Théorie globale des connexions et des groupes d'holonomie*. Edizioni Cremonese (1957).
87. M. Lübke *Chernklassen von Hermite–Einstein Vektorbündeln*. Math. Ann. 260 (1982), 133–141.
88. M. Lübke, A. Teleman *The Kobayashi–Hitchin correspondence*. World Scientific (1995).
89. Y. Manin *Frobenius manifolds, quantum cohomology, and moduli spaces*. AMS Coll. Publ. 47 (1999).
90. S. A. Merkulov *Formality of canonical symplectic complexes and Frobenius manifolds*. Internat. Math. Res. Notices 4 (1998), 727–733.
91. J. Milnor, J. Stasheff *Characteristic classes*. Annals Math. Stud. 76. Princeton University Press (1974).
92. D. Mumford, J. Fogarty, F. Kirwan *Geometric invariant theory*. Third edition. Erg. Math. Grenzgebiete (2) 34. Springer (1994).
93. R. Narasimhan *Several Complex Variables*. Chicago Lectures in Math. (1971).
94. M. Narasimhan, C. Seshadri *Stable and unitary vector bundles on a compact Riemann surface*. Ann. Math. 82 (1965), 540–567.
95. A. Newlander, L. Nierenberg *Complex analytic coordinates in almost complex manifolds*. Ann. Math. 65 (1957), 391–404.
96. N. O'Brian, D. Toledo, Y. Tong *Grothendieck–Riemann–Roch for complex manifolds*. Bull. Amer. Math. Soc. 5 (1981), 182–184.

97. R. Remmert *Meromorphe Funktionen in kompakten komplexen Räumen.* Math. Ann. 132 (1956), 277–288.
98. R. Remmert *Theory of complex functions.* GTM 122. Readings in Math. Springer (1991).
99. I. Shafarevich *Basic Algebraic Geometry 2.* Springer (1996).
100. B. Shiffman, A. Sommese *Vanishing theorems on complex manifolds.* Progr. Math. 56. Birkhäuser (1985).
101. Y.-T. Siu *Every K3 surface is Kähler.* Invent. Math. 73 (1983), 139–150.
102. M. Spivak *A comprehensive introduction to differential geometry.* Publish or Perish. (1979).
103. D. Sullivan *Infinitesimal computations in topology.* Publ. Math. IHES 47 (1977), 269–331.
104. C. Taubes *The existence of anti-self-dual conformal structures.* J. Diff. Geom. 36 (1992), 163–253.
105. G. Tian *Extremal metrics and geometric stability.* Houston J. Math. 28 (2002), 411–432.
106. G. Tian *Canonical metrics in Kähler geometry.* Lect. Math. ETH. Birkhäuser (2000).
107. G. Tian *Smoothness of the universal deformation space of compact Calabi–Yau manifolds and its Petersson–Weil metric.* In: Math. aspects of string theory. Ed. S.-T. Yau. Adv. Series in Math. Phys. 1. (1987), 629–646.
108. A. Todorov *The Weil–Petersson geometry of the moduli space of* SU$(n \geq 3)$ *(Calabi–Yau) manifolds. I.* Comm. Math. Phys. 126 (1989), 325–346.
109. D. Toledo, Y. Tong *A parametrix for $\bar{\partial}$ and Riemann–Roch in Čech theory.* Topology 15 (1976), 273–301.
110. T. tom Dieck *Transformation Groups.* de Gruyter Studies in Math. (1987).
111. K. Ueno *Classification theory of algebraic varieties and compact complex spaces.* Lect. Notes Math. 439. Springer (1975).
112. K. Uhlenbeck, S.-T. Yau *On the existence of Hermitian–Yang–Mills connections in stable vector bundles.* Comm. Pure Appl. Math. 39 (1986), S257–S293.
113. C. Voisin *Hodge theory and complex algebraic geometry.* Cambridge Stud. Adv. Math. 76. (2002).
114. A. Weil *Introduction à l'étude des variétés kählériennes.* Hermann (1958).
115. A. Weil *Final report on contract AF 18(603)-57.* Coll. Works II. Springer.
116. R. O. Jr. Wells *Differential analysis on complex manifolds.* Second edition. GTM 65. Springer (1980).
117. E. Witten *Supersymmetry and Morse theory.* J. Diff. Geom. 17 (1982), 661–692.
118. F. Zheng *Complex Differential Geometry.* AMS/IP Studies in Advanced Math. 18 (2000).

Index

Universitext

Aksoy, A.; Khamsi, M. A.: Methods in Fixed Point Theory

Alevras, D.; Padberg M. W.: Linear Optimization and Extensions

Andersson, M.: Topics in Complex Analysis

Aoki, M.: State Space Modeling of Time Series

Audin, M.: Geometry

Aupetit, B.: A Primer on Spectral Theory

Bachem, A.; Kern, W.: Linear Programming Duality

Bachmann, G.; Narici, L.; Beckenstein, E.: Fourier and Wavelet Analysis

Badescu, L.: Algebraic Surfaces

Balakrishnan, R.; Ranganathan, K.: A Textbook of Graph Theory

Balser, W.: Formal Power Series and Linear Systems of Meromorphic Ordinary Differential Equations

Bapat, R.B.: Linear Algebra and Linear Models

Benedetti, R.; Petronio, C.: Lectures on Hyperbolic Geometry

Benth, F. E.: Option Theory with Stochastic Analysis

Berberian, S. K.: Fundamentals of Real Analysis

Berger, M.: Geometry I, and II

Bliedtner, J.; Hansen, W.: Potential Theory

Blowey, J. F.; Coleman, J. P.; Craig, A. W. (Eds.): Theory and Numerics of Differential Equations

Börger, E.; Grädel, E.; Gurevich, Y.: The Classical Decision Problem

Böttcher, A; Silbermann, B.: Introduction to Large Truncated Toeplitz Matrices

Boltyanski, V.; Martini, H.; Soltan, P. S.: Excursions into Combinatorial Geometry

Boltyanskii, V. G.; Efremovich, V. A.: Intuitive Combinatorial Topology

Booss, B.; Bleecker, D. D.: Topology and Analysis

Borkar, V. S.: Probability Theory

Carleson, L.; Gamelin, T. W.: Complex Dynamics

Cecil, T. E.: Lie Sphere Geometry: With Applications of Submanifolds

Chae, S. B.: Lebesgue Integration

Chandrasekharan, K.: Classical Fourier Transform

Charlap, L. S.: Bieberbach Groups and Flat Manifolds

Chern, S.: Complex Manifolds without Potential Theory

Chorin, A. J.; Marsden, J. E.: Mathematical Introduction to Fluid Mechanics

Cohn, H.: A Classical Invitation to Algebraic Numbers and Class Fields

Curtis, M. L.: Abstract Linear Algebra

Curtis, M. L.: Matrix Groups

Cyganowski, S.; Kloeden, P.; Ombach, J.: From Elementary Probability to Stochastic Differential Equations with MAPLE

Dalen, D. van: Logic and Structure

Das, A.: The Special Theory of Relativity: A Mathematical Exposition

Debarre, O.: Higher-Dimensional Algebraic Geometry

Deitmar, A.: A First Course in Harmonic Analysis

Demazure, M.: Bifurcations and Catastrophes

Devlin, K. J.: Fundamentals of Contemporary Set Theory

DiBenedetto, E.: Degenerate Parabolic Equations

Diener, F.; Diener, M.(Eds.): Nonstandard Analysis in Practice

Dimca, A.: Sheaves in Topology

Dimca, A.: Singularities and Topology of Hypersurfaces

DoCarmo, M. P.: Differential Forms and Applications

Duistermaat, J. J.; Kolk, J. A. C.: Lie Groups

Edwards, R. E.: A Formal Background to Higher Mathematics Ia, and Ib

Edwards, R. E.: A Formal Background to Higher Mathematics IIa, and IIb

Emery, M.: Stochastic Calculus in Manifolds

Endler, O.: Valuation Theory

Erez, B.: Galois Modules in Arithmetic

Everest, G.; Ward, T.: Heights of Polynomials and Entropy in Algebraic Dynamics

Farenick, D. R.: Algebras of Linear Transformations

Foulds, L. R.: Graph Theory Applications

Frauenthal, J. C.: Mathematical Modeling in Epidemiology

Friedman, R.: Algebraic Surfaces and Holomorphic Vector Bundles

Fuks, D. B.; Rokhlin, V. A.: Beginner's Course in Topology

Fuhrmann, P. A.: A Polynomial Approach to Linear Algebra

Gallot, S.; Hulin, D.; Lafontaine, J.: Riemannian Geometry

Gardiner, C. F.: A First Course in Group Theory

Gårding, L.; Tambour, T.: Algebra for Computer Science

Godbillon, C.: Dynamical Systems on Surfaces

Godement, R.: Analysis I

Goldblatt, R.: Orthogonality and Spacetime Geometry

Gouvêa, F. Q.: p-Adic Numbers

Gustafson, K. E.; Rao, D. K. M.: Numerical Range. The Field of Values of Linear Operators and Matrices

Gustafson, S. J.; Sigal, I. M.: Mathematical Concepts of Quantum Mechanics

Hahn, A. J.: Quadratic Algebras, Clifford Algebras, and Arithmetic Witt Groups

Hájek, P.; Havránek, T.: Mechanizing Hypothesis Formation

Heinonen, J.: Lectures on Analysis on Metric Spaces

Hlawka, E.; Schoißengeier, J.; Taschner, R.: Geometric and Analytic Number Theory

Holmgren, R. A.: A First Course in Discrete Dynamical Systems

Howe, R., Tan, E. Ch.: Non-Abelian Harmonic Analysis

Howes, N. R.: Modern Analysis and Topology

Hsieh, P.-F.; Sibuya, Y. (Eds.): Basic Theory of Ordinary Differential Equations

Humi, M., Miller, W.: Second Course in Ordinary Differential Equations for Scientists and Engineers

Hurwitz, A.; Kritikos, N.: Lectures on Number Theory

Huybrechts, D.: Complex Geometry: An Introduction

Isaev, A.: Introduction to Mathematical Methods in Bioinformatics

Iversen, B.: Cohomology of Sheaves

Jacod, J.; Protter, P.: Probability Essentials

Jennings, G. A.: Modern Geometry with Applications

Jones, A.; Morris, S. A.; Pearson, K. R.: Abstract Algebra and Famous Inpossibilities

Jost, J.: Compact Riemann Surfaces

Jost, J.: Postmodern Analysis

Jost, J.: Riemannian Geometry and Geometric Analysis

Kac, V.; Cheung, P.: Quantum Calculus

Kannan, R.; Krueger, C. K.: Advanced Analysis on the Real Line

Kelly, P.; Matthews, G.: The Non-Euclidean Hyperbolic Plane

Kempf, G.: Complex Abelian Varieties and Theta Functions

Kitchens, B. P.: Symbolic Dynamics

Kloeden, P.; Ombach, J.; Cyganowski, S.: From Elementary Probability to Stochastic Differential Equations with MAPLE